HIGH ENERGY ASTROPHYSICS AND COSMOLOGY

Proceedings of Academia Sinica — Max-Planck Society Workshop on High Energy Astrophysics

HIGH ENERGY ASTROPHYSICS AND COSMOLOGY

Proceedings of Academia Sinica — Max-Planck Society
Workshop on High Energy Astrophysics Held
in Nanjing, China, April 9—17, 1982

Edited by

YANG Jian (杨 建)
ZHU Cisheng (朱慈墭)

Science Press, Beijing, China, 1983
Gordon and Breach Science Publishers S. A.

Responsible Editor Fang Kaiwen

Published by Science Press, Beijing

Distribution rights throughout the world, excluding the People's Republic of China, granted to Gordon and Breach Science Publishers SA, Switzerland

Printed by C & C Joint Printing Co., (H.K.) Ltd.

First published 1983

ISBN 0-677-3134-03

Science Press Book No. 3358-36

JOINT SCIENTIFIC COUNCIL

Chairman:

GONG Shumo (龚树模) (CHINA)

Co-Chairman:

F. Meyer (F.R.G.)

Members:

G.Börner (F.R.G.)
W. Hillebrandt (F.R.G.)
LU Tan (陆 埮)(CHINA)
H.-C. Thomas (F.R.G.)
FANG Lizhi (方励之) (CHINA)
LI Qibin (李启斌)(CHINA)
QU Qinyue (曲钦岳) (CHINA)
ZHANG Heqi (张和祺) (CHINA)

PREFACE

A workshop on high energy astrophysics was held under the auspices of the Academia Sinica and the Max-Planck Society on April 9—17, 1982 in Nanjing, with the aim of strengthening the scientific exchange between Chinese and West German astrophysicists, present at the workshop were 38 Chinese and 9 West German astrophysicists, as well as 4 astrophysicists from Japan and the United States of America.

The following subjects in the field of high energy astrophysics and in related fields were discussed at the workshop: pulsars, neutron stars, supernova explosions, quark stars, cataclysmic binaries, active galactic nuclei, X-ray and γ-ray sources, neutrino processes, cosmology, and others. This Proceedings has been compiled in the hope that it will be of some help in promoting the scientific exchange with astrophysicists in other countries.

Joint Scientific Council
of the Workshop on High Energy
Astrophysics
Nanjing, China
July 1982

CONTENTS

I. PULSARS

II. COMPACT STARS

III. SUPERNOVAE

IV. BINARY SYSTEMS

V. ACCRETION DISKS

VI. BLACK HOLES

VII QUASARS AND GALAXIES

VIII. MASSIVE NEUTRINO AND COSMOLOGY

IX. MISCELLANEOUS

All the cartoons in this book are drawn by Fang Lizhi.

I. PULSARS

MASS SPECTRUM OF PULSARS

C. Y. FAN* (范章云), HANG Hengrong (杭恒荣)

(Purple Mountain Observatory,
Academia Sinica, Nanjing)

I. INTRODUCTION

The resemblance between Jupiter and a pulsar was first suggested by Burbidge and Strittmater[1] from the consideration of the radio noise emitted by the two bodies. The interest in the analogy was rekindled by the observation on the structure of Jupiter's outer magnetosphere on Pioneer 10 and 11 and subsequently on Voyager 1 and 2. In this paper, we make use of the information on the energetic particles of the magnetosphere for a quantitative comparison of the energy generation by Jupiter with that by a pulsar and then explore the consequence of the scaling.

The feature of the radiation environment relevant to the present paper is the observation of a large flux of high energy particles escaping from the Jovian magnetosphere[2]. An order of magnitude estimate indicates that the total energy needed to maintain the intensity level is about 2×10^{20} erg s^{-1}. Assuming a reasonable conversion efficiency, the solar wind can supply only a fraction of this energy. It was, thus, concluded that the energy must be deduced from the rotational energy of Jupiter. In this

*Permanent address: Department of physics, University of Arizona, Tucson, AZ, 85721, U.S.A.

respect, again, Jupiter resembles a pulsar.

II. SCALING FROM JUPITER TO PULSARS

In order to scale the results on Jupiter to pulsars, one is confronted with two tasks: the first is to derive an expression for the rate of energy generation by a rotating magnetic body and the second is to obtain an expression for the magnetic dipole moment of the body. In view of the complexity of the problems, we have chosen to use dimensional analysis guided by the observational facts on Jupiter and other celestial bodies. It turns out that the result obtained agrees remarkably well with the observational facts on the Crab pulsar.

The rate of energy generation in the vicinity of a spinning celestial body, whose magnetic dipole axis is aligned with the rotational axis, would depend on four physical quantities: the magnetic dipole moment M, the angular velocity ω, the radius of the body R and an empirical quantity K, characterizing the physical nature of the plasma medium in its vicinity. From dimensional argument, it can be verified readily that the rate of the energy generation is given by the following expression:

$$\frac{dT}{dt} = K\,\frac{M^2\omega^2}{Rc^2}, \qquad (1)$$

where c is the velocity of light. For Jupiter, $M=1.46\times10^{30}$ Gauss cm, $\omega = 1.8\times10^{-4}$ rad s^{-1}, $R=7.1\times10^{9}$ cm and $dT/dt = 2\times10^{20}$ erg s^{-1}, we find

$$K = 2\times10^{-2}\ s^{-1}.$$

The coefficient K is regarded as a scaling constant.

It is important to note that Eq. (1) is different from the energy loss from magnetic dipole radiation

$$\frac{dT}{dt} = \frac{2}{3c^3} M^2 \omega^4 , \qquad (2)$$

which was proposed by Pacini and others[3] as the mechanism for the energy loss by a pulsar. The difference basically comes from the fact that the latter neglects the effect of the plasma medium surrounding the pulsar. Thus, the physical quantities K and R do not appear in the expression. When using Eq.(2) to calculate the energy generation by Jupiter, one finds that it is seven orders of magnitude lower than what was observed.

The magnetic field of a celestial body must be produced by an internal current system which is presumably excited by its rotation. Therefore, it seems reasonable to assume that, as a first approximation, the dipole moment depends linearly on its angular momentum with a numerical coefficient characterizing the excitation mechanism. That is, if L is the angular momentum,

$$M = \alpha L, \qquad (3)$$

where α is regarded as a scaling constant.

This proposition seems to be supported by the so called magnetic Bode's law for the magnetic dipole moments of seven bodies in the solar system as shown in Fig. 1, where the data for Mercury (My), Venus (V), Mars (Ms), Earth (E), (Earth and Moon, EM), Jupiter (J), Saturn (S) and the Sun do lie more or less along a straight line.[4] For the angular momentum of Jupiter as 6.56×10^{45} g cm² s^{-1}, we find

$$\alpha = 2\times10^{-16} \text{ gauss cm s g}^{-1} .$$

This linear relationship is also consistent with the estimates of the values of M and L of 24 magnetic stars whose magnetic fields were reported by Babcock[5] and rotational periods by Wolff[6], and that of one white dwarf (L795-7)[7] and that of a neutron star[8] as shown in Fig. 1. For comparison, the values for seven magnetic stars and nine white dwarfs with less reliable M and L values are plotted.

By combining Eq.(3) with Eq.(1), we have for the rate of energy generation

$$\frac{dT}{dt} = \frac{K\alpha^2}{c^2 R} I^2 \omega^4, \tag{4}$$

where I is the moment of inertia of the body and the scaling constant $K\alpha^2$ deduced from the results on Jupiter is

$$K\alpha^2 = 8\times10^{-34} \text{ cm s}^{-1} \text{ g}^{-1}.$$

The physical meaning of scaling is to show that the value of $K\alpha^2$ deduced from the observation on the Crab pulsar has the same value as that deduced from Jupiter. In order to avoid the dependence on a specific neutron star

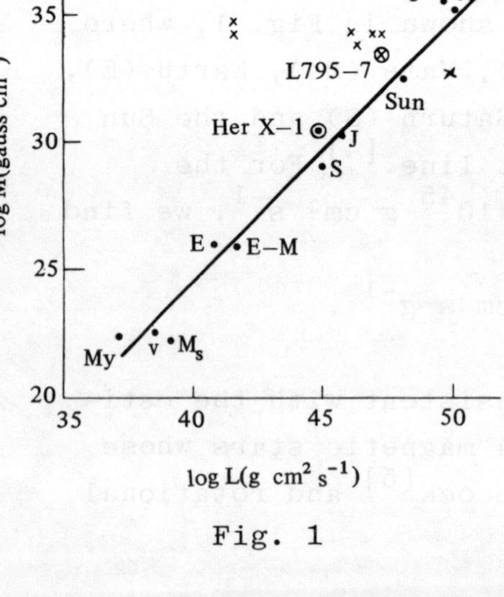

Fig. 1

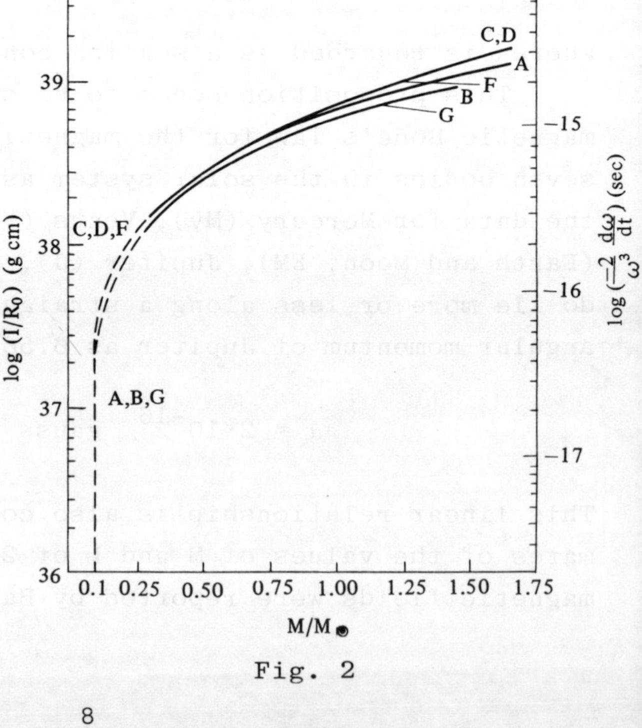

Fig. 2

model, we use 6×10^{44} g cm² for the value of I of the Crab pulsar, deduced from the observed dT/dt by Trimble and Rees[9] and 10^{6} cm for the radius R which can be estimated from the argument of stability of a fast spinning body. For the Crab pulsar, $\omega = 190$ rad s^{-1}, and $-d\,\omega/dt = 2.43\times10^{-9}\ s^{-2}$, we find

$$K\alpha^2 = 5.3\times10^{-34}\ \text{cm s}^{-1}\ \text{g}^{-1}.$$

The agreement is rather astonishing. We thus have demonstrated the justification for the scaling.

III. MASS SPECTRUM OF PULSARS

Having demonstrated the observational evidence for the scaling, we now proceed to link observational facts on pulsars to neutron star models, from which we obtain the mass spectrum of pulsars.

Eq.(4) can be rewritten as

$$-\frac{2}{\omega^3}\frac{d\omega}{dt} = \frac{2K\alpha^2}{c^2}\frac{I}{R}\ . \qquad (5)$$

For a pulsar, the quantity of the left-hand side can be precisely measured, whereas that on the right-hand side is determined by the mass of the pulsar according to neutron star models. Fig. 2 shows the relation to six different models and, for our purpose, they do not make much difference. We determined the masses of 194 pulsars. The remarkable feature is that, without adjustment of the scaling constant, the masses thus determined are confined within the neutron star mass limits (0.1 - 3.0 $M_\odot$) in compliance with the current neutron star model.

Fig. 3 shows the distribution of the star masses in 0.1 $M_\odot$ in five energy bins. A least square fit yields a spectrum

$$\frac{dN}{dm} \simeq \frac{1}{m^x} \tag{5}$$

with x = 1.8. Eight stars are excluded in the plot: they are P0153+61 and 0820+02 whose I/R values are $\sim 10^{40}$, and 0254-53, 1541-52, 1804-08, 1913+16, 1944+17 and 1952+20 whose I/R values are < 10^{36} g cm, which are beyond the limits of the theoretical curve in Fig. 2 to allow the determination.

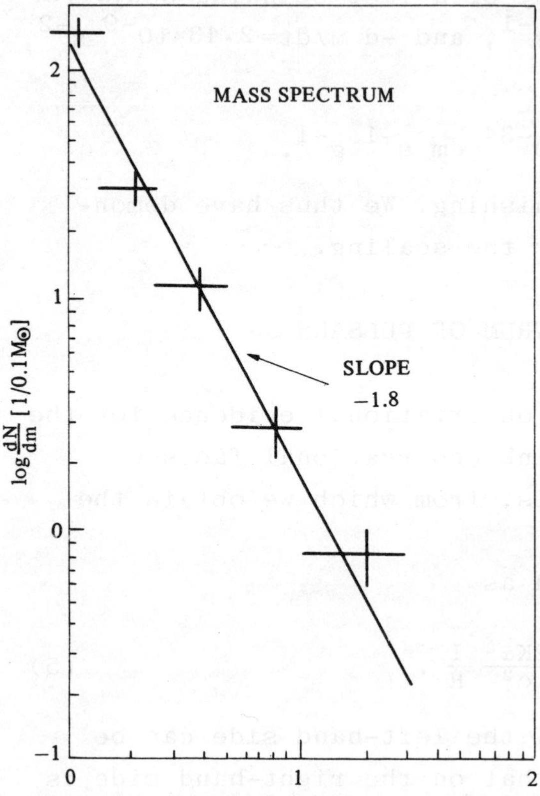

Fig. 3

It should be noted that P1913+16 is a binary pulsar and its mass was estimated as $1.4M_\odot$[10]. This estimate contradicts the value which one may conclude from Fig.2. A plausible explanation for this disagreement is the coupling with its companion star which renders the assumption used in deriving Eq. (1) invalid. One interesting question arises: whether the other five stars may also be binaries. In the region where P0153+61 and 0820+02 lie the stars would collapse into black holes. One possibility is that the collapse will occur when their spin velocities are sufficiently reduced.

The most significant result of this paper seems to be that one can now determine the mass of a single star. Using pure Newtonian mechanics, one can only estimate the mass of one member of a binary. Through the mechanical and electromagnetic coupling the determination is made possible.

REFERENCES

[1] Burbidge, G.R., Strittmater, P.A., 1968, Nature, Vol. 218, 433.

[2] Krimigis, S.M. et al., 1981, J. Geophys. Res., Vol. 86, 8227.

[3] Pacini, F., 1968, Nature, Vol. 219, 145.

[4] Russell, C.T., 1978, Letters to Nature, Vol. 272, 147.

[5] Babcock, H.W., 1960, "Stellar Atmosphere", J.L. Greenstein (ed.), Univ. Chicago Press, 282.

[6] Wolff, S.C., 1975, IAU Colloquium, No. 29, p. 55.

[7] Angel, J.R.P., 1978, Ann. Rev. Astron. Astrophys., Vol. 16, 487.

[8] Trumper, J., 1979, "X-ray Astronomy", Giacconi, R., Setti, G. (eds.), Reidel, Dordrecht.

[9] Trimble, V., Rees, M., 1970, Astrophys. Letters, Vol. 5, 93.

[10] Taylor, L.A. et al., 1979, Nature, Vol. 277, 437.

THERMAL X-RAY EMISSION FROM ISOLATED OLDER PULSARS: A NEW HEATING MECHANISM

HUANG Jiehao (黄介浩)

(Department of Astronomy, Nanjing University, Nanjing, China)

R. E. Lingenfelter

(Center for Astrophysics and Space Sciences, University of california, San Diego La Jolla, California 92093, U.S.A.)

PENG Qiuhe(彭秋和), HUANG Keliang (黄克谅)

(Department of Astronomy, Nanjing University, Nanjing, China)

A new pulsar heating mechanism, magnetic dipole radiation of superfluid neutrons, is proposed. It is also a new mechanism for pulsar spindown, $\dot{P} \propto P^2$.

As is well known, eletroneutral neutrons have an anomalous magnetic moment, thus there is an electromagnetic interaction between neutrons and electromagnetic field. This interaction leads to dipole radiation of the neutron magnetic moment for neutrons in circular motion, such as those in superfluid vortexes.

For neutron Cooper pairs coupled by the $^1S_0$ wave interaction, the spins of the two neutrons are anti-parallel, so that the total spin is zero and there is no net magnetic moment. But for neutron pairs coupled by the $^3P_2$ wave interaction the spins are parallel, so that the total spin is equal to 1 and there is a net magnetic moment. If there is also a magnetic field in the interior, the $^3P_2$ superfluid neutrons should lose energy and angular momentum by magnetic dipole radiation, contributing to both the spindown and the heating of the star.

The total power of the magnetic dipole radiation of superfluid neutrons in the neutron star is

$$W_m = 3.67\times10^3(\overline{\Delta^2B^2})R_p^3\frac{\overline{n^3}}{\pi}P^{-1}, \qquad (1)$$

where the bar denotes an average of some physical quantities for all the vortex lines, Δ the energy gap, R_p the radius of $^3P_2$ region, B the magnetic field strength, n the circulation quantum number of superfluid vortex, and P the pulsar period.

The W_m should exceed W_d(magnetic dipole radiation of the pulsar) for periods greater than some transition period P_m at which $W_m = W_d$. For periods $P < P_m$ the rotational energy loss rate $W_{rot} \propto P^{-4}$ for dipole radiation of the pulsar leads to $\dot{P} \propto P^{-1}$, while for $P > P_m$ the rate $W_{rot} \propto P^{-1}$ for magnetic dipole radiation of superfluid neutrons leads to $\dot{P} \propto P^2$.

In fact, such a change is suggested by a statistical study of the P and $\dot{P}$ measured for 269 pulsars. It shows that a statistically significant dependence of $\dot{P} \propto P^2$ for those pulsars with $P \gtrsim 1.25$ sec at the 1% level, and a $\dot{P} \propto P^{-1}$ dependence for those with $P \lesssim 0.5$ sec at the same level. Those with intermediate periods show no significant correlation between $\dot{P}$ and P at the 1% level, suggesting a range of transition periods, 0.5 sec<P<1.25 sec, for the observed pulsars.

As a heating mechanism the heating rate is

$$W_m = 4\pi^2 \, fI\dot{P}P^{-3}, \qquad (2)$$

where

$$f = \frac{W_m}{W_{rot}} = [1+(P_m/P)^3]^{-1} . \tag{3}$$

For older pulsars without pion condensates, where the internal temperature is $< 8\times10^8$K, we can then calculate the expected surface temperatures from heating rate W_m, assuming that it is equal to the cooling rate by black body radiation from the surface, L_r, and the ordinary neutrino bremsstrahlung emission in the interior, L_ν^B[1];

$$W_m = L_r + L_\nu^B , \tag{4}$$

where

$$L_\nu^B(C) \simeq 3\times10^{39} \left\{\int_{R_i}^{R} (\frac{\rho}{\rho_N}) r^2 dr\right\} (\frac{z^2}{A}) T^6 . \tag{5}$$

Where the radii are in units of 10^6 cm, R the stellar radius, R_i the inner boundary radius of the heavy ion crust, $\rho_N \simeq 3\times10^{14}$ g cm^{-3} the nuclear density, T the internal temperature in units of 10^9 K, and we set $(z^2/A) = 1$. Finally, from the surface temperature, T_{se}, we calculate the expected temperature, T_∞, seen by an observer infinitely far from the surface of the neutron star,

$$T_\infty \simeq T_{s_e} (1 + z)^{-1} , \tag{6}$$

where z the redshift on the surface of pulsar.

In order to see the effects of equations of state, we choose two extreme cases: the stiffest equation of state and the very soft equation of state. For comparison we set the stellar mass at $M = 1.3\ M_\odot$. On the other hand, we choose another stellar mass for Model B (soft equation of state) to estimate the effect of mass.

The effects of equations of state on the thermal X-ray radiation could be seen in Table 1. Also listed in column 8 are the preliminary temperatures reported for the three detected pulsars and upper limits on six others observed by the Einstein Observatory, assuming an interstellar hydrogen density of 0.3 cm^{-3}[2-4].

From the resultant surface temperatures, we see that the two models make no difference indeed. The interesting thing is the quite consistent surface temperatures between the two models are just caused by the very differences of these two models.

In the last column of Table 1 we list the resultant surface temperatures for Model B with mass of 0.5 $M_\odot$, from which we can see the effect of stellar masses on the thermal X-ray radiation. Obviously, the temperatures are decreased with decreasing masses. Although the mass of 1.33 $M_\odot$ seems to be more probable from the comparison with the temperatures reported for the three detected pulsars, the mass of 0.5 $M_\odot$ still could not be rejected according to our mechanism.

Using the magnetic dipole radiation of superfluid vortexes, we can estimate the internal magnetic field strength. The ratio of internal to surface magnetic field

$$\left(\frac{B_p}{B_s}\right)^2 = \frac{32\pi^3}{11c} \frac{\hbar^2}{r^4 m_n^2 \Delta_p^2 n^2} \frac{R^6}{R_p^3} (P^3 + P_m^3)^{-1}. \qquad (7)$$

Taking $n = 1$ and $P_m = 0.5$ sec, we can estimate the upper limit of internal magnetic field and the ratio for each pulsar in different models. The results are listed in Table 2. 10^{11}-10^{12} gauss is the order of magnitude in which the internal fields could be no matter what model we take.

REFERENCES

[1] Tsuruta, S., 1979, Phys. Rep., Vol. 56, 237.

[2] Einstein (HEAO 2) X-Ray Observatory Consortium Observing Program (CFA/HEA 79-285, 1979), 224.

[3] Helfand, D.J., 1981, In "I.A.U. Symposium,No. 95, Pulsars", D. Reidel, Dordrecht, 343.

[4] Helfand, D.J., Chanan, G.A., Novik, R., 1980, Nature, Vol. 283, 337.

TABLE 1. ESTIMATED PULSAR SURFACE TEMFERATURES FOR MAGNETIC DIPOLE RADIATION HEATING

PSR	D kpc	F (Pm=5)	MODEL Wm erg s^{-1}	A T 10^6K	MODEL Wm erg s^{-1}	B T 10^6K	T_{obs} 10^6K	MUDEL B T 10^6K
0031-07	.44	.87	3.3E31	.32	1.0E31	.28	$<$.4	.25
0149-16	.56	.82	1.5E32	.46	4.5E31	.40	$<$.5	.37
0355+54	1.5	.03	2.7E33	.83	8.3E32	.83	$\simeq$.8	.76
0402+61	.87	.63	1.3E33	.70	4.1E32	.69		.64
0450+55	.55	.24	1.1E33	.67	3.5E32	.67		.61
0540+23	2.8	.11	8.7E33	1.1	2.7E33	1.1		1.0
0611+22	3.5	.23	2.9E34	1.5	9.0E33	1.5		1.4
0727-18	2.0	.52	5.8E33	1.0	1.8E33	1.0		.92
0740-28	2.0	.36	1.0E34	1.2	3.2E33	1.2		1.1
0743-53	.39	.07	1.6E33	.73	5.0E32	.73		.67
0809+74	.20	.95	5.5E30	.20	1.7E30	.18		.16
0826-34	.32	.98	1.2E31	.25	3.8E30	.22		.20
0840-48	.42	.68	1.9E33	.76	5.9E32	.76		.70
0905-52	.40	.12	1.0E33	.65	3.2E32	.65		.60
0919+06	1.3	.39	5.3E33	.99	1.6E33	.98		.90
0922-52	.45	.77	5.2E33	.98	1.6E33	.98		.90
0940-55	.52	.70	4.3E33	.94	1.3E33	.93		.86
0941-56	.51	.81	4.8E33	.96	1.5E33	.96		.88
0950+08	.10	.12	1.3E32	.45	4.0E31	.39		.36
1055-52	1.1	.06	3.5E33	.89	1.1E33	.88	$\simeq$1.0	.81
1133+16	.18	.93	1.6E32	.48	5.1E31	.41		.38
1221-63	.99	.08	2.9E33	.85	9.0E32	.84		.78
1449-64	2.6	.04	1.7E33	.74	5.1E32	.73		.68
1451-68	.29	.13	5.3E31	.36	1.6E31	.31		.29
1508+55	.92	.76	7.5E32	.61	2.3E32	.60		.55
1530+27	.63	.92	2.8E31	.31	8.5E30	.26	$<$.5	.24
1641-45	4.9	.43	7.2E33	1.1	2.2E33	1.1		.98
1642-03	.16	.32	7.7E32	.61	2.4E32	.61	$\simeq$.3	.56
1648-17	.16	.88	2.3E32	.52	7.1E31	.45		.41
1706-16	.16	.69	1.2E33	.69	3.9E32	.68	$<$.3	.63
1727-47	5.9	.82	1.9E34	1.4	5.8E33	1.3		1.2
1742-30	2.9	.28	4.8E33	.96	1.5E33	.96		.88
1747-46	.74	.77	1.9E32	.50	5.9E31	.43		.39
1749-28	1.0	.59	2.1E33	.79	6.6E32	.78		.72
1822-09	.65	.28	7.1E33	1.1	2.2E33	1.1		.97
1844-04	4.9	.63	1.2E34	1.2	3.8E33	1.2		1.1
1916+14	1.0	.93	9.4E33	1.1	2.9E33	1.1		1.0
1930+22	8.1	.02	3.6E34	1.6	1.1E34	1.6		1.5
1952+29	.67	.38	6.2E29	.12	1.9E29	.10	$<$.3	.09
2151-56	.18	.95	1.2E32	.44	3.8E31	.38		.35
2327-20	.32	.97	8.0E31	.40	2.5E31	.34	$<$.4	.32

TABLE 2. ESTIMATION OF INTERNAL MAGNETIC FIELDS FOR DIFFERENT MODELS

PSR	MODEL A Bp (gauss)	MODEL A Bp/Bs	MODEL B Bp (gauss)	MODEL B Bp/Bs
0031-07	1.8E11	.82	1.0E11	.10
0149-16	3.5E11	.96	2.0E11	.11
0355+54	6.5E11	2.2	3.8E11	.27
0402+61	8.9E11	1.4	5.2E11	.17
0450+55	6.3E11	2.0	3.6E11	.24
0540+23	1.5E12	2.2	8.5E11	.26
0611+22	3.1E12	2.0	1.8E12	.24
0727-18	1.7E12	1.6	1.0E12	.19
0740-28	1.3E12	2.2	7.6E11	.27
0743-53	5.9E11	2.2	3.4E11	.26
0809+74	8.5E10	.53	4.9E10	.06
0826-34	1.5E11	.32	8.8E10	.04
0840-48	1.1E12	1.3	6.5E11	.15
0905-52	5.1E11	2.1	3.1E11	.26
0919+06	1.5E12	1.8	8.8E11	.21
0922-52	2.0E12	1.1	1.2E12	.13
0940-55	1.7E12	1.2	9.8E11	.15
0941-56	2.0E12	.99	1.2E12	.12
0950+08	1.8E11	2.1	1.1E11	.26
1055-52	8.3E11	2.2	4.8E11	.26
1133+16	4.4E11	.60	2.6E11	.07
1221-63	8.0E11	2.2	4.6E11	.26
1449-64	5.5E11	2.2	3.2E11	.27
1451-68	1.2E11	2.1	6.9E10	.25
1508+55	7.5E11	1.1	4.3E11	.13
1530+27	1.8E11	.64	1.0E11	.08
1641-45	1.8E12	1.7	1.1E12	.21
1642-03	5.5E11	1.9	3.2E11	.22
1648-17	4.7E11	.78	2.8E11	.09
1706-16	9.1E11	1.3	5.3E11	.15
1727-47	4.0E12	.96	2.3E12	.12
1742-30	1.3E12	1.9	7.8E11	.23
1747-46	3.8E11	1.1	2.2E11	.13
1749-28	1.1E12	1.5	6.4E11	.17
1822-09	2.4E12	1.1	1.4E12	.13
1844-04	2.7E12	1.4	1.6E12	.17
1916+14	3.4E12	.60	1.9E12	.07
1930+22	2.3E12	2.2	1.3E12	.27
1952+29	1.6E10	1.8	9.4E9	.21
2151-56	4.1E11	.49	2.4E11	.06
2327-20	3.7E11	.38	2.1E11	.05

THE DISTRIBUTION OF THE MAGNETIC INCLINATION IN A PULSAR'S POLAR CAP MODEL

WU Xinji(吴鑫基), WU Fei(邬 飞), QIAO Guojun(乔国俊),
CHENG Jiuheng(程久恒), DENG Guoxiang(邓国祥)

(Department of Geophysics, Beijing University, Beijing)

Magnetic inclination ϕ is an important parameter of the pulsar. The investigation of the distribution and the evolution of ϕ is very significant in determining the magnetosphere structure and emission mechanism. These problems are discussed in this paper.

I. THE CONDITIONED DISTRIBUTION DENSITY OF THE APPARENT BEAMWIDTH

In the polar cap model, there exists a functional dependence[1] of α on β, ϕ and θ

$$\alpha = 2\cos^{-1}\left(\frac{\cos\beta/2-\cos\phi\cos\theta}{\sin\phi\sin\theta}\right), \tag{1}$$

where the apparent beamwidth $\alpha = 360\frac{W}{P}$ (W is the pulse width and P is the period), ϕ is the magnetic inclination and θ is the observer's angle with respect to the rotational axis (see Fig. 1).

By symmetry, we need only to discuss the hemisphere above equator ($\phi,\theta \leqslant \pi/2$). From Eq.(1) we find that

$$\theta_{1,2}(\alpha)=\cos^{-1}\left(\frac{(\cos\phi)^{-1}\cos\beta/2\pm\sqrt{tg^4\phi\cos^4\alpha/2-tg^2\phi\cos^2\alpha/2\,(\cos^2\beta/2\,(\cos\phi)^{-2}-1)}}{1+tg^2\phi\cos^2\alpha/2}\right). \tag{2}$$

Under the assumptions that the axis of rotation is oriented randomly and β is constant, the formula of the conditioned distribution density of α is

$$\rho(\alpha/\phi)=\sin(\theta_1(\alpha))\frac{d\theta_1}{d\alpha}-\sin(\theta_2(\alpha))\frac{d\theta_2}{d\alpha}, \tag{3}$$

$$\frac{d\theta_{1,2}}{d\alpha}=-\frac{1}{2}\,\frac{\sin\alpha/2\,\sin\phi\,(1-\cos^2\theta_{1,2})}{\cos\phi-\cos\beta/2\,\cos\theta_{1,2}}. \tag{4}$$

From Eq.(1), for each ϕ, there is a maximum value of α:

$$\alpha_{max}=2\cos^{-1}\left(\frac{\sqrt{\cos^2\beta/2-\cos^2\phi}}{\sin\phi}\right), \tag{5}$$

When $\alpha=\alpha_{max}$, $\rho(\alpha/\phi)$ reaches its peak value so that $\alpha_{peak}=\alpha_{max}$.

When $\phi=\frac{\beta}{2}$, we have

$$\rho(\alpha/\phi)=\rho_\theta(\theta(\alpha))\left|\frac{d\theta}{d\alpha}\right|=\frac{tg^2\beta/2\,\sin\alpha}{(1+tg^2\beta/2\cos^2\alpha/2)^2}, \tag{6}$$

$$\alpha_{peak}=\cos^{-1}\left(\frac{1+2tg^2\beta/2-\sqrt{(1+2ctg^2\beta/2)^2+8}}{2}\right) \tag{7}$$

When $\phi<\frac{\beta}{2}$, we have

$$\rho(\alpha/\phi)=\begin{cases}-\rho_\theta(\theta_2(\alpha))\,\dfrac{d\theta_2}{d\alpha} & 0\leqslant\alpha\leqslant\pi,\\[2ex] -\rho_\theta(\theta_1(\alpha))\,\dfrac{d\theta_1}{d\alpha} & \pi\leqslant\alpha\leqslant 2\pi.\end{cases} \tag{8}$$

II. THE FORMULA OF DISTRIBUTION DENSITY OF ϕ

If $\rho(\alpha)$ and $\rho(\phi)$ represent the distribution densities of α and ϕ respectively, we have

$$\rho(\alpha)=\int\rho(\alpha/\phi)\,\rho(\phi)d\phi, \tag{9}$$

$$\rho(\alpha)=\sum_j \rho(\alpha/\phi_j)\rho(\phi_j)\Delta_j\phi, \tag{10}$$

since

$$f_i \simeq \int_{\alpha_i}^{\alpha_i+\Delta\alpha} \rho(\alpha)d\alpha = p_i = p\{\alpha_i \leqslant \xi < (\alpha_i+\Delta\alpha)\}, \tag{11}$$

where P_i is the probability of the pulsars falling into the interval $\alpha_i \rightarrow (\alpha_i + \Delta\alpha)$, f_i is the relative frequency in that interval ($f_i = \frac{\Delta N_i}{N}$, ΔN_i is pulsar's number in that interval and N is the total number).

Substituting Eq. (10) into Eq. (11) and then multiplying Eq. (11) by N, we obtain

$$\Delta N_i = \sum_j x_j \int_{\alpha_i}^{\alpha_i + \Delta\alpha} \rho(\alpha/\phi_j)d\alpha, \tag{12}$$

$$x_j = N\,\rho(\phi_j)\,\Delta_j\phi, \tag{13}$$

where x_j is the number of pulsars falling into the interval $\Delta_j\phi$. Since the value of ΔN_i can be obtained from the observational data of 299 pulsars and the values of $\int_{\alpha_i}^{\alpha_i+\Delta\alpha_i} \rho(\alpha/\phi_j)d\alpha$ calculated by numerical integration, we can obtain values of x_j from Eq. (12).

III. THE DISTRIBUTION OF Φ

We analyse the distribution of α for a sample of 299 pulsars with known period P and pulse equivalent width W_E. Since it has been suggested that β may be strongly dependent upon the period, we make an estimation for β according to the peak value of the distribution density of period and the relation $\beta \propto P^{-\frac{1}{2}}$[2] and obtain that $\beta=3°$.

The results of numerical calculations are listed in Table 1. It is obvious that there is a preference for small angles in the distribution of ϕ. Pulsars with angles $\phi \leqslant 30°$ make up about 77%. Two possibilities for $\rho(\phi)$ are considered: random orientation in space[3] ($\rho(\phi) = \frac{1}{2}\sin\phi$) and random alignment[4] ($\rho(\phi) = 1/\pi$). Our conclusion is different from the above ideas and consistent with the point of view in paper[1].

Table 1

Range of ϕ	<5°	5°-10°	10°-20°	20°-30°	>30°
Percentage	7%	18%	31%	21%	23%

In addition, because β is dependent on the age[4], a sample of 195 pulsars with known ages is subdivided into four groups in terms of their ages. β has different value for each group. (The method of estimating β has been described above.) Some informations about the evolution of the magnetic inclinations obtained. The results of calculations are shown in Table 2.

Table 2

Group	Number of Pulsars	β	Range of ϕ				
			<5°	5°-10°	10°-20°	20°-30°	>30°
A($t<1.5\times10^6$yr)	49	7°	0	0	16%	29%	55%
B ($1.5\times10^6 \leqslant t < 4\times10^6$)	47	3°	2%	17%	20%	19%	42%
C($4\times10^6 \leqslant t < 15\times10^6$)	55	2°.5	3%	31%	45%	11%	10%
D($t \geqslant 15\times10^6$yr)	44	2°	13%	56%	28%	2%	1%
Total	195		4%	26%	28%	15%	27%

From Table 2 we know that pulsars with $\phi \lesssim 30°$ make up about 73% of 195 pulsars.[5] The distribution of ϕ in the sample of 299 pulsars is not very different from the sample of 195 pulsars. We conclude that ϕ decreases with increasing the age of pulsars.

REFERENCES

[1] Prószyński, M., 1979, Astron. Astrophys., Vol. 79, 8.

[2] Gunn, J.E., Ostriker, J.P., 1970, Ap. J., Vol. 160, 979.

[3] Roberts, D.H., 1976, Ap.J., Vol. 207, 949.

[4] Manchester, R.N., Lyne, A.G., 1977, M.N.R.A.S., Vol. 181, 761.

[5] Taylor, J.H., Manchester, R.N., 1980, Five college radio astronomy observatory pulsar list.

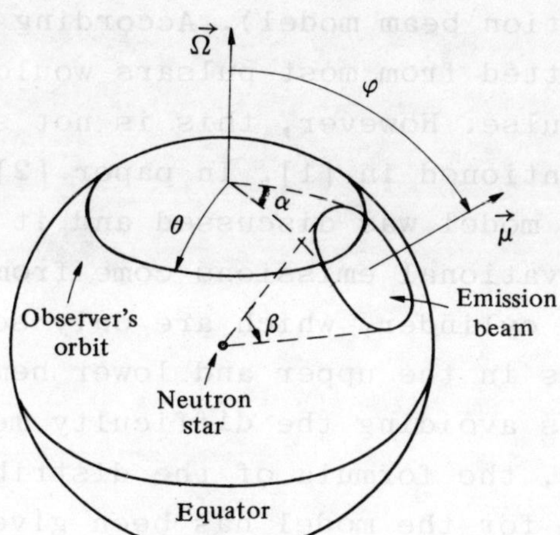

Fig. 1 Geometry of a pulsar for polar Cap model (after Prószyński, M., 1979[1]).

THE DISTRIBUTIONS OF APPARENT BEAMWIDTH FOR THE SHEET – SHAPE RADIATION BEAM MODEL OF PULSARS

QIAO Guojun(乔国俊), WU Xinji(吴鑫基)

(Department of Geophysics, Beijing University, Beijing)

In the early stage of the discovery of pulsars, it was discussed that the observational radiations of pulsars come from the regions close to the magnetic equator and the radiation beam is in the shape of a sheet of ring structure (sheet-shape radiation beam model). According to this model, the radiations emitted from most pulsars would give rise to a strong interpulse. However, this is not supported by observations as mentioned in [1]. In paper [2] a possible structure for this model was discussed and it was pointed out that the observational emissions come from two regions close to the light cylinder, which are only seen respectively by observers in the upper and lower hemispheres of neutron stars, thus avoiding the difficulty mentioned in[1].

In this paper, the formula of the distributions of apparent beamwidth for the model has been given and the theoretical distributions have been calculated by numerical integration. A fit between the theoretical and the observational distributions has been performed. The result shows that the theory is in agreement with observations.

I. THE DISTRIBUTION OF APPARENT BEAMWIDTH AND ITS OBSERVATIONAL FIT

In Fig. 1, β is the angular width of the sheet-shape radiation beam, θ the angle which the direction of rotational axis makes with the observers, W the pulse width, P the period, $\alpha = \frac{W}{P}\times 360°$, the apparent beamwidth, and ϕ the magnetic inclination.

From Fig. 1 we obtain

$$\alpha = \cos^{-1}(-1)\frac{\sin\frac{\beta}{2}+\cos\phi\cos\theta}{\sin\phi\sin\theta} - \cos^{-1}\frac{\sin\frac{\beta}{2}-\cos\phi\cos\theta}{\sin\phi\sin\theta} \tag{1}$$

$$\text{for } \left(\frac{\pi}{2}-\phi+\frac{\beta}{2}\right)<\theta\leqslant\frac{\pi}{2},$$

$$\alpha = 2\pi - 2\cos^{-1}\frac{\sin\frac{\beta}{2}-\cos\phi\cos\theta}{\sin\phi\sin\theta} \tag{2}$$

$$\text{for } \left(\frac{\pi}{2}-\phi-\frac{\beta}{2}\right)<\theta\leqslant\left(\frac{\pi}{2}-\phi+\frac{\beta}{2}\right).$$

If $\rho(\alpha)$ and $\rho(\theta)$ represent the distribution density of α and θ respectively, we have

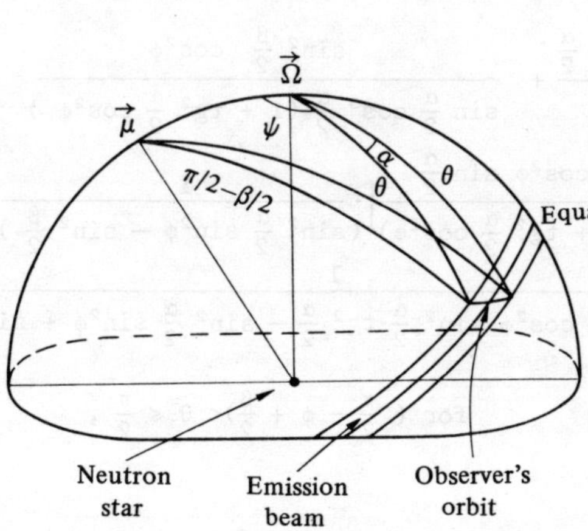

Fig. 1 The apparent beamwidth in the sheet-shape beam radiation model.

$$\rho(\alpha) = \rho(\theta)\,\theta'(\alpha). \tag{3}$$

Under the assumption that the directions of rotational axes of neutron stars are distributed randomly. We get $\rho(\theta) = \frac{1}{2}\sin\theta$. From Eqs. (1) and (2), we have

$$\rho\left(\frac{\alpha}{\phi}\right) = \frac{1}{2}\sin\left\{\cos^{-1}\left[\frac{\sin\frac{\beta}{2}\cos\phi + \sin\phi\cos\frac{\alpha}{2}\left(\cos^2\phi + \cos^2\frac{\alpha}{2}\sin^2\phi - \sin^2\frac{\beta}{2}\right)^{\frac{1}{2}}}{\cos^2\phi + \cos^2\frac{\alpha}{2}\sin^2\phi}\right]\right\}$$

$$\cdot\left\{\frac{1}{2}\sin\phi\sin\frac{\alpha}{2}\left(\cos^2\phi + \cos^2\frac{\alpha}{2}\sin^2\phi - \sin^2\frac{\beta}{2}\right)^{\frac{1}{2}} + \frac{\cos^2\frac{\alpha}{2}\sin\frac{\alpha}{2}\sin^3\phi}{2\left(\cos^2\phi + \cos^2\frac{\alpha}{2}\sin^2\phi - \sin^2\frac{\beta}{2}\right)^{\frac{1}{2}}}\right.$$

$$\left. - \frac{\sin\frac{\beta}{2}\cos\phi\sin^2\phi\cos\frac{\alpha}{2}\sin\frac{\alpha}{2} + \cos^2\frac{\alpha}{2}\sin\frac{\alpha}{2}\cos^3\phi\left(\cos^2\phi + \cos^2\frac{\alpha}{2}\sin^2\phi - \sin^2\frac{\beta}{2}\right)^{\frac{1}{2}}}{\cos^2\phi + \cos^2\frac{\alpha}{2}\sin^2\phi}\right\}$$

$$\cdot\frac{1}{\left\{\left(\cos^2\phi + \cos^2\frac{\alpha}{2}\sin^2\phi\right)^2 - \left[\sin\frac{\beta}{2}\cos\phi + \sin\phi\cos\frac{\alpha}{2}\left(\cos^2\phi + \cos^2\frac{\alpha}{2}\sin^2\phi - \sin^2\frac{\beta}{2}\right)^{\frac{1}{2}}\right]^2\right\}^{\frac{1}{2}}}$$

$$\text{for } \left(\frac{\pi}{2} - \phi - \frac{\beta}{2}\right) < \theta \leqslant \left(\frac{\pi}{2} - \phi + \frac{\beta}{2}\right), \tag{4}$$

$$\rho\left(\frac{\alpha}{\phi}\right) = -\frac{1}{2}\sin\left\{\cos^{-1}\left[\frac{\sin^2\phi\sin^2\frac{\alpha}{2} - \sin^2\frac{\beta}{2}}{\sin^2\frac{\alpha}{2}\left(1 + tg^2\alpha\cos^2\phi\right)}\right]^{\frac{1}{2}}\right\}$$

$$\cdot\left[\frac{\sin^2\frac{\beta}{2}\cos\frac{\alpha}{2}}{\sin^3\frac{\alpha}{2}} + \frac{\sin^2\frac{\beta}{2}\cos^2\phi}{\sin\frac{\alpha}{2}\cos^3\frac{\alpha}{2}\left(1 + tg^2\frac{\alpha}{2}\cos^2\phi\right)}\right.$$

$$\left. - \frac{\sin^2\phi\cos^2\phi\sin\frac{\alpha}{2}}{\cos^3\frac{\alpha}{2}\left(1 + tg^2\frac{\alpha}{2}\cos^2\phi\right)}\right]\cdot\frac{1}{\left(\sin^2\frac{\alpha}{2}\sin^2\phi - \sin^2\frac{\beta}{2}\right)^{\frac{1}{2}}}$$

$$\cdot\frac{1}{2\left[\sin^2\frac{\alpha}{2} + \cos^2\phi\sin^2\frac{\alpha}{2}tg^2\frac{\alpha}{2} - \sin^2\frac{\alpha}{2}\sin^2\phi + \sin^2\frac{\beta}{2}\right]^{\frac{1}{2}}}$$

$$\text{for } \left(\frac{\pi}{2} - \phi + \frac{\beta}{2}\right) < \theta \leqslant \frac{\pi}{2}, \tag{5}$$

where $\rho(\alpha/\phi)$ represents the conditional probability. The theoretical distributions of apparent beamwidth can fit observations by the method mentioned in[3], as is shown in Fig. 2, where the dashed lines represent the theoretical values and the solid lines represent the observational values

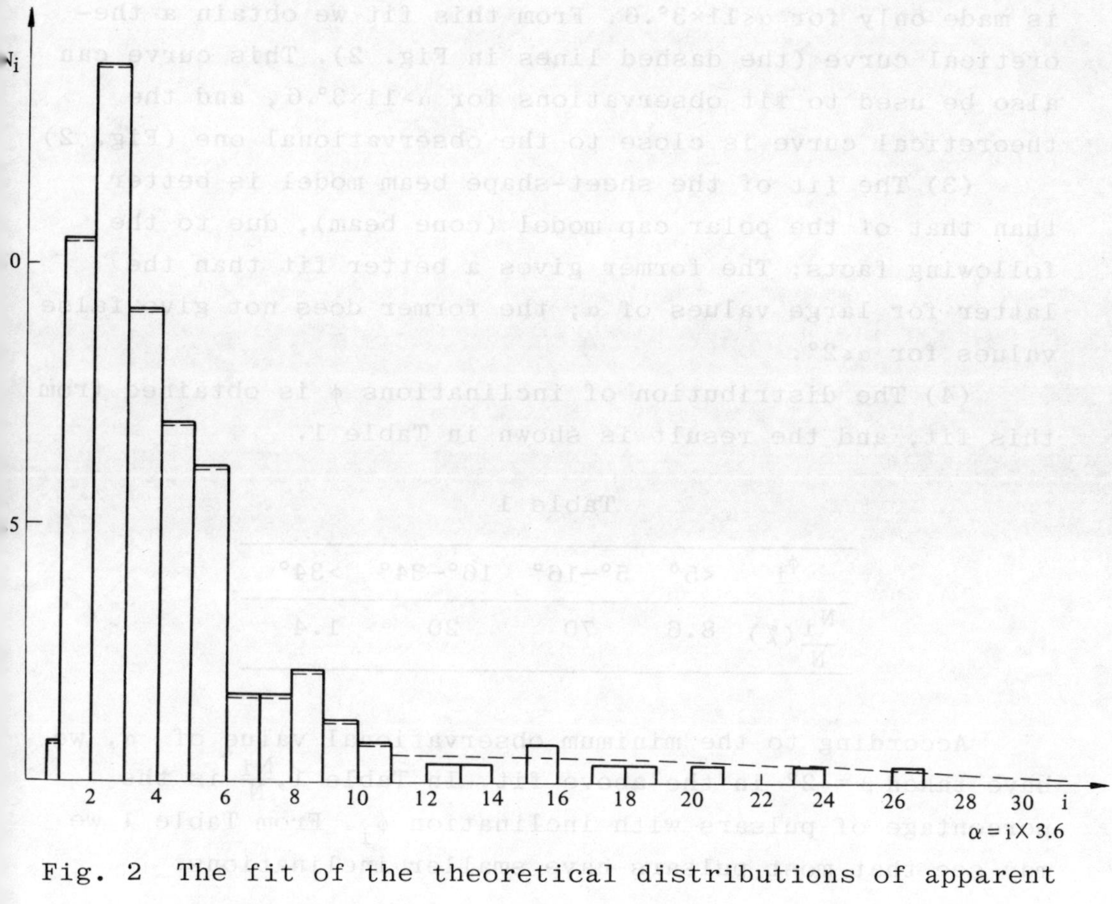

Fig. 2 The fit of the theoretical distributions of apparent beamwidth for the sheet-shape model with observations.

II. DISCUSSION

(1) In this paper the fit between the theoretical distributions and the observations is carried out for two cases of 195 pulsars ($\dot{P}$ known) and 299 pulsars[4], respectively. The results for both cases are very similar. This shows that the number of samples is enough in our study.

(2) The fit of the distributions of apparent beamwidth is made only for $\alpha \leqslant 11 \times 3^{\circ}.6$. From this fit we obtain a theoretical curve (the dashed lines in Fig. 2). This curve can also be used to fit observations for $\alpha > 11 \times 3^{\circ}.6$, and the theoretical curve is close to the observational one (Fig. 2)

(3) The fit of the sheet-shape beam model is better than that of the polar cap model (cone beam), due to the following facts: The former gives a better fit than the latter for large values of α; the former does not give false values for $\alpha \leqslant 2^{\circ}$.

(4) The distribution of inclinations ϕ is obtained from this fit, and the result is shown in Table 1.

Table 1

ϕ_i	<5°	5°–16°	16°–34°	>34°
$\frac{N_i}{N}$(%)	8.6	70	20	1.4

According to the minimum observational value of α, we have taken $\beta = 2^{\circ}$ in the above fit. In Table 1, $\frac{N_i}{N}$ is the percentage of pulsars with inclination ϕ_i. From Table 1 we may see that most pulsars have smaller inclinations.

REFERENCES

[1] Papaginnis, M.P., 1969, Nature, Vol. 222, 1261.

[2] Qiao Guojun, Yang Haishou, Wu Xinji, Lin Jianxiang, Deng Guoxiang, 1980, Chinese Astronomy, Vol. 4, 372.

[3] Wu Xinji, Wu Fei, Qiao Guojun, Cheng Jiuheng, Deng Guoxiang, in this Proceedings p.19.

[4] Taylor, J.H., Manchester, R.N., 1980, Five College Radio Astronomy Observatory Pulsar List.

A STATISTICAL STUDY OF PULSAR ROTATIONAL BRAKING

QIAO Guojun(乔国俊), WU Xinji(吴鑫基)

(Department of Geophysics, Beijing University, Beijing)

I. INTRODUCTION

Pacini (1973)[1] described the basic theoretical expectations of the magnetic dipole radiation and unipolar induction of pulsars and presented observational results. This paper further discusses the difference between the two braking mechanisms and analyses the results of a statistical study through observed data of apparent beamwidth's distributions and a geometrical method[2]. The theory of magnetic dipole radiation of pulsars agrees with the observations. Unipolar induction as a braking mechanism is not consistent with the observations. If the braking mechanism of pulsars is not unipolar inductions, then the associated radiation mechanism (RS model[3,4]) must have a high transformation efficiency ($\eta_u = L_{radio}/W_u$ = Radio radiation luminosity/Power of unipolar radiation). Such a high radio transformation efficiencies ($\eta_R = L_{radio}/\dot{E}_{rot}$) is a serious requirement for pulsars.

II. THE THEORY OF TWO BRAKING MECHANISMS

Pacini[1] have pointed out that the rate of the rotational energy loss of pulsars, $\dot{E}_{rot}$, and the period generally satisfy

$$|\dot{E}_{rot}| = I\Omega\dot{\Omega} \simeq \frac{B_{01}^2 R_0^6 (2\pi)^4}{c^3\ P^4}, \quad (1)$$

where I is the moment of inertia of the star, Ω the angular velocity, $\dot{\Omega} = d\Omega/dt$, B_{01} the magnetic field at the star's surface, R_0 the radius of neutron star, C the velocity of light and P the period of rotation. For the magnetic dipole radiation, $B_{01}^2 R_0^6 = \mu^2 \sin^2\phi$; for the unipole induction radiation, $B_{01}^2 R_0^6 = \mu^2\cos^2\phi$, where ϕ is the magnetic inclination. When the perpendicular magnetic dipole radiation plays a main role in pulsar's braking, we have

$$\mu^2 \sin^2\phi = \frac{C^3}{4\pi^2} I\dot{P}P. \quad (2)$$

It has a difference of a factor of 2/3 with magnetic dipole radiation in vacuum. When the unipolar induction radiation plays a main role in pulsar's braking, we have

$$\mu^2 \cos^2\phi = \frac{C^3}{4\pi^2} I\dot{P}P. \quad (3)$$

For a magnetic torque parallel with the rotation axis $\mu_{//}$, between the equator and a point of latitude λ, the induction potential is

$$\varepsilon = \frac{B_{01}\Omega R_0^2}{2C} (\cos^2\lambda - 1). \quad (4)$$

For a oblique rotator with magnetic inclination ϕ, taking $B_{01} = \mu\cos\phi/R_0^3$, if $|\varepsilon|$ is the maximum potential, then we have

$$|\varepsilon| = \frac{\mu\Omega}{2CR_0} \cos\phi. \quad (5)$$

In the RS model, ε varies in magnitude on a time scale of 10 μs, and it is supposed that the variances take place in equal interval of time, then we have

$$<\varepsilon> = \frac{\Omega\mu}{4CR_0}\cos\phi. \tag{6}$$

The total current flowing from the polar region has been estimated by Kennel et al. (1979)[4]:

$$I_t = \frac{C^2}{2\Omega}\left(\frac{B\phi}{B_r}\right)_c B_{rc}, \tag{7}$$

where

$$\left(\frac{B\phi}{B_r}\right) \simeq 1, \quad B_{rc} = \frac{B_0 R_c}{2\ R_A}\left(\frac{R_0\Omega}{C}\right). \tag{8}$$

According to [5]

$$R_A = \frac{V_A}{\Omega} = \frac{(2\pi)^{3/5} B_0^{\ 2/5} R_0^{3/5}}{(4\pi\rho_0)^{1/5}\Omega p^{3/5}}, \tag{9}$$

where V_A is the Alfv́en velocity, ρ_0 the mass density in the surface of the neutron star. So that the total power of the unipolar induction is

$$W_u = <\varepsilon>I_t = \frac{(2\pi)^{0.6}(2\rho_0)^{0.2} B_0^{0.6} R_0^{1.4}\ \mu}{16C} P^{-2.4}\cos\phi, \tag{10}$$

where μ is the magnetic moment. If the unipolar induction radiation is the major mechanism of the pulsar braking, we have

$$\cos^{1.6}\phi = \frac{16C\ I\ R_0^{0.4}}{(2\pi)^{0.6}(2\rho_0)^{0.2}\mu^{1.6}}\dot{P}P^{-0.6}. \tag{11}$$

Eqs.(2) (3) and (11) show the major difference between the magnetic dipole radiation and the unipolar induction radiation.

III. A STATISTICAL STUDY FOR TWO BRAKING MECHANISMS

A statistical study is carried out to compare these

two braking mechanisms with observations. The observational data for the distribution of apparent beamwidth and a geometrical method are used in this study. We divide the 195 pulsars into three groups according to $\dot{P}P$ and $\dot{P}P^{-0.6}$ respectively, as can be seen in the following two cases.

a) case 1

Group	$\dot{P}P$	Number	The percentage of the PSR Number Observed in an Order of Magnitude(%
A	$5 \leqslant \dot{P}P \times 10^{15} < 500$	50	82
B	$0.5 \leqslant \dot{P}P \times 10^{15} < 5$	99	100
C	$10^{-3} \leqslant \dot{P}P \times 10^{15} < 0.5$	46*	65

* There are 41 pulsars (89%) in the range of 0.025–0.5 of $\dot{P}P \times 10^{15}$.

b) case 2

Group	$\dot{P}P^{-0.6}$	Number	The Percentage of the PSR Number Observed in an Order of Magnitude(%)
A	$8 \leqslant \dot{P}P^{-0.6} \times 10^{15} \leqslant 3 \times 10^{3}$	54	87
B	$1 \leqslant \dot{P}P^{-0.6} \times 10^{15} < 8$	100	100
C	$2 \times 10^{-3} < \dot{P}P^{-0.6} \times 10^{15} < 1$	41	85

Using the method as we did in paper[2], the statistics is carried out for every group. The results are shown in Table 1.

Table 1

Group	A	B	C
For Case 1, the percentage of pulsars with $\phi>10^{\circ}$ (%)	54	51	17
For Case 2, the Percentage of pulsars with $\phi>10^{\circ}$ (%)	72	53	4.7

IV. CONCLUSIONS

a) Table 1 shows that in both of the cases mentioned above, a large number of pulsars with larger ϕ are included in group A, while a large number of pulsars with smaller ϕ are in group C. This agrees with the theory of magnetic dipole radiation. If the middle value of $\dot{P}P$ and the peak value of ϕ are taken, the theoretical relation is in agreement with observations quantitatively.

b) According to the theory of unipolar induction radiation, the magnetic inclination ϕ decreases with increasing the value of $\dot{P}P^{-0.6}$ or $\dot{P}P$ of pulsars, but we cannot conclude this from the observations (see Table 1).

c) For the 195 pulsars[6], $\dot{P}P$ and $\frac{P}{\dot{P}}$ have a close relationship (R = -0.7). The statistical study mentioned above shows that older pulsars have smaller ϕ.

REFERENCES

[1] Pacini, F., 1973, "Astrophysics and Gravitation ", 1.

[2] Wu Xinji, Wu Fei, Qiao Guojun, Cheng Jiuheng, Deng Guoxiang, in this Proceedings p.19 .

[3] Ruderman, M.A., Sutherland, P.G., 1975, Ap. J.,Vol. 196, 51.

[4] Kennel, C.F., Fujimura, F.S., Pellat, R., 1979, Space Science Reviews, Vol. 24, 407.

[5] Qiao Guojun, Wu Xinji, Xia Xiaoyang, 1982, Acta Astrophysica Sinica,Vol. 2, 35.

[6] Taylor, J. H., Manchester, R. N., 1980, Five College Radio Astronomy Observatory Pulsar List.

A EXTERNAL FACTOR AFFECTING PULSAR'S $\dot{P}$ — THE DIFFERENTIAL ROTATION OF THE GALAXY

RONG Jianxiang(容建湘), XIA Yifei(夏一飞)

(Department of Astronomy, Nanjing University, Nanjing)

I. INTRODUCTION

In this paper the effect of the differential rotation of the Galaxy on pulsar's $\dot{P}$ is analysed. This subject is involved with an external cause, which is independent of intrinsic mechanisms, affecting the change of radio pulse period. Owing to the presence of differential rotation of the Galaxy[1,2], for each pulsar there exists a factor affecting the rate of change, $\dot{P}$, of pulsar period. Therefore this effect is of general significance.

Let the pulse period of a pulsar in its proper coordinate system be P_0. Then when the pulsar has the radial velocity v_r relative to an observer, the observed pulse period P and P_0 have the following relation (without taking account of the effect of relativity):

$$\frac{P-P_0}{P} = \frac{v_r}{c}\ , \tag{1}$$

where c is the velocity of light. The time derivative of the above equation is

$$\frac{1}{P}\left(\frac{P_0}{P}\frac{dP}{dt} - \frac{dP_0}{dt}\right) \simeq \frac{1}{P}(\dot{P}-\dot{P}_0) \simeq \frac{1}{c}\frac{dv_r}{dt}\ . \tag{2}$$

In general, because of the presence of differential rotation of the Galaxy, not only $v_r \neq 0$ but also $\frac{dv_r}{dt} \neq 0$. Therefore, the problem of the effect of differential rotation on the rate of change of pulsar period is raised.

II. COMPUTATIVE FORMULAE

(1) For the Pulsars near the Galactic Plane

Let the distances of the Sun and a pulsar (whose galactic coordinates are l and b) from the galactic center be R_0 and R respectively, and let the angular velocities of the galactic rotation at the Sun and the pulsar be ω_0 and ω respectively. We suppose that the differential rotation of the Galaxy for pulsars is similar to that of Population I stars, thus the radial velocity of the pulsar relative to the Sun is

$$v_r = R_0(\omega-\omega_0)\cos b \sin l. \tag{3}$$

Since b is small, $\cos b \simeq 1$. So we can derive

$$\frac{1}{P}(\dot{P}-\dot{P}_0) = \frac{RR_0(\omega-\omega_0)^2}{c} \frac{\cos l \cos(\phi+l)}{R_0 \cos l - R\cos(\phi+l)}, \tag{4}$$

where ϕ is the angle made by the Sun and the pulsar at the galactic center. Eq. (4) shows that when $\omega=\omega_0$, $\dot{P}-\dot{P}_0=0$; when $l = \pm 90°$ or $\phi+l = \pm 90°$, $\dot{P}-\dot{P}_0 = 0$; when $l = 0°$ and $\phi+l = 0°$ or $l = 180°$, $\dot{P}-\dot{P}_0$ has maximum, and Eq. (4) is written as

$$\frac{1}{P}(\dot{P}-\dot{P}_0) = \frac{RR_0(\omega-\omega_0)^2}{C|R_0-R|} \tag{5}$$

(2). For the Pulsars with Higher Galactic Latitude

Let the orbit of a pulsar revolving round the galactic center be an ellipse with eccentricity e and semi-major axis a. We set up a polar coordinate system with the origin at the galactic center, and polar angle θ is measured from the pericenter. Let the distance of the pulsar from the galactic center be R and its velocity be v. For the convenience of computation, let the orbital plane of the pulsar be perpendicular to the galactic plane and pass through the Sun. We denote the distance of the Sun from the galactic center by R_0 and the polar angle by θ_0. Owing to differential rotation, the radial velocity of the pulsar relative to the Sun then is $v_r = -v \cos\phi$. ϕ is the angle between velocity direction of the pulsar and the straight line drawn through the pulsar and the Sun. In terms of areal velocity theorem, we obtain an expression of $\frac{dv_r}{dt}$, and then substitution of the expression into Eq. (2) gives further the expression for $\frac{1}{P}(\dot{P}-\dot{P}_0)$.

III. COMPUTATIVE RESULTS AND DISCUSSION

(1). For the Pulsar near the Galactic Plane

$R_0 = 3.08\times10^{22}$ cm and $\dot{\omega}_0 = 8\times10^{-16}$ rad $\sec^{-1}$ are known. The average velocities of the galactic rotation at different distances from the galactic center are substituted into Eq. (5) to obtain the numerical values of $\frac{1}{P}(\dot{P}-\dot{P}_0)$ for various R which are listed on the second line of Table 1 and written as $\frac{1}{P}(\dot{P}-\dot{P}_0)_I$.

This average velocity represents the velocity of centroid of stars revolving round the galactic center. The velocity of every star always deviates more or less from the average velocity at the point where the star lies. Therefore if it deviates from the average velocity by 50%, from Eq. (5) we perform the calculation once again and the results marked with $\frac{1}{P}(\dot{P}-\dot{P}_0)_{II}$ are listed on the third line of Table 1.

Table 1. The effect of differential rotation of the galaxy on $\dot{P}$ for the pulsars with low galactic latitude

R/R_0	0.03	0.07	0.1	0.2	0.3	0.4	0.5	0.6	0.7	0.8
$\frac{1}{P}(\dot{P}-\dot{P}_0)_I \times 10^{18}\ sec^{-1}$	26.0	17.0	9.7	3.3	1.5	1.4	1.0	1.0	0.8	0.4
$\frac{1}{P}(\dot{P}-\dot{P}_0)_{II} \times 10^{18}\ sec^{-1}$	59.0	39.0	23.0	8.8	4.5	4.2	3.5	4.0	3.9	3.2

R/R_0	0.9	0.93	0.96	0.99	1.01	1.05	1.1	1.2	1.3
$\frac{1}{P}(\dot{P}-\dot{P}_0)_I \times 10^{18}\ sec^{-1}$	0.2	0.2	0.1	0.03	0.01	0.1	0.1	0.2	0.2
$\frac{1}{P}(\dot{P}-\dot{P}_0)_{II} \times 10^{-18} sec^{-1}$	3.8	4.4	6.1	18.4	17.3	3.9	2.3	1.4	1.1

(2) For the Pulsars with Higher Galactic Latitude

Let $\theta_0 = 225°$, $a = R_0/3.3$, $e = 0.7$ and the velocity of a pulsar at apocenter be $v_d = 100 km\ sec^{-1}$. Then the calculated results are given in Table 2.

Table 2. The effect of differential rotation of the galaxy on $\dot{P}$ for the pulsars with high galactic latitude

θ	70°-100°	100°-200°	200°-340°	340°-70°
$\frac{1}{P}(\dot{P}-\dot{P}_0) \times 10^{18}\ sec^{-1}$	1.8	0.1	3.4	11.0

We can draw the following conclusions from the above results:

a) Because of the differential rotation of the Galaxy, even if the intrinsic period of a pulsar is constant, the observed radio pulse period would still change. The rate of change of radio pulse period caused by this effect has positive or negative value, depending on pulsar distance from the galactic center and pulsar direction.

b) For most pulsars discovered so far, this effect influnces only the last one or two digits of effective number. In general, it can be negligible.

c) For the pulsars with less $\dot{P}$ (for example $\dot{P}_0 \simeq 10^{-17}$ sec sec^{-1} or less), the effect caused by differential rotation must be considered. For some pulsars this effect may be the main reason of the change of period. This effect will be more significant with the continuous improvement of observational precision and discovering more pulsars with less $\dot{P}$.

d) It is well known that basic model for pulsars is the rotating neutron star model and the radio pulse period is equal to the rotation period of the star. Therefore, while a pulsar moves relative to the Sun, the observed radio pulse period will also change because of the continuous variation in observed direction. It is another external cause affecting the change of P. The further computation shows that the change of P caused by this effect can be negligible when it compares with that caused by the differential rotation of the Galaxy.

The authors wish to thank Chen Fusheng for his helpful discussion.

REFERENCES

[1] "The Chinese Encyclopaedia, Astronomy", 1980, 506.

[2] Куликовский П.Г., 1978, Звездная Астрономия,《Наука》, Москва, 90.

II. COMPACT STARS

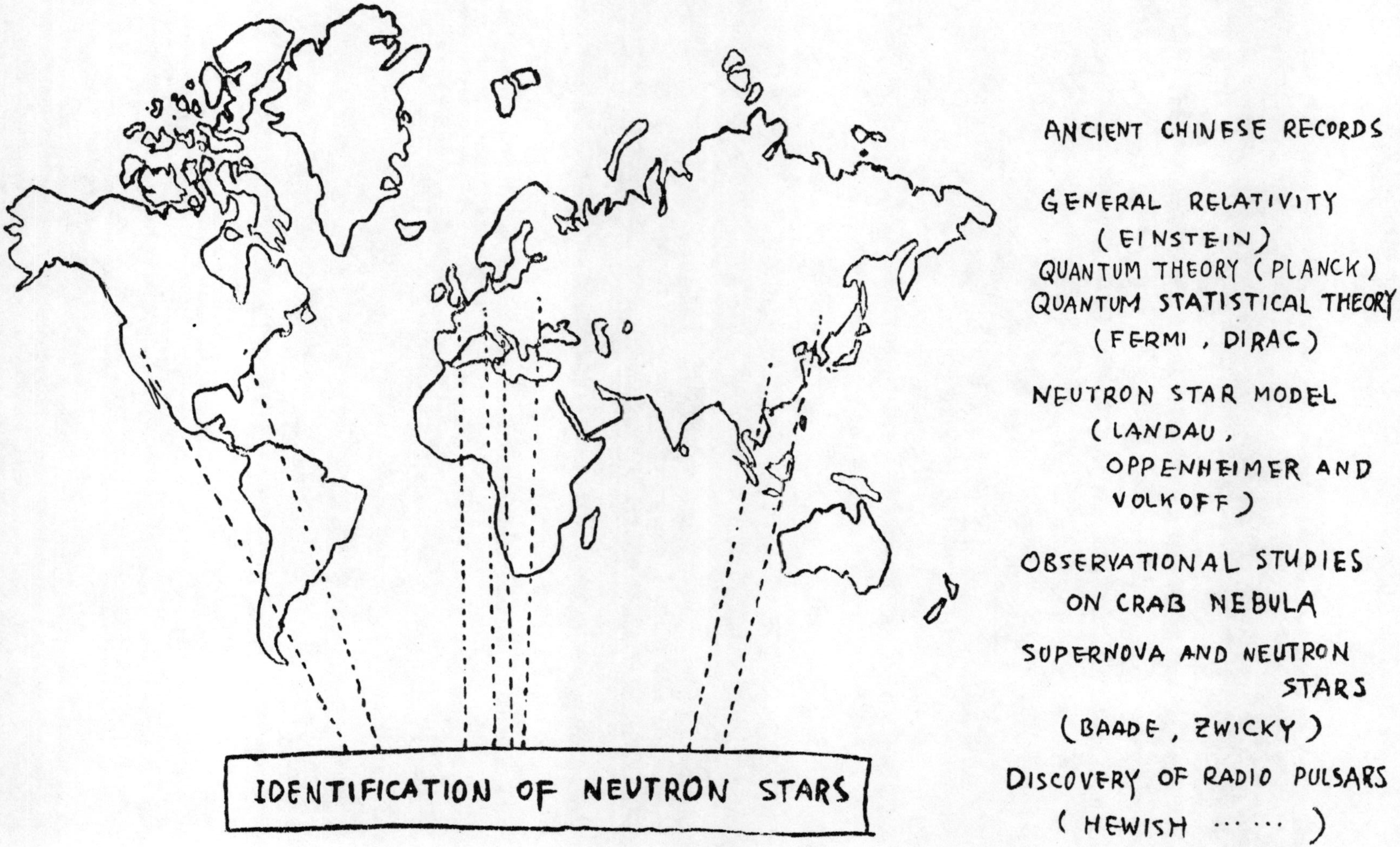
ANCIENT CHINESE RECORDS
GENERAL RELATIVITY (EINSTEIN)
QUANTUM THEORY (PLANCK)
QUANTUM STATISTICAL THEORY (FERMI, DIRAC)
NEUTRON STAR MODEL (LANDAU, OPPENHEIMER AND VOLKOFF)
OBSERVATIONAL STUDIES ON CRAB NEBULA
SUPERNOVA AND NEUTRON STARS (BAADE, ZWICKY)
DISCOVERY OF RADIO PULSARS (HEWISH ......)
IDENTIFICATION OF NEUTRON STARS

QUARK STARS AND ABNORMAL NEUTRON STARS

QU Qinyue (曲钦岳)

(Department of Astronomy, Nanjing University, Nanjing)

The stars of late stage occupy an important position in the theory of stellar evolution. The modern theory of stellar evolution regards them as some kinds of compact stars, such as white dwarfs, neutron stars and black holes. The discoveries of radio pulsars, pulsating X-ray binaries, transient X-ray sources, X-ray bursts and γ-ray bursts confirm the existence of compact stars and give challenges to the theory of stellar structure. Are there only the three categories mentioned above of compact stars? Are there any other new categories? We think it is quite important to investigate these questions. Two new categories of compact stars which may exist—quark stars and abnormal neutron stars will be discussed in this paper.

I. QUARK PHASE AND QUARK STARS

When the density of matter is much greater than the density within ordinary nuclei ($\rho_n \simeq 3\times10^{14}$ g cm^{-3}), the hadrons overlap strongly. The state of matter should be discribed by quark soup or quark gas rather than nucleons. In the interior of neutron stars, the density may be much greater than ρ_n. It is necessary to ask the following questions: Has the phase transition from nuclear matter into quark matter taken place in the interior of neutron star? Is it possible to exist a quark phase in the interior of neutron stars?

In the middle of 1970s, these problems were discussed by Baym and Chin (1976)[1] and Chapline and Nauenberg (1976)[2]. They used the MIT bag model[3,4] with two sets of parameters: $B^{1/4} = 145$ MeV, $\alpha_c = 0.55$ and $B^{1/4} = 125$ MeV, $\alpha_c = 0.75$ (in the unit of $\hbar = c = 1$). As an example, the work of Baym and Chin is introduced as follows: Let us consider N nucleons in a volume V. The corresponding quark phase has N_i quarks of flavor i (= u,d) with energy density $\rho = E/V$ given by

$$\rho = B + \sum_i \left(\frac{3}{4\pi^2} + \frac{2\alpha_c}{\pi^3}\right) P_{f_i}^4, \tag{1}$$

where $P_{fi} = (\pi^2 f_i \frac{N}{V})^{1/3}$, $f_i = N_i/N$ and $\sum_i f_i = 3$. The three terms above are the bag energy, the massless quarks' kinetic energy and the lowest order color interaction (or excharge) energy. For the quark phase of neutron matter: $f_u = 1$, $f_d = 2$. The energy per baryon in the quark phase is thus

$$E/N = BV/N + D(N/V)^{1/3} \tag{2}$$

with $D \equiv \frac{3}{4}\pi^{2/3}(1 + \frac{8\alpha_c}{3\pi}) \sum_i f_i^{4/3}$. The corresponding equation of state is

$$P \equiv -\frac{\partial E}{\partial V} = \frac{1}{3}(\rho - 4B). \tag{3}$$

In order to compare the energy per baryon in the quark matter with that in neutron matter, Baym and Chin have selected four calculations as being representative: a) Pandharipande's variational calculation employing Reid's potential (it is called PR Model for short)[5], b) Bethe and Johnson's calculation using their own model V neutron potential (BJVN Model)[6,7], c) Pandharipande and Smith's neutron solid calculation with tensor interaction (PSS

Model)[8], d) Walecka's calculation based on a relativistic mean field theory (WMF Model)[9]. The results of PR Model and WMF Model are shown in Fig. 1. The results of the remaining two, BJVN Model and PSS Model lying between PR and WMF, are not shown in Fig. 1. The curve calculated by Eq. (2) is also shown in Fig. 1, labelled by bag. As we see in Fig. 1 the quark phase has lower energy than the neutron phase at higher density. Hence, the matter should be in quark phase at higher density. The "double tangent" constructions are shown for the case of curve bag and curve PR and the case of curve bag and cure WMF in Fig. 1 as dashed lines. At the tangent points, the pressures are equal in the two phase but densities are different. The density for phase transition is obtained from that and given in Table 1.

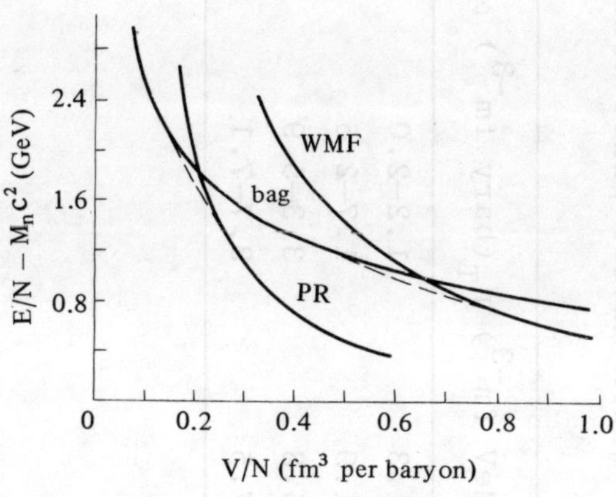

Fig. 1

In Table 1, column 1 refers to the equation of state, column 2 gives the pressure at the phase transition, column 3 and 4 delineate the transition region in terms of baryon and mass density respectively, column 5 and 6 give the maximum neutron star mass and its central density when the corresponding equation of state of column 1 is used. It can be seen by comparing column 4 with 6 that in all cases considered the neutron-quark phase transition density lies

Table 1. Neutron matter-quark matter phase transition for various models of pure neutron matter

Eqs. of state	P_T(10^3MeV fm^{-3})	n_T(bary fm^{-3})	ρ_T(10^{15}g cm^{-3})	$M_{max}/M_\odot$	ρ_c(10^{15}g cm^{-3})
WMF [8]	1.3	1.2–2.0	3.5–7.5	2.57	1.6
PSS [9]	1.9	1.7–2.6	5.4–11	2.28	1.1
BJVN [6,7]	5.8	3.3–5.9	13–31	1.76	3.3
PR [5]	7.5	3.7–7.1	14–40	1.66	4.1

above the maximum density obtainable in a stable neutron star. Hence quark matter is unlikely to be expected within these dense stellar objects. But this conclusion is not accepted generally. Collins and Perry (1975)[10] point out that the parameter α_c which characterizes the strength of quark-quark interaction will decrease in a very high density because of the screened effect of many-body on the long range interaction of quark-quark. Using the screened value of α_c (so called screened charge) Freedman and McLerran (1978)[11], Fechner and Joss (1978)[12] obtain that quark matter might occur in the interior of neutron stars. The quark model used by Fechner and Joss has two parameters: μ, which describes the strength of quark-quark interaction and B, which is the parameter of MIT bag model. They select two values of B for calculation: a) B = 56 MeV fm^{-3}, i.e. $B^{1/4}$ = 145 MeV; b) B = 0. Their approach is as follows: As the density is greater than the phase transition density ρ_T, they use the equation of state for quark matter. While the density is less than ρ_T, the equation of state for nuclear matter is used. The two equations of state are connected at the phase transition density ρ_T under the following condition: for the first order phase transition, the pressure must be continuous and for the second order phase transition, both the pressure and the density must be continuous. Three nuclear-matter equations of state, i.e. PR Model[5], BJVH Model[6,7] and TI3 Model[8] calculated by Pandharipande and Smith with their own tensor interaction potential 3, are used. The results of their calculation are given in Table 2. Column 1 and 2 are the two parameters of quark matter, column 3 is the interaction potential used in the nuclear matter model, column 4 is the type of phase transition, column 5 and 6 are the minimum and maximum densities across the phase transition respectively, column 7, 8, 9 and 10 are the maximum of the central density, mass, inertia moment and surface redshift of a stable star respectively.

It can be seen from Table 2 that there may be a kind of compact stars that has a core of quark phase by proper selecting the parameters of B and μ.

However, it is necessary to point out that above results have been obtained by means of the equation of state for quark matter given by Freedman and McLerran who use the perturbation theory of quantum colordynamics (QCD). Recently, Alvarez et al. (1981)[13] use Hartree-Fock approximation rather than the perturbation theory to calculate the equation of state for quark matter and estimate the phase transition density ρ_T from nuclear matter to quark matter. They obtain a phase transition density ρ_T greater than the maximum center density of stable stars and arrive at the conclusion that there could not be quark matter in the interior of neutron stars.

In summary, it can be seen that the phase transition density from nuclear matter to quark matter can not now be exactly determined. The reason is that QCD theory is not perfect and experiments can not provide accurate parameters at present when one discusses the properties and state of matter at very high density. Therefore some contradictory conclusions still exist on this problem.

Now let us discuss the quark stars. Quark stars were first investigated by Itoh. He discusses the gas of massive quarks without interaction and the star whose gravity is balanced by the degenerate pressure of free quarks. He obtains a maximum mass of the order of 10^{-3} $M_\odot$ for stable quark stars[14]. At present, we know that quarks are always confined in the interior of hadrons and free massive quarks can not exist. Hence, the stars consisting of free quarks can not exist. Later, Brecher and Caporaso (1976)[15], Fechner and Joss (1978)[12] obtain the quark stars by use of the bag model of quarks. However, Bowers et al. (1977)[16],

Table 2. Properties of the stellar models

B ($MeV\ fm^{-3}$)	μ (MeV)	Nuclear Potential	Phase transition type	ρ_a (10^{14} g cm^{-3})	ρ_0 (10^{14} g cm^{-3})	$\rho_{c,max}$ (10^{14} g cm^{-3})	M_{max} ($M_\odot$)	I_{max} (10^{44} g cm^2)	z_{max}
0	273	R	I	2.9	4.0	20	1.92	22	0.40
0	275	R	II	4.4	4.4	20	1.89	21	0.39
0	273	BJVH	I	3.0	4.0	20	1.92	22	0.40
0	276	BJVH	II	4.5	4.5	20	1.89	21	0.40
0	295	TI3	I	3.2	6.0	25	1.73	18	0.34
0	315	TI3	I	4.8	8.7	4.7 20	1.83 1.71	26 23	0.22 0.28
56	85	R	I	3.8	5.0	25	1.69	15	0.39
56	95	R	II	5.4	5.4	25	1.69	15	0.38
56	85	BJVH	I	3.1	4.9	25	1.69	15	0.38
56	102	BJVH	II	5.8	5.8	25	1.68	15	0.38
56	170	TI3	I	2.8	5.9	25	1.61	14	0.33
56	230	TI3	I	4.2	8.2	8.2 25	2.24 1.60	44 19	0.30 0.28

Alvarez et al. (1981)[13] dicuss the stability of quark stars and conclude that quark stars are unstable, i.e. there are no stable quark stars.

The introduction of the research on quark phase and quark stars mentioned above is not complete. The reader may refer to the review written by Bahm and Pethick (1979)[17] and the papers of Baluni[18] and Morley and Kislinger[19].

II. ABNORMAL NEUTRON STARS

(1) Abnormal Neutron Stars without Abnormal Protons

According to the theory suggested by T.D. Lee and others (1974)[18–21], the coupling of nucleon with 0^+ meson which may probably exist, will change its rest mass from m_n into $m_{eff} = m_n + g\phi$ where ϕ is the meson field and g is a coupling constant. When the nucleonic number density is low, the ground state of the meson-nucleon system would have $\phi \simeq 0$, i.e. $m_{eff} = m_n$. This is the normal nuclear state. Once the nucleonic number density $n > n_c$, the system would have $\phi \simeq -m_n/g$, i.e. $m_{eff} \simeq 0$. This is the abnormal nuclear state. There would be phase transition between normal neutron state and abnormal neutron state at the critical density n_c. When $n < n_c$ the normal neutron state is stable and the abnormal neutron state is stable when $n > n_c$. The value of n_c is a bit greater than the nucleonic number density within the normal nucleus, therefore the nucleons observed seems to remain always in the normal nuclear state. However, the nucleon density in the interior of neutron stars may be much greater than n_c, so there would be abnormal neutrons in neutron stars.

The possibility of the appearance of abnormal neutron state in dense matter was first discussed by Källman (1975)[22], Källman and Moszkowski (1975)[23], but the conclusion they drew is that there are no abnormal neutron stars[24]. In 1977, Fang Lizhi et al. cooperated with the High Energy Astrophysics Research Group of Nanjing University in the investigation on the structure and stability of the abnormal neutron star and concluded that the abnormal neutron star might be a new kind of compact stars[25]. It is well known that there are two mass limits for compact stars, one is the maximum mass of white dwarfs $\sim 1.4\ M_\odot$, another is the maximum mass of neutron stars $\sim 2.0\ M_\odot$. In paper[25], we use the homogeneous stellar model and obtain the maximum mass of about 4 $M_\odot$ for abnormal neutron stars. In 1979, Fang Lizhi and Hsieh Kuangchung discussed the inhomogeneous stellar model and revised the maximum mass to about 3.2 $M_\odot$[26]. This should affect the judgement of the existence of black hole in terms of the stellar mass. The other results obtained in paper [25] are as follows: There may exist metastable compact stars. The energy gap between stable and metastable stars is of the order of 1/100 $M_\odot$. Therefore, when phase transition of a star occurs as a whole, the released energy is of the order of 10^{52} erg. Just near the released energy of a supernovae outburst. It provides a mechanism for outburst energy sources.

(2) Abnormal Neutron Stars with Abnormal Protons

In paper [25], we assume that the abnormal neutron stars consist solely of abnormal neutrons. However, because of the difference in electromagnetic property between abnormal neutron and abnormal proton and the same value of ϕ field, there should be mass difference between abnormal neutron and abnormal proton, hence the following process of weak decay could occur.

$$N_a \longrightarrow P_a + e^- + \bar{\nu}, \qquad (4)$$

where N_a and P_a represent abnormal neutron and abnormal proton respectively. Only when the Fermi energy of electrons E_{Fe} equals the difference between the Fermi energy of abnormal neutrons E_{FN} and that of abnormal protons E_{FP}, i.e.

$$E_{FN} - E_{FP} = E_{Fe}, \tag{5}$$

the weak decay of abnormal neutrons will be forbidden, and abnormal neutrons become stable.

Because the effective mass of either abnormal neutron or abnormal proton is nearly zero, they will be relativistic when nucleonic number density is very high*. From Eq. (5) we can conclude that the number of abnormal protons can not be neglected. So it is necessary to extend the previous research and to discuss the abnormal neutron star with abnormal protons.

This problem has been investigated by the High Energy Astrophysics Research Group of Nanjing University[28–30] and the following results have been obtained.

a) There are lots of abnormal protons in an abnormal neutron star. n_P/n_N will increase with n and its maximum is about 0.35.

b) The maximum mass of the abnormal neutron star with abnormal protons are 3.0 $M_\odot$ and 3.1 $M_\odot$ for $\alpha = 0.60$ f and $\alpha = 0.62$ f respectively.

c) There might be abnormal neutron stars which entirely consist of abnormal neutrons (including abnormal protons and electrons), but there might be no abnormal neutron stars whose cores consist of abnormal neutrons while their outer layers consist of normal neutrons. Hence the change must be global rather than local when a phase transition takes place between a normal neutron star and an abnormal neutron star. This process may probably provide an explosive release of

---

* Similar result has been obtained for relativistic case in normal neutron stars[27].

energy of $10^{52} - 10^{53}$ erg. This mechanism may probably explain the reexplosion of supernovae[31]. There would be a tremendous emission of neutrinos and electromagnetic radiation when a normal neutron star changes into an abnormal neutron star.

Some authors have different views on the abnormal nuclear state. Our opinion is that it is worthwhile to investigate the abnormal neutron star when there is no full evidence to show that abnormal nuclear state can not exist. Futhermore, if abnormal nuclear state does exist, there may be many abnormal neutron stars among several hundreds of pulsars and X-ray binaries.

III. QUARK - ABNORMAL NEUTRON STARS

Recently, the High Energy Astrophysics Research Group of Nanjing University has discussed another interesting problem[32]: Is it possible for quark phase to exist in the interior of abnormal neutron stars? Is it possible to exist the quark-abnormal neutron star of which the interior is in the quark state and the outer layer is in the abnormal neutron state?

Using the screened value of α_c, Freedman and McLerran (1978)[11] obtain the relation among pressure, energy density and baryon number density for two models of quark matter. The parameters of Model I are: B = 56 MeV fm^{-3}, μ_0 = 100 MeV (screened charge $\alpha_c = 1$ at μ_0) and the parameters of Model II are: B = 0, μ_0 = 275 MeV. There is only a very little difference between the results of two models. So it is good enough to only discuss Model I in this paper.

Table 3 is taken from paper [11], where column 1 is the baryon number density, column 2 is the energy per baryon, ρ is the energy density and column 3 is the pressure.

Table 3. The pressure, energy density and number density relations

$n(fm^{-3})$	ρ/n(MeV)	P(MeV fm^{-3})	$n(fm^{-3})$	ρ/n(MeV)	P(MeV fm^{-3})
0.100	952	0.7	0.888	1117	207
0.150	956	1.8	0.985	1142	246
0.200	961	3.7	1.20	1194	338
0.250	966	6.8	1.38	1234	420
0.300	972	11.4	1.57	1276	514
0.347	978	20.0	1.79	1318	620
0.402	988	35.5	2.02	1362	740
0.494	1010	63.7	2.27	1405	876
0.597	1037	98.1	2.53	1449	1028
0.713	1070	139	2.72	1479	1138
0.796	1093	171	3.02	1523	1320

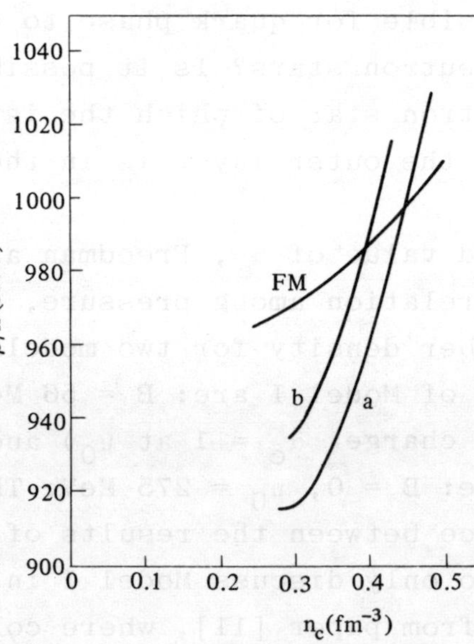

Fig. 2 The relation of ρ/n-n.

a—the case of α=0.60 f

b—the case of α=0.62 f

Fig. 2 shows ρ/n against the baryon number density n. Curve FM is drawn on the basis of Table 3. Curve a and curve b refer to α= 0.60 f and α = 0.62 f respectively for abnormal neutron state with abnormal protons according to the Eqs. (7)–(9) in paper[29]. It can be seen from Fig. 2:

a) Curve a crosses with curve FM at n_1 = 0.443 fm^{-3} and curve b crosses with curve FM at n_2 = 0.39 fm^{-3}.

b) For the case of α = 0.60, f, when $n< n_1$ the energy per baryon of abnormal neutron state is less than that of quark state at the same number density and vice versa when $n>n_1$. This result means that from the view of energy matter should be in abnormal neutron state when $n<n_1$ and in quark state when $n>n_1$.

c) For the case of α = 0.62 f, the result is similar but changes n_1 into n_2.

The fundamental differential equations determining the structure of a static and spherically symmetric star are

$$\frac{dP(r)}{dr} = - \frac{GM(r)\rho(r)}{r^2}\left[1+\frac{P(r)}{\rho(r)}\right]\left[1+\frac{4\pi r^3 P(r)}{M(r)}\right]\left[1-\frac{2GM(r)}{r}\right]^{-1} \tag{6}$$

$$\frac{dM(r)}{dr} = 4\pi r^2 \rho(r) \tag{7}$$

In order to construct a stellar model with interior in quark state and outer layer in abnormal neutron state, we use the following method: Taking a certain central number density (greater than n_1 or n_2) and using the relation (p, ρ) given in Table 3, we integrate the differential Eqs. (6) and (7) by numerical method. The number density will continuously decrease as r increases. As soon as the number density $n<n_1$ (or n_2) for α = 0.60 f (or α = 0.62 f), we use the equation of state for abnormal neutron state to replace that of quark state and keep the pressure continuous. After changing the equation of state, we continue to integrate the differential Eqs. (6) and (7) by numerical method until

$p = 0$. The R and M corresponding to $p = 0$ are the radius and mass of the whole star. 80 models are calculated for the two cases of α(α=0.60 f, 0.62 f) by the similar numerical method mentioned in paper[32]. A part of the results is shown in Tables 4 and 5 and Figs. 3 – 6. In Tables 4 and 5, column 1 is the center number density n_c, columns 2 and 3 are the mass M and radius R of a star, column 4 is the ratio of the mass of quark matter M_q to the total mass M, column 5 is the ratio of the radius of quark matter R_q to the radius of the whole star R and column 6 is the gravitational redshift at stellar surface. The relations between M and n_c, M and R, Mq/M and n_c, and z and M are shown in Figs. 3–6, respectively.

Table 4. The parameters of quark-abnormal neutron stars (α = 0.60 f)

$n_c(fm^{-3})$	$M(M_\odot)$	R(km)	M_q/M	R_q/R	z
0.50	1.52	11.2	0.09	0.38	0.29
0.55	1.58	11.2	0.17	0.48	0.31
0.60	1.63	11.2	0.25	0.54	0.33
0.70	1.70	11.0	0.36	0.61	0.36
0.80	1.73	10.8	0.44	0.65	0.38
0.90	1.75	10.7	0.50	0.68	0.40
1.00	1.76	10.5	0.54	0.70	0.41
1.10	1.76	10.4	0.57	0.71	0.42
1.20	1.75	10.2	0.60	0.72	0.42
1.30	1.74	10.1	0.62	0.73	0.43
1.40	1.73	9.97	0.64	0.73	0.44
1.50	1.72	9.84	0.65	0.74	0.44
1.60	1.71	9.73	0.66	0.74	0.44

Table 5. The parameters of quark-abnormal neutron stars (α = 0.62 f)

$n_c(fm^{-3})$	$M(M_\odot)$	R(km)	M_q/M	R_q/R	z
0.40	1.08	10.3	0.02	0.24	0.21
0.45	1.22	10.5	0.16	0.48	0.24
0.50	1.34	10.6	0.28	0.58	0.26
0.60	1.50	10.7	0.44	0.68	0.31
0.70	1.60	10.6	0.53	0.72	0.34
0.80	1.65	10.5	0.59	0.75	0.37
0.90	1.68	10.4	0.64	0.77	0.39
1.00	1.70	10.2	0.67	0.78	0.40
1.10	1.70	10.1	0.69	0.79	0.41
1.20	1.70	9.97	0.71	0.80	0.42
1.30	1.70	9.85	0.73	0.80	0.43
1.40	1.69	9.73	0.74	0.81	0.44
1.50	1.68	9.61	0.75	0.81	0.44
1.60	1.67	9.51	0.76	0.82	0.44

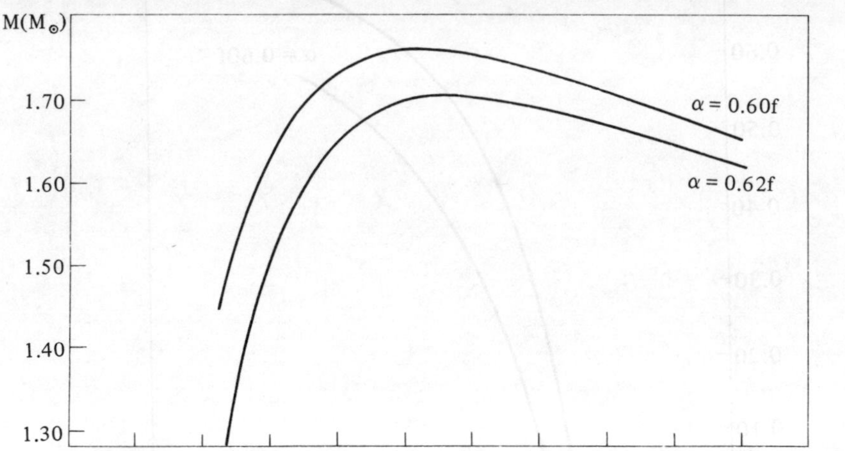

Fig. 3 The relation of M-n_c.

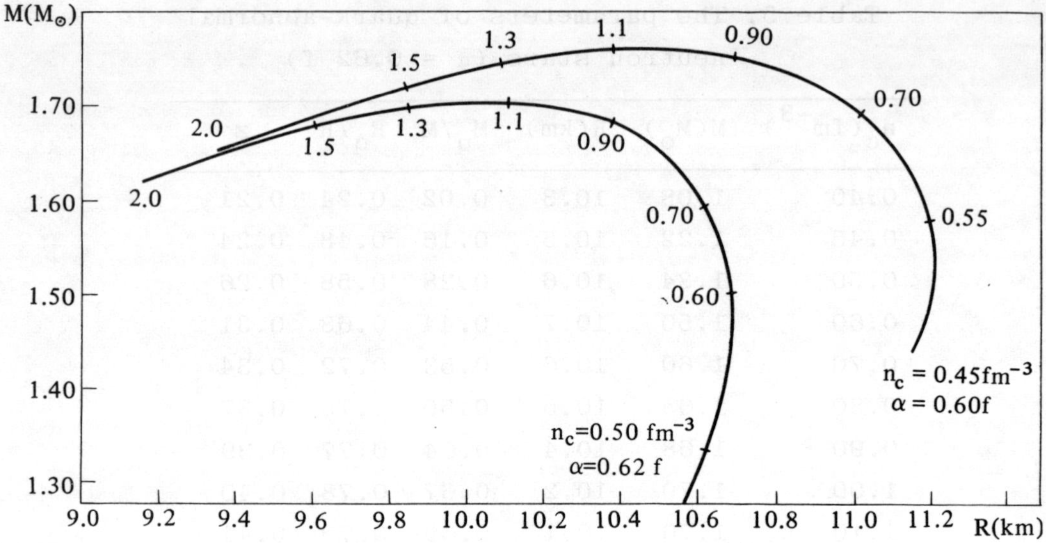

Fig. 4 The relation of M—R

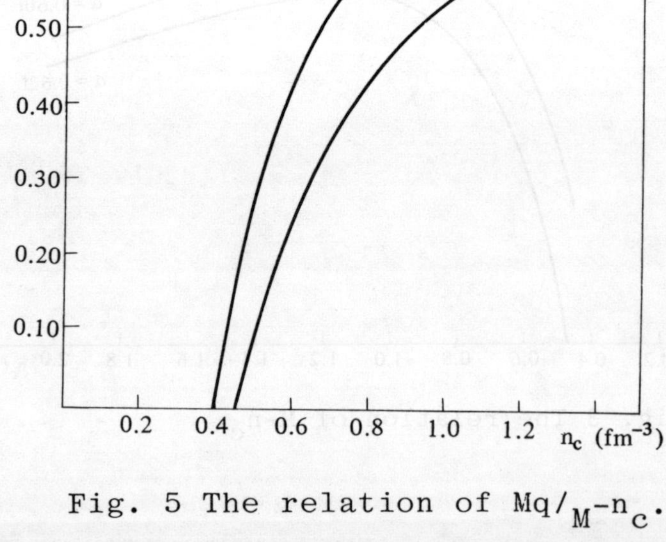

Fig. 5 The relation of $Mq/_M-n_c$.

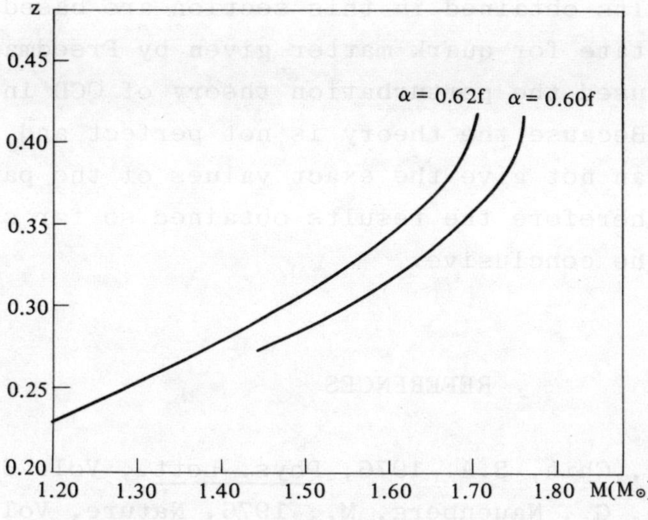

Fig. 6 The relations of Z—M.

It can be seen from Fig. 3 that the mass of quark-abnormal neutron stars increases with the center number density until the maximum mass and then decreases. This kind of curves generally means that the stars are stable before the maximum and unstable against gravitation after the maximum. It can be seen from Fig. 4 that the direction of increase of center number density is counter-clockwise in the M-R curve. According to paper[33], the stars change from stable to unstable as M(R) passes through the maximum in a counter-clockwise direction. It is pointed out further that the stars before maximum mass are stable. The maximum masses of quark-abnormal neutron stars are 1.76 $M_{\odot}$ and 1.70 $M_{\odot}$ for α = 0.60 f and α = 0.62 f respectively. It can be seen from Tables 4 and 5 that the ratio M_q/M increases with center number density and tends to 60—70% as M approaches to the maximum. Besides, the ratio Rq/R increases with center number density and tends to 70—80% as M approaches to the maximum.

The results obtained in this section are based on the equation of state for quark matter given by Freedman and McLerran who used the perturbation theory of QCD in their calculation. Because the theory is not perfect and the experiments can not give the exact values of the parameters at present, therefore the results obtained so far can not be regarded as the conclusive.

REFERENCES

[1] Baym, G., Chin, S.A., 1976, Phys. Lett., Vol. 62B 241.

[2] Chapline, G., Nauenberg, M., 1976, Nature, Vol. 264, 235.

[3] Chodos, A. et al., 1974, Phys Rev., Vol. D9, 3471.

[4] De Grand, T. et al., 1975, Phys. Rev., Vol. D12, 2060.

[5] Pandharipande, V. R., 1971, Nucl. Phys.,Vol. A178, 123.

[6] Bethe, M. A., Johnson, M. B., 1974, Nucl. Phys., Vol. A230, 1.

[7] Malone, R. C. et al., 1973, Ap. J., Vol. 199, 741.

[8] Pandharipande, V. R., Smith, R. A., 1975, Nucl. Phys., Vol. A237, 507.

[9] Walecka, J.D.,1974, Ann Phys., Vol. 83, 491.

[10] Collins, J.C., Perry, M.J., 1975, Phys. Rev. Lett., Vol. 34, 1353.

[11] Freedman, B., McLerran, L., 1978, Phys. Rev., Vol. D17, 1109.

[12] Fechner, W.B., Joss, P.C., 1978, Nature, Vol 274, 347.

[13] Alvarez, E., Ibáñez, J., 1981, Astron. Astrophys., Vol. 98, 390.

[14] Itoh, N., 1970, Prog. Theor. Phys., Vol. 44, 291.

[15] Brecher, K., Caporaso, G., 1976, Nature, Vol. 259, 377.

[16] Bowers, R.L. et al., 1977, Ap. J., Vol. 213, 840.

[17] Baym, G., Pethick, C., 1979, Ann. Rev. Astron. Astrophys. Vol. 17, 415.

[18] Baluni, V., 1978, Phys. Rev., Vol. D17, 2092.

[19] Morley, P.D., Kislinger, M.B., 1979, Phys. Rep., Vol. 51, 64.

[20] Lee, T.D., Wick, G.C., 1974, Phys. Rev., Vol. D9, 2291.

[21] Lee, T.D., 1975, Rev. Mod. Phys., Vol. 47, 267.

[22] Källman, C.G., 1975, Phys. Lett., Vol. 55B, 178.

[23] Källman, C.G., Moszkowski, S.A., 1975, Phys. Lett., Vol. 57B, 183.

[24] Moszhowski, S.A., Källman, C.G., 1977, Nucl. Phys., Vol. A287, 495.

[25] Fang Lizhi, Qu Qinyue, Wang Zhenru, Lu Tan, Luo Liaofu, 1978, Scientia Sinica, No. 4, 406.

[26] Fang Lizhi, Hsieh Kuangchung, 1979, Kexue Tongbao, Vol. 24, 167.

[27] Zeldovich, Y.B., Novikov, I.D., 1971, "Relativistic Astrophysics", Vol. 1, University of Chicago Press, Chicago, 189.

[28] Qu Qinyue, Wang Zhenru, Chu Yi, Lu Tan, 1981, Kexue Tongbao, Vol. 26, 1436.

[29] Wang Zhenru, Qu Qinyue, Lu Tan, Chu Yi, Commun. Theor. Phys.(to be published).

[30] Qu Qinyue, Wang Zhenru, Commun.Theor. Phys., (to be published).

[31] Wang Jianmin, 1979, Publ. Beijing Astron. Obs., No. 1, 69.

[32] Qu Qinyue, Wang Zhenru, Zhao Jiabing, (to be published).

[33] Bardeen, J.M., Thorne, K.S., Meltzer, D.W., 1966, Ap. J., Vol. 145, 505.

X-RAY OBSERVATIONS BY THE EINSTEIN SATELLITE AND NEUTRON STAR COOLING

Sachiko Tsuruta

(Department of Physics, Montana State University, Bozeman, Montana, U.S.A.)

I. INTRODUCTION

Among the various important contributions by the Einstein observatory is the effort to detect X-ray point sources in supernova remnants and radio pulsars. Comparison of the observed temperatures (or their upper limits) with neutron star cooling and heating theories may give valuable insight to the structure of neutron stars. It may also give profound effect on the current theories of stellar evolution. After the recent successful completion of the Einstein mission, therefore, it may be timely to give a brief review of where this important problem stands today. A more detailed account will be found elsewhere[1,2].

In the next section we discuss observations of these point sources. Their possible theoretical implications are considered in Sect. III.

II. NEUTRON STAR DATA FROM EINSTEIN OBSERVATORY

In this report, we confine our subject to the observing programs whose aim is the attempt to measure (or set upper limits on) the surface temperature of isolated neutron stars

Among various instruments on board the Einstein, those which are especially important for the present problem are IPC (Imaging Proportional Counter) and HRI (High Resolution Imaging Detector). The IPC has a spatial resolution of a few arc minutes and moderate spectral resolution, while the HRI has a spatial resolution of a few arc seconds and no spectral resolution. These values should be compared with the best spatial resolutions which were available before the Einstein, that is, ~20 minutes or so. Moreover, sensitivity of the Einstein is ~1000 times that of the earlier detectors. The energy window is 0.1—4 keV. From the above information, it is clear that the Einstein Observatory was the most ideal telescope to date, for the investigation of very weak, soft X-ray point sources, such as neutron stars and quasars. We refer to other appropriate references[3] for detailed discussion of the Einstein instrumentation and other exciting observations.

The results for point source observations in supernova remnants (SNR's)are summarized in Table 1. The names of SNR's are given in the 2nd column. The next three columns give the estimated distances, diameters and ages whenever known[2]. We note that they are in the average at a distance of ~0.4—8 kpc (kiloparsec), with a diameter of ~1—60 pc (1 pc = 3× 10^{18} cm), and about ~300—10000 years old.

Point sources are detected in five of these SNR's, marked as (*). For these sources, the last column gives surface temperatures (observed at infinity), assuming that all of the non-pulsed portion of the point source radiation comes from a neutron star surface[4-7]. For the other sources, the last column gives the 3σ level upper limits to the surface temperatures, assuming the presence of a neutron star in these remnants[8-11]. These values are obtained by assuming blackbody radiation and by using "apparent" stellar radius of ~10 – 18 km, which corresponds to a variety of "realistic" equations of state and mass range of neutron stars. Measured values of n_0, interstellar density, are used whenever possible, but otherwise, it is assumed that $n_0 = 0.3–1\ cm^{-3}$.

Table 1. Some characteristics of supernova remnants observed by the Einstein Observatory, for possible detection of the thermal X-ray emission from neutron stars

	Supernova Remnant	Distance (kpc)	Diameter (pc)	Age (years)	Temperature (10^6K)
(1)	Cas A	2.8	2	300	1.5–1.7
(2)	Kepler	8.0	–	375	1.8–2.2
(3)	Tycho	3.0	6	407	1.2–1.5
(4)	CTB 80(*)	3.0	–	(575?)	1.2–1.6
(5)	3C 58 (*)	8.0	10–15	(800?)	2.0–2.4
(6)	Crab (*)	1.7–2	1	925	2.0–2.4
(7)	SN 1006	1.2	4.4	973	0.65–0.8
(8)	RCW 103(*)	3.3	–	{~1000 - 2000}	1.7–2.2
(9)	RCW 86	2.5	–	1794	1.4–1.8
(10)	W 28	2.3	10	3400	1.6–2.0
(11)	G 350.0–18	4.0	35	8000	1.8–2.2
(12)	G 32.7–0.2	4.8	35	~1,000	2.0–2.4
(13)	Vela X (*)	0.4	20–40	1,200	0.8–1.5

The origin of the point source emission remains unclear, especially for the Crab, where > ~99% of the point source X-rays are pulsed[5]. No X-ray pulsations are detected in other remnants. However, their point source flux is generally too low to detect period modulation[4,7]. In the absence of spatially resolved spectral data for the point source emission and with lack of information for temporal variation, we are yet to prove that the point source detection indeed means measurement of stellar temperatures. In this sense, the above "detection" should be, strictly speaking, still regarded as "upper limits". However, it is already important in the sense that it strongly suggests the existence of an active neutron star in these remnants (marked as (*)). Indeed, whenever detailed information was available, it was shown that the extended nebula region of these SNR's generally exhibit brightness distribution centrally peaked toward the point source, power-law spectra, and significant degree of polarization[4]. In this sense, these remnants may be called "Crab-like". On the other hand, the nebula distribution of the other remnants in Table 1 is "shell-like", and no point sources are detected. For these remnants, the last column gives true upper limits, in the sense that a point source should have been detected if a neutron star of that temperature or higher were present.

The Columbia group investigated by the IPC the possible existence of point thermal X-ray sources at the location of ~35 radio pulsars[8]. They are chosen as favorable candidates for detection, due to their relatively short distance, younger age and clearer environment. Their distance and period are in the range of ~150–170 pc and ~0.03–1.7 sec, respectively. The age of these sources is somewhat uncertain. If we estimate from their period and rate of increase of period, they are in the range of $\sim 10^3$–10^7 years. The results[8] are summarized below. The upper limits are in the range of $\sim(2–3) \times 10^5$ K to a few million degrees. A general trend of decreasing temperature with increasing pulsar period is noted. Positive detection was obtained for six sources, the Crab, Vela, and PSR's 0355, 1055, 1642, and 0031. The detected temperatures for the faster pulsar, Crab and Vela, are a few million degrees (Table 1). Those for medium rate pulsars ($\sim$0.1 sec range), PSR's 0355 and 1055, are around a million degrees. For PSR 1642 with the 0.388 sec period, it is only $\sim 2 \times 10^5$ K, while an older pulsar PSR 0031 with about 1 sec period is as hot as $\sim(3–7) \times 10^5$ K. In order to arrive at the above temperature values or upper limits, they used a typical neutron star of $1.3M_{\odot}$ and 16 km radius (18 km "apparent" radius), and interstellar density of $n_0 = 0.3–1\ cm^{-3}$ [8]. (The temperatures quoted above refer to local values at the star.)

In the next section, let us consider the possible significance of these observations by comparing them with current theories.

III. COMPARISON WITH THEORIES AND CONCLUSION

It is now generally considered that a neutron star is formed through a supernova explosion. In the absence of residual fuels, the collapsed core just keeps cooling after the explosion. The time history of the stellar temperature is found by solving simultaneously the basic stellar structure and evolution equations (hydrostatic and thermodynamic) together with the appropriate supplementary equations (equations of state, opacities, and composition). (See, e.g., reference[1] for details.)

We have carried out such calculations by using the best physical input currently available[1], as follows: The fully general relativistic equations[12] for stellar structure and evolution are solved by using a Henyey-type stellar evolution code[13]. The effects of crystallization[14] are taken into account. Neutrino emissivities included are: modified Urca, neutron-neutron and neutron-proton bremsstrahlung[15], electron-ion bremsstrahlung[16], and plasmon, pair and photoneutrino emissivities[17]. The effects of superfluid protons and neutrons on neutrino emissivities and specific heats are taken into account by using the work of Takatsuka[18], Maxwell[19], and Chao, Clark and Yang[20]. We set the effective mass of neutrons and protons to be 0.8[19]. We use the latest Los Alamos table[21] for iron radiative opacities and the work of Flowers-Itoh and Yakovlev-Urpin[22] for conductive opacities. The effect of magnetic fields on opacities is taken into account by adopting the approach of reference [23].

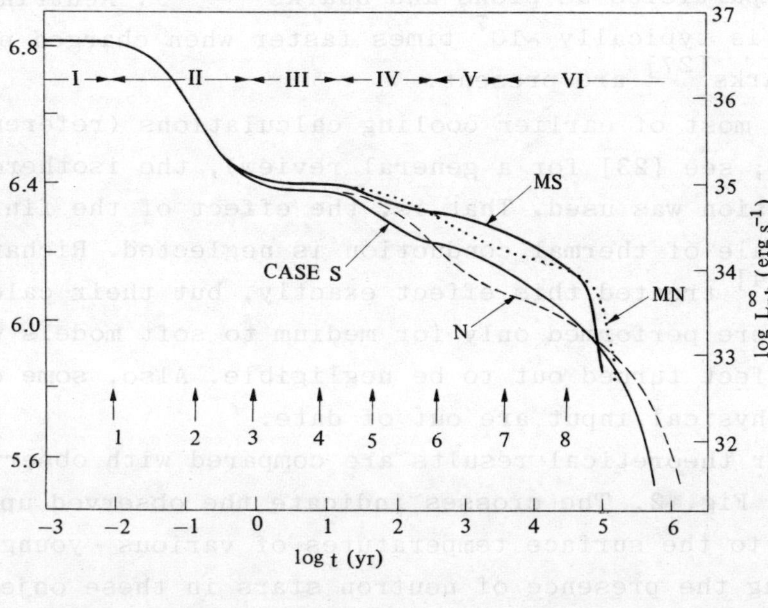

Fig. 1 The surface temperature T_e and photon luminosity L^{∞}_{ph} (both observed at infinity) are shown as a function of age (measured at infinity) for the PS model with $1.3M_{\odot}$.

Our results are summarized in Fig. 1 for the model with a stiff equation of state by pandharipande and Smith[24], generally called the PS model, where: the gravitational mass is $M_G = 1.3M_{\odot}$, the stellar radius $R = 16$ km (the "apparent" radius $R^{\infty} = 18.5$ km), and the central density $\rho^c = 4.0 \times 10^{14}$ g cm^{-3}. The "standard" cooling curves are shown for four cases: Case S with superfluid nucleons and Case N with non-superfluid (normal) nucleons, both for the zero magnetic field; and Cases MS(superfluid) and MN(normal) for the magnetic field of $H = 4.4 \times 10^{12}$ gauss. Similar curves are obtained for the soft model of the Reid potential type, generally called the BPS model[25], with $M_G = 1.3M_\odot$, R=8.0 km (R^∞ = 11 km), and $\rho^c = 3.2 \times 10^{15}$g cm^{-3}. They are essentially the same as our earlier results[23,26]. The "standard" cooling means cooling in the absence of such exotic particles as pions and quarks[23,26]. Neutrino emissivity is typically $\sim 10^7$ times faster when charged pions[23] and quarks[27] are present.

In most of earlier cooling calculations (references [26-28]; see [23] for a general review), the isothermal approximation was used. That is, the effect of the finite time scale of thermal conduction is neglected. Richardson et al[14] treated this effect exactly, but their calculations were performed only for medium to soft models where such effect turned out to be negligible. Also, some of their physical input are out of date.

Our theoretical results are compared with observational data in Fig. 2. The crosses indicate the observed upper limits to the surface temperatures of various young SNR's assuming the presence of neutron stars in these objects. The circles refer to possible "detection" of point sources (as marked (*) in Table 1). These points are accompanied by the estimated error bars which are mostly due to uncertainties in interstellar absorption and source distance.

Our conclusion is summarized below.

a) Some of the observed upper limits (for Cas A, SN 1006, and probably Tycho) are below the "standard" theoretical curves. They are all safely above the "non-standard" cooling curves with charged pions[23] or quarks[27]. A natural interpretation is that there are no neutron stars in these SNR's, or that these stars contain such exotic particles as charged pion condensates and quarks. Since Tycho is thought to be the remnant of a Type I supernova, the present result is consistent with the detonation/ deflagration-type explosion model for Type I supernova which leaves no neutron stars[29].

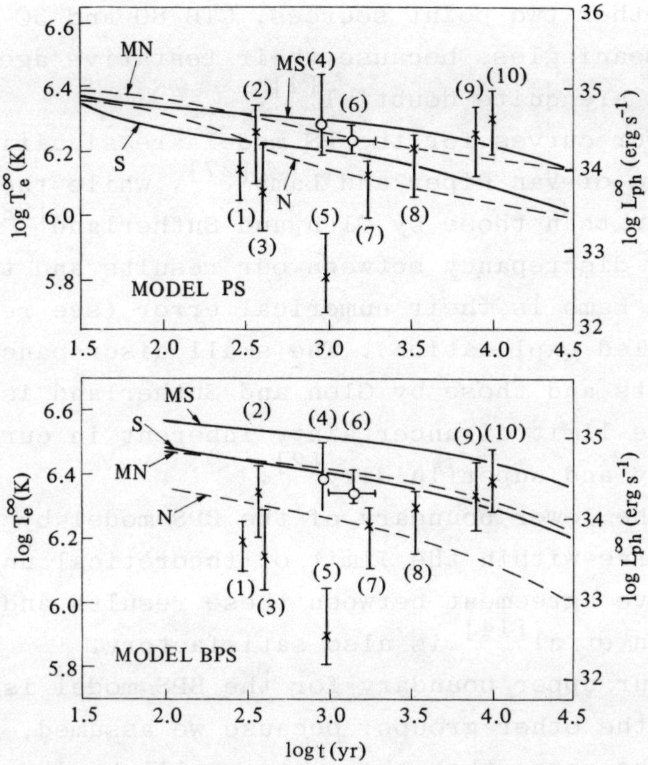

Fig. 2 Comparison between the observational results from the Einstein Observatory and the theoretical curves for "standard" cooling of the PS model (upper half) and the BPS model (lower half). T_e^{∞} (left) and L_{ph}^{∞} (right) are shown as a function of age (measured at infinity). The numbers refer to the observational data of: (1) Cas A, (2) Kepler, (3) Tycho, (4) Crab, (5) SN 1006, (6) RCW 103, (7) RCW 86, (8) W 28, (9) G350.0–18, and (10) G22.7–0.2.

b) The "detection" points for RCW 103, vela, and the crab are consistent with these "standard" theoretical curves.

For the other two point sources, CTB 80 and 3C 58, comparison is meaningless because their tentative ages suggested (Table 1) are quite doubtful[4].

c) Our curves for the PS model are significantly higher than those of Van Riper and Lamb[27], while they are somewhat lower than those by Glen and Sutherland[28]. A major source of discrepancy between our results and those by Van Riper and Lamb is their numerical error (see reference[2] for detailed explanation). The small discrepancy between our results and those by Glen and Sutherland is acceptable within the limit of uncertainty inherent in current theories of opacity and superfluidity[2].

d) The lower boundary of the BPS model by these three groups agree within the limit of theoretical uncertainties. Qualitative agreement between these results and those by Richardson et al[14] is also satisfactory.

e) Our upper boundary for the BPS model is higher than those of the other groups, because we assumed, as may very well be the case, that the protons will be in a state of p-wave superfluidity in the high density region where $\rho = \sim 10^{15}\text{-}5 \times 10^{15}$ g cm^{-3} [23].

f) Van Riper and Lamb reached essentially the same conclusion as ours for SN 1006 and Tycho.

g) For the Crab and Cas A, however, the conclusion Van Riper and Lamb presented contradicts ours, mainly because observed points for these sources lie above their "standard" curves. The major reason for this discrepancy is that they misplaced these observed points (from the positions given in the references they quote)(see Fig. 3 in reference[27]. In the above comparison, we used the luminosity scale because the BPS temperature scale of Van Riper and Lamb is in error.

h) For medium to soft models, the effects of a finite time scale of thermal conduction are negligible, during the period of our interest ($t > \sim 100$ years)[1, 14].

i) For the stiff model these effects are important. For instance, for the PS model of mass $1.3M_{\odot}$, it takes 4000 years for Case S and ~1500 years for Case N before the isothermal state is reached[1]. Many young SNR's (Cas A, Kepler, Tycho, the Crab, SN 1006, RCW 103, and W 28) have an age of a few thousand years or less. Therefore, for the stiff model, exact stellar evolution calculations are desirable if we are to make relevant quantitative comparisons between theory and the greatly improved observational data.

j) The observed temperature of some of the older radio pulsars, e.g. PSR 0031, is far above the theoretical cooling curves[8]. Some heating mechanisms[8,23] should be working in these sources.

k) It is desirable that in our future theoretical work more realistic equations of state and better treatment of magnetic fields become possible. The equations of state recently developed by the Illinois group[30] may be one step toward that direction. Further careful studies of "non-standard" cooling are also desirable. Such investigations are currently in progress by the Kyoto group[31].

l) If future observational progress leads to further lowering of the upper limits, decrease of error bars, and measurement of neutron star surface temperatures (by e.g. better spectral, spatial and temporal studies of point sources), the effect will be far reaching.

ACKNOWLEDGEMENTS

We thank many of our colleagues for interesting and valuable discussion. Our special thanks are due to Drs. G. Fabbiano, F.R. Harnden, D.J. Helfand, S.S. Murray, P. Gorenstein, and F.D. Seward, for the carefully checked observational information. The manuscript was prepared in ISAS (the Institute of Space and Astronautical Science). We are grateful to Professors M. Oda, Y. Tanaka, and other members of ISAS, for hospitality and help.

REFERENCES

[1] Nomoto, K., Tsuruta, S., 1981, Ap. J., Vol. 250, L19. Nomoto, K., Tsuruta, S. 1982 (in preparation).

[2] Tsuruta, S., 1982, to appear in Comments on Astrophysics.

[3] Giacconi, R. et al., 1979, Ap. J., Vol. 230, 540. Giacconi, R., 1981, "Proceedings of the 15th ESLAB Symposium on X-Ray Astronomy".

[4] Becker, R.H., Helfand, D.J., Szymkowiak, A. E., 1981, CAL Preprint, No. 206.

[5] Harnden, Jr., F.R., Hertz, P., Gorenstein, R., Grindley, J., Schreier, E., Seward, F.D., 1979, B.A.A.S., Vol. 11, 424. Harnden Jr., F. R., Buchler, B., Glacconi, R., Grindley, J., Hertz, P., Schreier, E., Seward, F. D., Tananbaum, H., Van Speybroeck, L., 1979, B.A.A.S., Vol. 11, 789.

[6] Fabbiano, G., 1980, "X-Ray Astronomy", Giacconi, R., Setti, G. (eds.), Reidel, Dordrecht; Fabbiano, G., 1981, private communication.

[7] Tuoy, I., Garmire, G., 1980, Ap. J., Vol. 239, L107.

[8] Helfand, D. J., Chanan, C. A., Novick, R., 1980, Nature, Vol. 283, 337; Helfand, D. J., 1981, private communication.

[9] Pye, J. P., Pounds, K. A., Rolf, D. P., Seward, F. D., Smith, A., Willingale, R., 1981, M.N.R.A.S., Vol. 194, 569.

[10] Murray, S. S., Fabbiano, G., Fabian, A. C., Epstein, A., Giacconi, R. 1979, Ap. J., Vol. 234, L69; Murray, S. S., 1981, private communication.

[11] Seward, F. D., 1981, private communication.

[12] Thorne, K. S., 1977, Ap. J., Vol. 212, 825.

[13] Sugimoto, D., 1970, Ap. J., Vol. 159, 619.

[14] Richardson, M.B, Van Horn, H. M., Ratcliff, K. F., Malone, R.C., 1981, preprint.

[15] Friman, B. L., Maxwell, O.V., 1979, Ap. J., Vol. 232, 541.

[16] Soyeur, M., Brown, G. E., 1979, Nucl. Phys., Vol. A 324, 464.

[17] Beaudet, G., Petrosian, V., Salpeter, E. E., 1967, Ap. J., Vol. 150, 9791.

[18] Takatsuka, T., 1972, Prog. Theor. Phys., Vol. 48, 1517.

[19] Maxwell, O. V., 1979, Ap. J., Vol. 231, 201.

[20] Chao, N. C., Clark, J. W., Yang, C. H., 1972, Nucl. Phys.,Vol. A 179, 320.

[21] Huebner, W. F., Merts, A. L., Magee Jr., N.H., Argo, M. F., 1977 , Los Alamos National Laboratory Report AL–6760–M.

[22] Flowers, E. G., Itoh, N., 1976, Ap. J., Vol. 206, 218; Flowers, E. G. 1979, Ap. J., Vol. 230, 847; Yakovlev, D. G., Urpin, V. A., 1980, Sov. Astr ., Vol. 24, 303.

[23] Tsuruta, S., 1979, Phys. Rept., Vol. 56, 237.

[24] Pandharipande, V.R., Pines, D., Smith, R.A., 1976, Ap. J.,Vol. 208, 550. (PS Model corresponds to Model (B) in references [23].)

[25] See, for instance, Baym, G., Pethick, C. J., Sutherland, P., 1971, Ap. J., Vol. 170, 299. (BPS Model corresponds to Model (A) in reference [23].)

[26] Tsuruta, S., 1980, "X-Ray Astronomy", Giacconi, R., Setti, G., (eds.), Reidel, Dordrecht, 73; Tsuruta, S., 1981, "Pulsars", Wielebinski, R., Sieber, W., (eds.), Reidel, Dordrecht, 331.

[27] Van Riper, K. A., Lamb, D. Q., 1981, Ap. J., Vol. 244, L13.

Note: Recently, discovery of two more pulsars in SNR has been reported[32,33].

[28] Glen, G., Sutherland, P. G., 1980, Ap. J., Vol. 239, 671.

[29] Nomoto, K., 1981, "Fundamental Problems in the Theory of Stellar Evolution", Sugimoto, D., et al., (eds.), Reidel, Dordrecht, 210.

[30] Friedman, B., Pandharipande, V. R., 1981, preprint.

[31] Tamagaki, R., 1981, "Proceedings of the 2nd International Conference on Recent Progress in Many Body Theories",

[32] Fahlman, G. G., Gregory, P. C., 1981, Nature, Vol. 293, 202.

[33] Seward, F. D., Harnden Jr., F.R., 1982, CFA preprint, No. 1604, submitted to Ap. J. Letters, March 1982.

THE X-RAY BURST SOURCES

LI Qibin(李启斌)

(Beijing Observatory, Academia Sinica, Beijing)

I. BRIEF HISTORY OF DISCOVERY

In December 1975, Grindley and Heise[1,2] discovered the first set of X-ray bursts which drew the attention of astronomers as an important class of celestial phenomena. Those X-ray bursts were detected from NGC 6624 on 28 September 1975 with the Astronomical Netherlands Satellite. Subsequently, using existing SAS-3 data from 1975, Clark et al. found a series of ten bursts from the source in NGC 6624.[3] Next year, two additional bursts sources were discovered by Lewin and his coworkers[4] within a few degrees of the Galactic centre. In the process the Rapid Burster (MXB 1730 - 335) with its extraordinary behaviour was discovered. Within a year over twenty burst sources were found, largely due to observations made with SAS-3 and OSO-8. Lewin[5] compiled the first catalog of burst sources in September 1976, containing 15 ones. By the end of 1976 the number of burst sources was about 22. The burst MXB 1735 - 44, with simultaneous optical and X-ray bursts, was observed in the summer of 1978.[6] In February 1979, a Japanese small X-ray astronomical satellite named Hakucho was launched. From Hakucho observations[7], eight X-ray bursters have been newly established and some interesting phenomena such as a trapezoid burster have been detected. We now know of 31 burst sources altogether.[8]

II. MAIN OBSERVATIONAL CHARACTERISTICS

(1) Distribution

The distribution of burst sources shows that the X-ray burst sources concentrate towards the galactic center (See Fig. 1). There are 23 bursters within 40° from the galactic

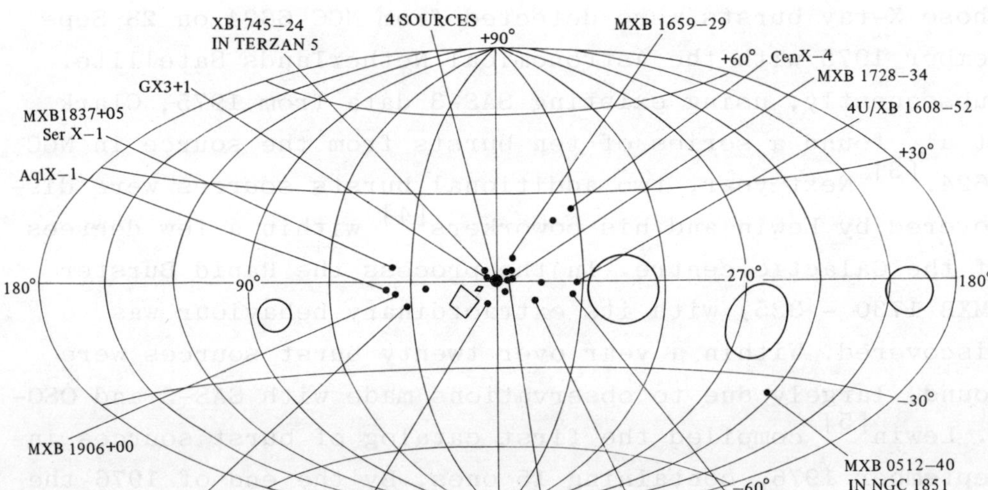

Fig. 1 The distribution of X-ray burst sources. The sky map shows the position (in galactic coordinates) of 31 X-ray burst sources. This figure is from Lewin & Joss[8].

center, among them 13 sources are located within 10° from the galactic center. The strong concentration of bursts towards the galactic center is remarkable in comparison with the distribution of general galactic X-ray sources.

(2) Burst Profile

Bursts always have rise times of about 1 sec and decay times of 10 sec. Exactly speaking, the rise times range from 1-10 sec and the decay times range from seconds to minutes. (see Fig. 2)

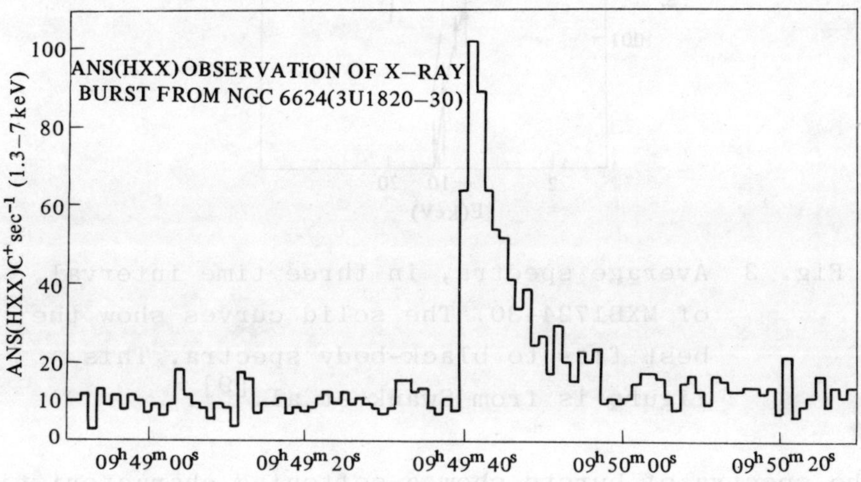

Fig. 2 The X-ray profile of the first X-ray burst.

The average peak luminosity of bursts is

$$L_{max}(obs) \simeq 3\times10^{38} \text{ erg sec}^{-1},$$

which is generally equal to the Eddington limit.

(3) Spectra

The burst spectra fit that of a black body reasonably well (see Fig. 3). Near burst maximum the black-body temperature is 3×10^{7}K.

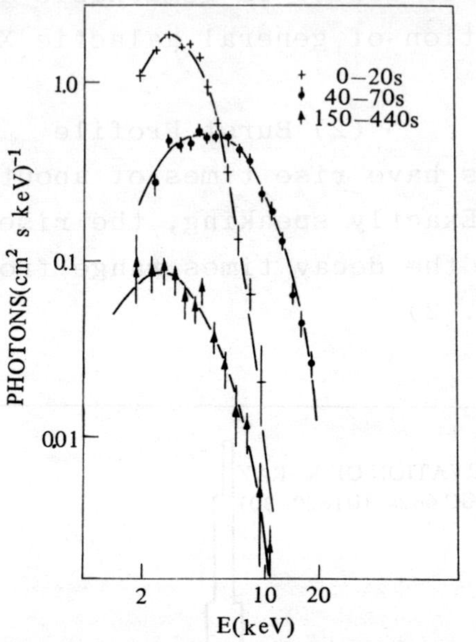

Fig. 3 Average spectra, in three time interval, of MXB1724-30. The solid curves show the best fits to black-body spectra. This figure is from Swank et al.[9].

The spectra of bursts show a softening characteristics while the luminosity decays[10].

(4) Burst Intervals

Burst intervals can be regular or irregular on time scales of hours to days. The burst occurrance rate is sometimes related to the level of persistent X-ray emission.[11]

(5) Idendification

Eight of the 31 sources are located in globular clusters. Six sources have been optically identified with faint stellar objects.

Simultaneous optical and X-ray bursts have been observed from some sources (i.e. MXB 1735-44, Ser X-1, MXB 1636-53).

Infrared bursts from the Rapid Bursters were observed.[12] Radio bursts from this source were also reported[13].

(6) Rapid Burster

The type II burst has own special charateristics. In this talk I shall confine myself to type I burst.

III. DEVELOPMENT OF THEORY

Obviously the short time scale structure of bursts suggests that they originate from a compact object. Shortly after the discovery of X-ray bursts Grindley and Gursky[14] argued that the X-ray burst source located in the globular cluster NGC 6624 is a very massive black hole. Their argument was largely based on the spectra hardening that was observed in two bursts. However, the discovery of more bursts sources showed that as a rule the spectra always soften during burst decay. In addition, the distribution of the observed burst sources, as a class, were not associated with globular clusters. The massive black hole idea disappeared.

Hoffman and coworkers[15] confirmed the consistency with black-body cooling for a series of bursts from two burst sources. They found values of radii of about 10 km for both sources if source distance of 10 kpc were assumed. The measured black-body radii strongly indicate that the collapsed object are neutron stars.

In 1976-1977, several models were proposed to explain the burst. Svestka[16], Henrikson[17], Joss and Rappaport[18], Baan[19], Lamb and cowork[20] cosidered various possible instabilities in the accretion flow at or near the magnetopause of a neutron star or a white dwarf. Wheeler[21], Liang [22] investigated potential instabilities in an accretion disk arounding a collapsed object. Grindley[23] suggested a thermal instability in accretion onto a massive black hole.

Woosley and Taam[24] proposed a different mechanism-neuclear flash in the surface layers of an accreting neutron star.

In September 1977, Hoffman, Marshall and Lewin[10] discovered Rapid Burster and introduced the classification of type I and type II burst and suggested that the type II bursts would be due to accretion instabilities but the type I bursts due to thermonuclear flashes.

IV. SURFACE EXPLOSIONS FROM THERMONUCLEAR FUSION ON ACCRETING NEUTRON STARS

(1) General Physical Picture

The accreting matter must pass through a series of nuclear burning shells as it is gradually compressed by the further accretion of still more material. If the core of the neutron star is sufficiently hot or the accretion rate is sufficiently high, the temperature in the surface layers will be high enough for the burning to proceed through thermonuclear reactions. These burning shells will tend to thermal runaway which is known as the "thin-shell instability" discovered by Schwarzschild and Härm.[25]

Hydrogen burning is not able to get a rise time of 1 sec. In fact, hydrogen cannot burn in less than 10^2 sec no matter what the temperature or density is. This is because in forming heavier elements from hydrogen, some protons must be converted to neutrons via $e^- + p \rightarrow n + \nu$ or $p \rightarrow n + \nu + e^+$ and such weak interactions are slow in the accessible temperature-density regime. The helium burning model, therefore, is the only choice which could explain the observations.

(2) Some General Properties of Surface Explosions from Thermonuclear Fusion on Accreting Neutron Stars

The following properties would be expected for helium burning:

1. Total energy and peak luminosity per flash. Helium flashes should occur after the accumulation of 10^{21} g of fuel and helium of one gram releases 10^{18} erg in the nuclear burning, so that a flash should release total energy of 10^{39} erg and result in a peak luminosity of 10^{38} erg.

2. α(the ratio of persistent luminosity to burst luminosity)

$$\alpha = \frac{L_s dt}{L_{burst} dt} = \frac{\text{gravitational energy per particle}}{\text{nuclear energy per particle}}$$

$$= \frac{GM/R}{E_n} = \frac{\eta_G\ c^2}{\eta_n\ c^2} = \frac{0.2}{\eta_n}\ ,$$

where η_n is the energy efficiency of the reaction:

$$\eta_n \begin{cases} \simeq 7\times10^{-3} & \text{for H burning} \\ \leqslant 1\times10^{-3} & \text{for He burning} \end{cases}$$

Then we have

$$\alpha \begin{cases} \simeq 30 & \text{for H burning} \\ \simeq 200 & \text{for He burning.} \end{cases}$$

As it can be seen from the above result that the value of α for helium burning is comparable with observations.

3. The burst duration τ. For a helium explosion we have

$$\tau \simeq \frac{E_{max}}{L_c} \sim \frac{10^{39}\text{ erg}}{10^{38}\text{ erg sec}^{-1}} \simeq 10\text{sec}$$

4. The time interval between bursts.

$$t = \frac{\text{total nergy emitted by a burst}}{\text{accretion rate nuclear energy per unit mass}}$$

$$= \frac{E}{\dot{m}\ 10^{18}\ \text{erg g}^{-1}}$$

Taking the typical accretion rate

$$\dot{m} = 10^{16} - 10^{18}\ \text{g sec}^{-1},$$

we have

$$t \simeq 10^{3} - 10^{5}\ \text{sec.}$$

5. The maximum temperature T_{max}. For a spherical accretion we have

$$L = E/4\pi R^2$$

and

$$L = \sigma T^4$$

so that

$$T = \left(\frac{E}{4\pi\sigma R^2}\right)^{\frac{1}{4}},$$

where $E = E_{b/\tau}$, E_b is the total energy in a burst:

$$E_b = mt,$$

then we obtain

$$kT = k\left(\frac{E_b}{4\pi\sigma\tau R^2}\right)^{\frac{1}{4}}$$

$$= 6\left(\frac{\eta_n}{10^{-3}}\right)^{\frac{1}{4}}\left(\frac{\dot{m}}{10^{17}\text{g s}^{-1}}\right)^{\frac{1}{4}}\left(\frac{t}{10^4\text{s}}\right)^{\frac{1}{4}}\left(\frac{\tau}{10\text{s}}\right)^{-\frac{1}{4}}\left(\frac{R}{10^6\text{cm}}\right)^{-\frac{2}{4}} \text{ keV}$$

$$= 6\text{keV}.$$

The emission should be in X-ray band.

6. The rise time to the burst maximum $\Delta.\Delta$ is limited by the radiation transfer time through the unexplored hydrogen layer above the burned helium layer.

$$\Delta \simeq \frac{\sigma_r \delta m\, h}{m_H\, c} \simeq \frac{h}{10^4} \text{ sec},$$

where σ_r is the Compton cross section, δm is the mass of the layer $\simeq 10^7$ g cm^{-2}, h is the thickness of the layer $\simeq 10^4$ cm. In this case, the rise time of the burst is about 1 sec.

(3) Numerical Calculation

The above estimates suggest that thermonuclear flashes on accreting neutron stars could account for the observed properties of the type I burst sources. With this encouragement, detailed numerical computations of the evolution of the surface layers of an accreting neutron stars have been carried out.

There are three groups actively working on the numerical calculation.

Joss and Li[26] explored the evolution of the helium burning shell. Their models contain two free parameters: the mass accretion rate $\dot{m}$ and the core temperature of the neutron star T_c. In Fig. 4 the hatched region denotes the

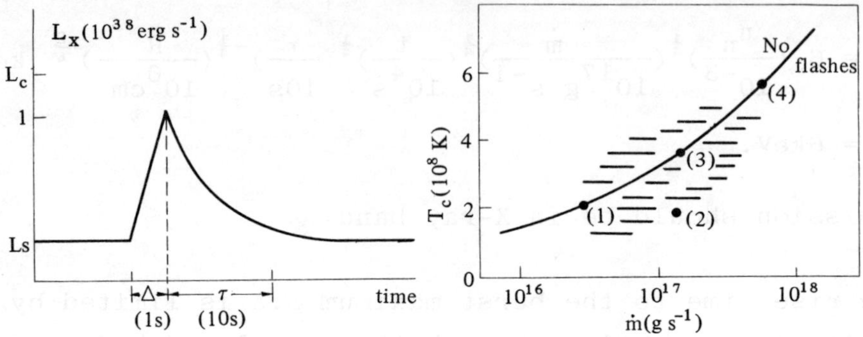

Fig. 4 (Left) Schematic representation of an X-ray burst from a thermonuclear explosion at an accreting neutron star surface.
(Right) Parameter range explored for thermonuclear explosions on accreting neutron stars by Joss.[27,28]

range of parameter values for which flashes may be expected. The helium-burning shell in model 4 does not display flashes. The shell become thermally stable at high values of T_c and $\dot{m}$. Flashing behaviour will also disapper at very low values of T_c and $\dot{m}$ ($T_c \leqslant 10^8$, $\dot{m} \leqslant 10^{15}$ g sec^{-1}). They further investgated the sensitivity of the flash properties to the assumed mass and radius of the neutron star.

Taam and Picklum[29] have showed that the hydrogen burning can have a major influence on the behaviour of the helium-burning shell. They have carried out the first fully time-dependent computations of the evolution of the surface layers of a neutron star, with a hydrogen-burning shell and found that the hydrogen and helium burning shell can indeed interact in a complex way.

Fujimoto and coworkers[30] have also studied the interaction between the hydrogen and helium burning shell. They argued that for neutron stars with low core temperature there would be three modes of thermonuclear flashes: helium flashes followed He-H simultaneous burning at high accretion rate, pure helium flashes at intermediate accretion rates, and hydrogen flashes ignite the helium burning shell at low accretion rates.

V. OPENING QUESTIONS AND FUTURE WORK

Further work has to be done to explain fully all the main observed facts. I would like to emphasis the following important special characteristics:

(1) Double or Multiple Peaks

Bursts with multiple peaks have been observed from MXB 1743-29, MXB 1728-34 and MXB 1850-08 with SAS-3[31] and from the burster in Terzan-2 with the Einstein Observatory[32]. Similar bursts of double peaks were observed with Hakucho: two from MXB 1608-522, one from MXB 1636-536, one from MXB 1905+000 and one from MXB 1728-337.[7]

In all these bursts double peaks are more distinct in high energy band than in the low energy band.

The existence of double peaks or multiple peaks structure of bursts have not been interpreted in terms of the nuclear flash model.

(2) Bursts of Very Short Intervals

A strong constraint to the burst mechanism is given by bursts of very short intervals. On 14 August 1980 two bursts from Terzan-5 occurred with an interval of 8 minutes (cf.[8]). Short intervals of 17 min and 4 min were observed for three bursts from MXB 1743-28.[33] An interval of 10 min was observed for two bursts on 6 July 1979 from MXB

1608-522.[34]

One might think that bursts of short intervals could be due to partial burning of nuclear fuel. If this were the case, the burst emitting area would fluctuate depending on the size and profile. However, the radii derived for all bursts are essentially the same(10km) so that the whole surface of a neutron star is responsible for burst emission. It is very difficult to understand why the area is constant despite that all the other features change.

(3) Super Luminosity

There are some indications that in some cases the Eddington limit is exceeded near the peak of some bursts. Höffman and coworkers found that in July 1976 the peak luminosity in burst from MXB 1728-34 were $(2.2\pm0.2)\times10^{39} \times(\frac{D}{10})^2$ erg sec^{-1}. If the distance is 5 kpc, the peak luminosity will exceed the Eddington limit for a 1.4 $M_\odot$ object by a factor of $\simeq$3.[35] Grindlay and coworkers reported that the peak luminosity of a burst from 4U 1722-30 may have exceeded the Eddington limit by a factor of$\simeq$10.[32]

(4) The Relation between the Burst Profile and the Level of Persistent Emission

Such relation was first found in 1976 by Clark and coworkers[11] for 3U/MXB 1820-30. They observed a gradual shortening in decay time (from $4^s.7$ to $<3^s$) while the level of persistent emission increased by a factor of 5. Hakucho team observed 4U 1608-52 in April 1979 and found that the bursts had a slower rise and a slower decay when the persistent flux had decreased by a factor of$\simeq$5.[36]

A lot of improvement in theory has been suggested to explain the observational characteristics. The following aspects should be mentioned here.

(1) Dynamical Effects

Since the luminosities of some bursts have exceeded the Eddington limit, the mass ejection probably occurs. The dynamical effects, therefore, must be considered in the problem.

(2) Non-spherical Symmetry

The numerical calculations have only been carried out to date relied on the assumption of spherical symmetry. Violations of spherical symmetry could be the key to the observed complexities in burst structure and recurrence patterns. A thorough investigation of such possibilities will eventually require two or three-dimensional numerical computations, which will be much more difficult than the computations of spherical symmetry models.

(3) Magnetic Field

A magnetic field will lead not only a violation of the spherical symmetry of accretion but also other effects, for instance, a change of opacity in the surface of neutron star which might lead to the super luminosity.

(4) Departure from Thermal Equilibrium of Core

If there is a change in the average accretion rate, the thermal inertia of the core is sufficient to require the elapse of 10^{2-3} years for thermal equilibrium to be reestablished. Thus departures from thermal equilibrium are entirely possible, and they could have substantial effects upon the behaviour of the nuclear burning shells.

REFERENCES

[1] Grindley, J.E. et al., 1976, Ap.J., Vol. 205, L 127.

[2] Grindley, J.E., Heise, J.,1976, IAU Circ., No. 2879.

[3] Clarc, G.W. et al., 1976, Ap.J., Vol. 207, L 105.

[4] Lewin, W.H.G., 1976, IAU Circ., No. 2911, 2918, 2922.

[5] Lewen, W.H.G., 1977, M.N.R.A.S., Vol. 179, 43.

[6] Grindley, J.E. et al., 1978, Nature, Vol. 274, 567.

[7] Hayakawa, S., 1981, Space Science Reviews, Vol. 29, No. 3,1.

[8] Lewin, W.H.G., Joss, P.C., 1981, Space Science Reviews, Vol. 28, No.1,1.

[9] Swank, J.H., et al., 1977, Ap.J., Vol. 212, L73.

[10] Hoffman, J.A., Marshall, H.L., Lewin, W.H.G., 1978, Nature, Vol. 271, 630.

[11] Clark, G.W. et al., 1977, M.N.R.A.S., Vol. 179, 651.

[12] Kulkarni, P.V., 1979, Nature,Vol. 280, 819.

[13] Canal, R., Schatzman, E., 1976, Astron. Astrophys., Vol. 46, 229.

[14] Grindley, J.E., Gursky, H., 1976, Ap.J., Vol. 205, L131.

[15] Hoffman, J.A., Lewin, W.H.G., Dofy, J., 1977, Ap.J., Vol. 217, L23; 1977, M.N.R.A.S., 179, 57.

[16] Svestka, J., 1976, Ap. Space Sci., Vol. 45, 21.

[17] Henrikson, R.N., 1976, Ap.J., Vol., 210, L19.

[18] Joss, P.C., Rappaport, S.A., 1979, Astron. Astrophys., Vol. 71, 217.

[19] Baan, W., 1977, Ap.J., Vol. 214, 245; 1979, Ap.J., Vol., 227, 987.

[20] Lamb, F.K. et al., 1977, Ap.J., Vol. 217, 197.

[21] Wheeler, J.C., 1977., Ap.J., Vol. 214, 560.

[22] Liang, E.P.T., 1977, Ap.J., Vol. 211, L67; Ap.J., Vol. 218, 243.

[23] Grindley, J.E., 1978, Ap.J., Vol. 224 L107.

[24] Woosley, S.E., Taam, R.E., 1976, Nature, Vol. 263, 101.

[25] Schwarzschild, M., Harm, R., 1965, Ap.J., Vol. 142, 855.

[26] Joss, P.C., Li, F.K., 1980, Ap.J., Vol. 238, 287.

[27] Joss, P.C., 1978, Ap.J., Vol. 225, L123.

[28] Joss, P.C., 1979, Comments on Astrophys., Vol,8, 109.

[29] Taam, R.E., Picklum, R.E., 1978, Ap.J., Vol. 224, 210, 1979, Ap.J., Vol. 223, 27.

[30] Fujimoto, M., Hanawa, T., Miyami, S., 1981, Ap.J., Vol. 246, 267.

[31] Hoffman, J.A. et al., 1980, Ap.J., Vol. 240, 27.

[32] Grindley, J. E. et al., 1980, Ap.J., Vol. 240, L21.

[33] Lewin, W,H.G. et al., 1976, M.N.R.A.S., Vol. 177, 83.

[34] Murakami, T., 1980, Inst. Spa. Astronau. Sci., Research Note, 105.

[35] Van Paradijs, J., 1978, Nature, Vol. 274, 650.

[36] Murakami, T. et al., 1980, Ap.J., Vol. 240, L143.

PHYSICAL PROCESSES IN THE STRONGLY MAGNETIZED ACCRETION PLASMA OF A NEUTRON STAR

J. G. Kirk

(Max-Planck-Institut für Physik und Astrophysik, Institut für Astrophysik, München, Federal Republic of Germany)

I. INTRODUCTION

The strongest magnetic field which can be maintained in a terrestial laboratory is about 10^6 gauss. Stronger fields can be produced, but only for short intervals of time, since the stresses they produce in the apparatus are sufficient to destroy it. With the help of a little TNT, it is possible to obtain magnetic fields of 10^7 gauss for a few microseconds[1]. In this paper, the magnetic fields of interest are about 10^{12} - 10^{13} gauss - one million times stronger.

The motivation for studying such a seemingly obscure and unlikely parameter range is twofold. First of all, it seems that the slow increase in the period of radio pulsars with time can best be explained by invoking a rotating neutron star with a surface magnetic field of about 10^{12} gauss[2]. Secondly, in the spectra of at least two and possibly three X-ray sources, line features have been observed which indicate a magnetic field strength of a few × 10^{12} gauss[3-5]. Similar features have also been seen in the spectra of γ-ray burst events[6]. Although the existence of such strong fields may at first seem surprising in view of the difficulties encountered in the laboratory, it is, in fact, the natural consequence of the formation process of neutron stars. 10^7 gauss can be produced by using chemical explosives, but the implosion of the core of a massive star generates pressures which are higher by many orders of magnitude. The magnetic field of the star is accordingly compressed, and held at its final value by the enormously strong gravitational force. A reduction in radius from 10^{11} cm to 10^6 cm for the resulting neutron star implies an increase in magnetic field strength by a factor 10^{10}. Thus the values which we shall consider can easily be obtained by the collapse of a star with a magnetic field of a few hundred gauss.

The theory of cyclotron and synchrotron (collectively referred to as gyromagnetic) radiation has been intensively investigated over a long period of time [7-9], for a review of the astrophysical aspects see[10]. These calculations are, however, not applicable to a plasma in a magnetic field as strong as 10^{12} gauss. Modifications are required essentially due to two kinds of effect, which may be loosely classified under the headings "quantum" and "relativistic".

II. QUANTUM EFFECTS

From the observations, it appears highly likely that the emission region of pulsating X-ray sources contains plasma at a temperature T such that

$$K_B \, T < \hbar \, \Omega, \tag{1}$$

where Ω(= eB/mc) is the gyrofrequency of the electron in the magnetic field B, and k_B is Boltzmann's constant. The thermal radiation from such a plasma cannot exceed that of a black body, so that the energy density of photons with frequency close to Ω must be lower than

$$u = \frac{\hbar \Omega^3}{2\pi^2 c^3} \left\{ \exp(\hbar\Omega/k_B T) - 1 \right\}^{-1}. \qquad (2)$$

The quantity $\left\{ \exp (\hbar\Omega/k_B T) - 1 \right\}^{-1}$ represents the occupation number of the photon states in the black body radiation field. According to Eq. (1), its value is less than or of order unity, so that the actual photon occupation number N_Ω fulfils the condition

$$N_\Omega < \left[\exp (\hbar\Omega/k_B T) - 1 \right]^{-1} < 1. \qquad (3)$$

This implies that a quantum description of gyromagnetic radiation is required. Classical treatments assume $N_\Omega >> 1$, and can be employed only when Ω lies in the Rayleigh-Jeans part of the Planck spectrum.

An electron in a magnetic field has, according to the Dirac equation, energy eigenvalues given by[11]

$$E = [m^2c^4 + c^2p_z^4 + (2_j + s + 1)\hbar\Omega]^{\frac{1}{2}}$$
$$(s = \pm 1,\ j = 0,1,2,\ldots). \qquad (4)$$

To lowest order in the quantity $(\frac{E - mc^2}{mc^2})$, these may be split into a continuous part, $E_{//}$, associated with translation in the direction of the field, and a discrete part, $E_\perp$, connected with motion in the plane perpendicular to the field:

$$E = E_{//} + E_\perp,$$
$$E_{//} = p_z^2 / 2m, \qquad (5)$$

$$E_\perp = n\hbar\Omega \ (n = 0,\ 1,\ 2,\ldots)$$ [12]

The spontaneous emission of a photon accompanied by a transition between the status n = 1 and n = 0 occurs at a rate

$$\Gamma = 4 \times 10^{15} \times B_{12}^2 \ \mathrm{sec}, \tag{6}$$

where $B_{12} = B \times 10^{-12}$. The rate of repopulation of this level due to collisions can be estimated under various assumptions[13,14]. The result is a rate which is somewhat slower than that given by Spitzer (1962)[15]

$$\Gamma_{\mathrm{Spitzer}} = 10^{10} \ n_{22} \ T_{\mathrm{keV}}^{-3/2} . \tag{7}$$

Comparing Eqs. (6) and (7) leads to the conclusion that the population of the Landau levels is determined not by collisions, but by radiative processes, unless the plasma is very dense ($n_{22} >> 10^5$). The radiation field, however, establishes the ratio of the population of the n = 1 (N_1) to the population of the n = 0 level (N_0) at the value

$$\frac{N_1}{N_0} = \frac{N_\Omega}{N_\Omega + 1} \tag{8}$$

$$<<1 \quad \text{for } k_B \ T << \hbar\Omega .$$

Thus the quantum effects enable the simplifying assumption to be made, that all electrons occupy the Landau ground state n = 0. This convenient property permits a treatment of the Coulomb cross-sections[16-18], the free-free absorption cross-sections[19] and the Compton cross-sections[20], and also lies at the heart of several calculations of the transfer of radiation[21-24].

III. RELATIVISTIC EFFECTS

For the present purpose, these effects include those interactions between photons and the positron sea which are forbidden in the absence of an external magnetic field. Under more normal circumstances such interactions may be neglected. However, in a magnetic field of $\sim 10^{12}$ gauss, the ratio of the energy of a cyclotron photon to the electron rest mass is about one tenth. Although this quantity may still be used as an expansion parameter, the low order terms can produce appreciable effects.

If we restrict our attention to photons of energy less than the electron rest mass, the dominant effect is that of vacuum polarization[25-27]. The propagation of electromagnetic waves is then controlled by the equation

$$\mathrm{Det}\ (\Lambda + \Psi + X) = 0, \tag{9}$$

where

$$\Lambda = 1 - n^2(1 - \hat{\underline{k}}\,\hat{\underline{k}}) \tag{10}$$

is the familiar tensor describing the propagation of transverse waves of wave vector $\hat{\underline{k}}$ and frequency ω ($n = \frac{ck}{\omega}$) according to the source-free Maxwell equations. The response of the vacuum (i.e. the virtual electron position pairs) is described by

$$\Psi = -2\delta\Lambda + n^2 p\ (\hat{\underline{k}}_\wedge \hat{\underline{B}})\ (\hat{\underline{k}}_\wedge \hat{\underline{B}}) + q\ \hat{\underline{B}}\ \hat{\underline{B}} \tag{11}$$

with

$$\delta = \frac{1}{45\pi}\left(\frac{e^2}{\hbar c}\right)\left(\frac{\hbar\Omega}{mc^2}\right) \tag{12}$$

$p = 4\delta$ and $q = 7\delta$. The response of the plasma is given by the tensor χ, related to the conductivity σ by

$$\chi = \frac{i\sigma}{\omega} \tag{13}$$

At frequencies below the cyclotron resonance, the largest elements of χ are of order ω_p^2/ω^2, where ω_p is the plasma frequency. When this is much smaller than the quantity δ, the behaviour of the refractive index is governed by the vacuum polarization effect. Conversely, the plasma contribution is dominant when δ is small compared to the terms contained in χ. The physical conditions which pertain in the emission region of pulsating X-ray sources seem to be just right for the observation of photons from each of these regimes, so that the interesting features produced in the scattering cross-sections at intermediate frequencies[28,29] may well be accessible to observation.

IV. COULOMB COLLISIONS

Most of the processes described above have been discussed in recent reviews[30,31]. Here we shall turn our attention to one particular process, and its implications for models of accreting magnetic neutron stars.

There are three ways in which matter falling towards the surface of a neutron star can be brought to rest. The first is by means of a radiative shock front[32]. However, most X-ray pulsars seem to have luminosities which would be too low for this purpose. The second is by means of a collisionless shock front[33,34]. This process requires a collective interaction between the accreting protons and the plasma. The indications are that the one dimensional nature to the plasma is unlikely to support a sufficiently strong instability[35,36]. The third possibility is that of a collisional shock front. No collective processes are involved here, but the details of Coulomb collisions between the energy carrying protons, and the electron component of the stellar atmosphere must be investigated.

Such a calculation has been performed in the cold plasma limit by Basko and Sunyaev (1975)[37] and Pavlov and Yakovlev (1976)[38] and more recently, for warm plasmas, by Kirk and Galloway (1981, 1982)[16,17]. The results can best be described in terms of the time evolution of the energy of an accreting proton as it penetrates an electron gas. Because of the anisotropy introduced by the magnetic field, the energy associated with motion along the field evolves in different manner to that connected with gyromotion. This effect can be approximately represented by two ordinary differential equations:

$$\frac{dE_{//}}{dr} + F_{//}(E_{\perp}, E_{//}, n, T), \qquad (14)$$

and

$$\frac{dE_{\perp}}{dr} = F_{\perp}(E_{\perp}, E_{//}, n, T), \qquad (15)$$

where $E_{//}$ ($E_{\perp}$) is the energy associated with the proton's motion parallel (perpendicular) to the magnetic field, and n and T are the density and temperature of the background electron gas, respectively. The quantities $F_{//}$ and $F_{\perp}$ are usually called "dynamic friction coefficients"[39]. In Fig. 1, $F_{//}$ is compared with the dynamic friction coefficient in an isotropic plasma. For velocities $v_{//}$ ($=\sqrt{2E_{//}/M}$) of order the electron thermal velocity s ($=\sqrt{T/m}$), the frictions is enhanced. For $v_{//} > s$ the effect is reversed.

In order to find the penetration depth of accreting protons one must integrate the Eqs. (14) and (15) in time, and make the simple conversion from the time taken to slow down to the distance travelled in slowing down. Given an initial speed for the proton, and the temperature and den-

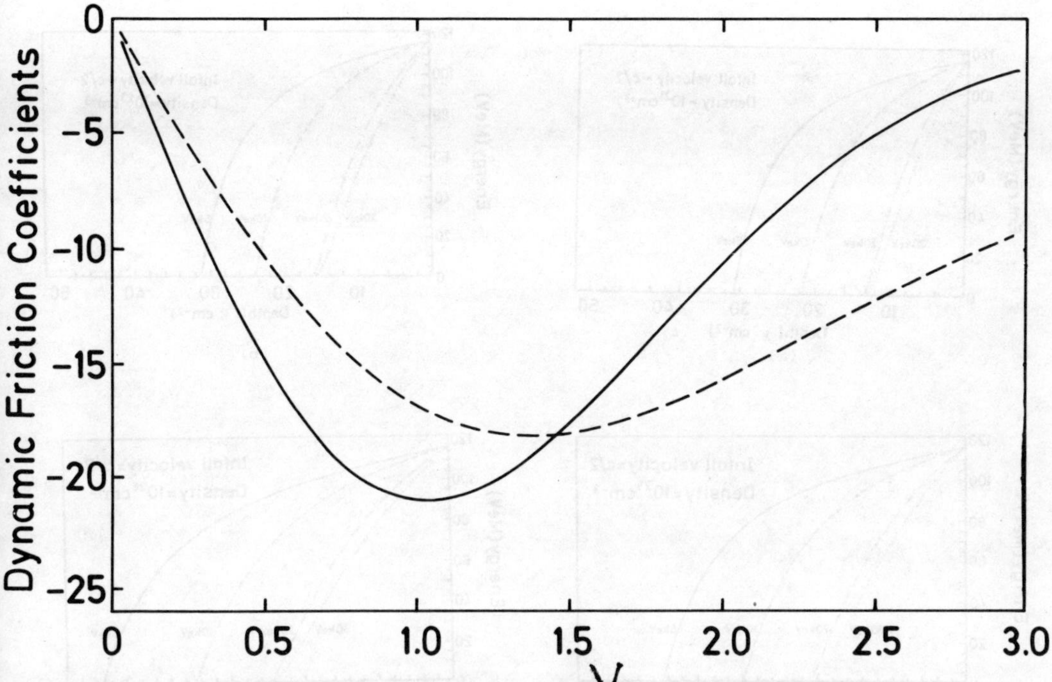

Fig. 1 A comparison of the dynamic friction coefficient in an isotropic plasma (dashed line) with the $F_{//}$ component of this coefficient in a strong magnetic field (solid line). For this purpose $v_{\perp}$ was fixed at 0.01. The density is 10^{23} cm^{-3} and the temperature is 5keV. v is expressed in units of the electron thermal speed $s = (T/m)^{1/2}$ and the coefficients are in units of $e^4 n/mMs^2$. The magnetic field is 5×10^{12} gauss.

sity of the plasma, the evolution of $E_{//}$ and $E_{\perp}$ with depth can be computed numerically. Figs. 2 and 3 show results for the total energy $E = E_{//} + E_{\perp}$ for different plasma densities and temperatures and for two initial velocities: c/2 and c/4.

These speeds are chosen to be approximately the free-fall speeds at the surface of a neutron star, which, neglecting a 10% relativistic correction, is given by

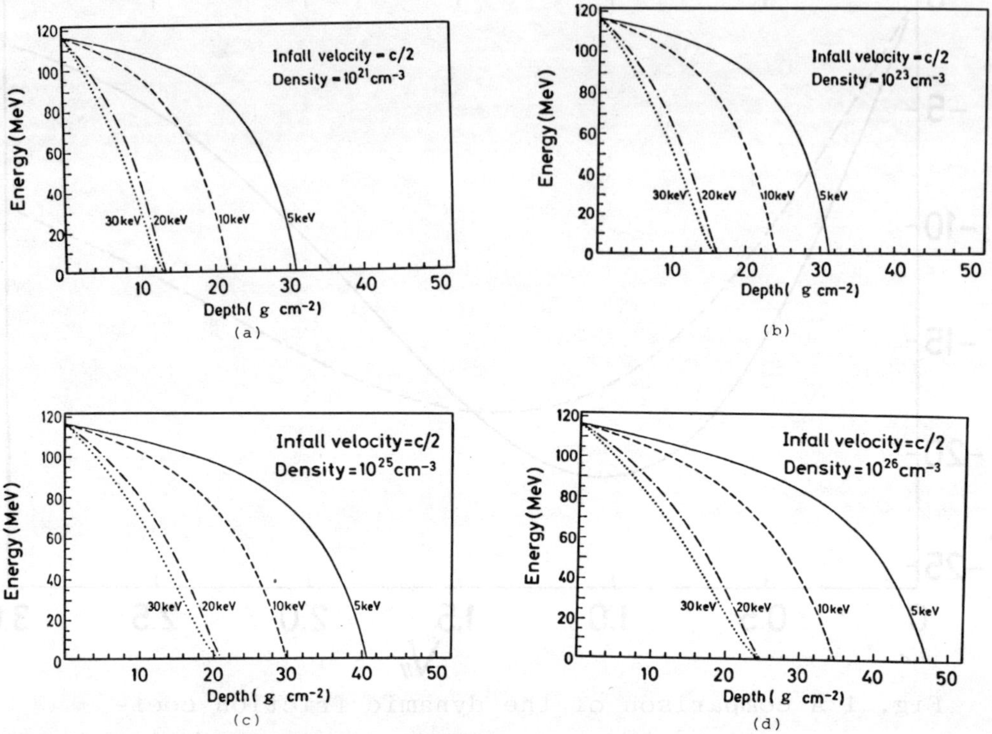

Fig. 2 The stopping length for plasma temperatures 5,10,20 and 30 keV and densities (a) 10^{21} cm^{-3}, (b) 10^{23} cm^{-3}, (c) 10^{25} cm^{-3} and (d) 10^{26} cm^{-3}. In each case the infall velocity is c/2 and the energy is normalised to the kinetic energy of a single proton. The magnetic field is 5×10^{12} gauss.

$$v = \left(\frac{2GM_*}{R_*}\right)^{1/2}$$

$$= 0.54c \quad \text{for } M_*=1M_\odot,\ R_*=10\text{km}$$

or $\qquad\qquad\qquad\qquad\qquad\qquad\qquad\qquad\qquad\qquad\qquad\qquad\qquad\qquad\qquad$ (16)

$$= 0.27c \quad \text{for } M_*=0.5M_\odot,\ R_*=20\text{km}.$$

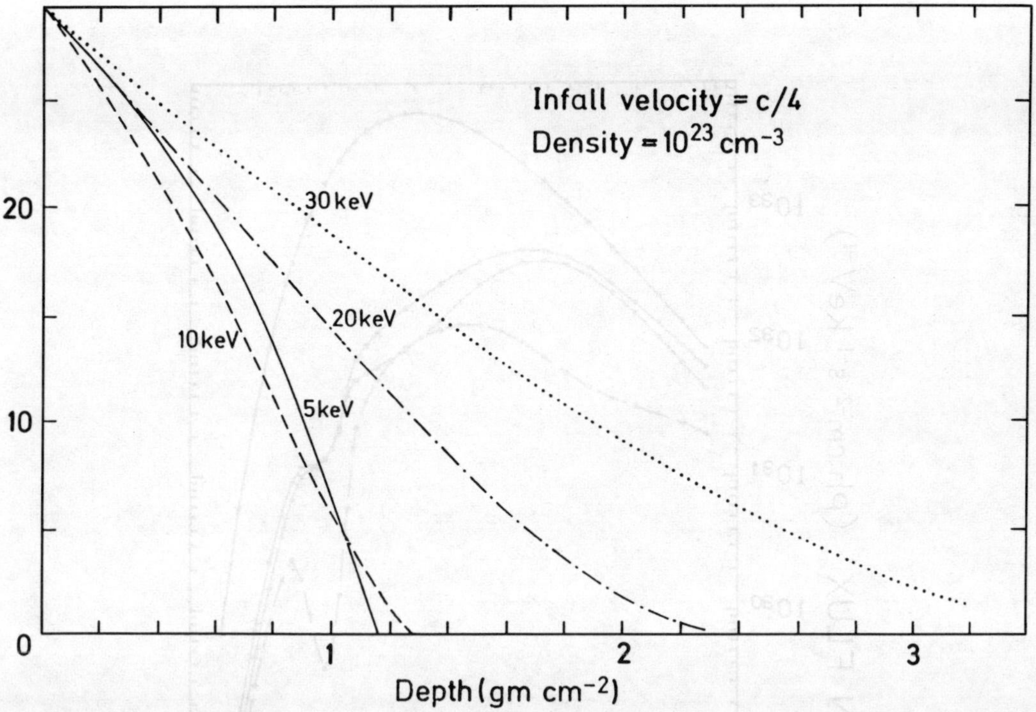

Fig. 3 As for Fig. 2(b) but with an infall velocity of c/4.

From Fig. 2 c, one sees that protons accreting onto a neutron star of mass $1M_\odot$ and radius 10 km would penetrate about 30 g cm^{-2} under the simplifying assumption that the atmosphere of the star had a uniform temperature of 10 keV, and a density of 10^{25} particles cm^{-3}. Fig. 4 shows results found by Nagel (1981)[21,22] for the spectrum of a uniform radiating slab having roughly this density and temperature, and with a thickness (about 2 cm) required to stop the infalling protons. Arguments concerning the structure of

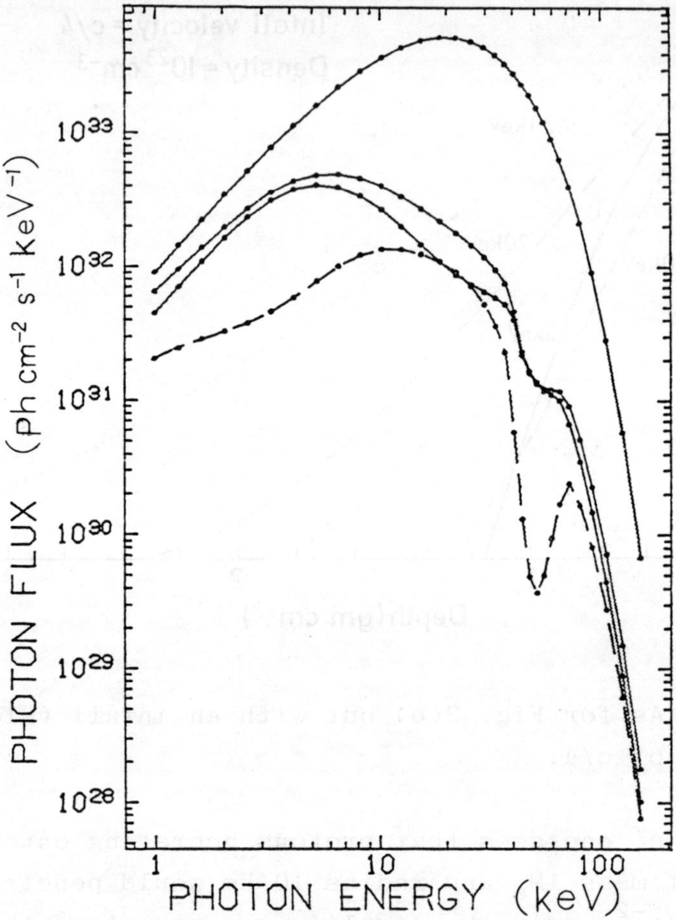

Fig. 4 Photon spectrum from a slab of hot strongly magnetized plasma. Solid lines: ordinary mode photons, dashed lines: extraordinary mode photons. The uppermost curve represents the Wien spectrum. The plasma parameters are $\rho = 10$ g cm^{-3}. T = 10 keV, $h\Omega = 50$keV, optical depth (for the Thomson scattering cross-section) $\tau = 8$. The luminosity is 3×10^{36} erg s^{-1} for an area of 10^{10} cm^{2}.

the accretion flow further away from the star indicate that such a slab should cover about 1% of the stellar surface. In this case the luminosity predicted by Nagel is 3×10^{36} erg s^{-1}, which is a value typical of X-ray pulsars.

Of course the assumption of a uniform atmosphere is a tremendous simplification, and a next step in such investigations would be to allow the density and temperature structure to be determined by a balance between the heating of the accreting protons and the cooling by radiation. Nevertheless, such calculations do indicate roughly the range of physical parameters which apply to X-ray pulsars.

V. γ-RAY BURST SOURCES AND UV EMISSION

Roughly three times per week, a flash of γ-rays reaches the earth from some location in the sky where previously nothing has been seen, and where afterwards no trace of the event is detectable. These mysterious γ-ray bursts were discovered in 1969. As of today there is no satisfactory explanation for them.

Nevertheless, progress has been forthcoming: In the spectra of three such bursts shown in Fig. 5, cyclotron resonance lines can be seen. The magnetic field strength indicated by these lines is about 5×10^{12} gauss, so the radiation must originate on the surface of a neutron star. Additional evidence for this is provided by the line feature at 450 keV, which can be identified as the electron-positron annihilation line, redshifted by the deep gravitational potential well of the neutron star.

Two different kinds of theoretical model have been proposed to account for these events. In the first, the sudden release of energy is due to the falling of a comet or asteroid sized body ($\sim 10^{19}$ g) onto the surface of the neutron star[40,41]. This is also the more difficult model

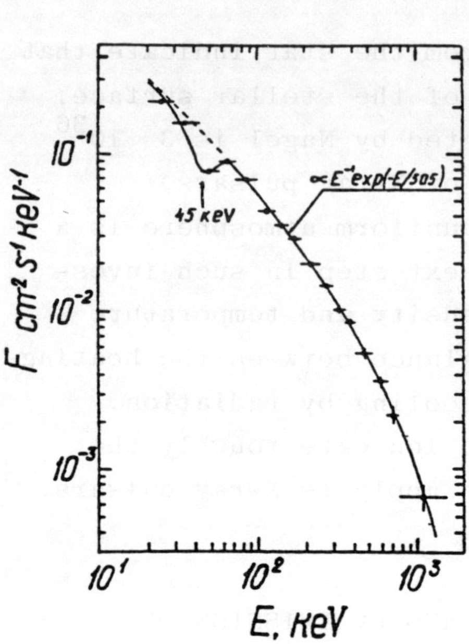

(a)

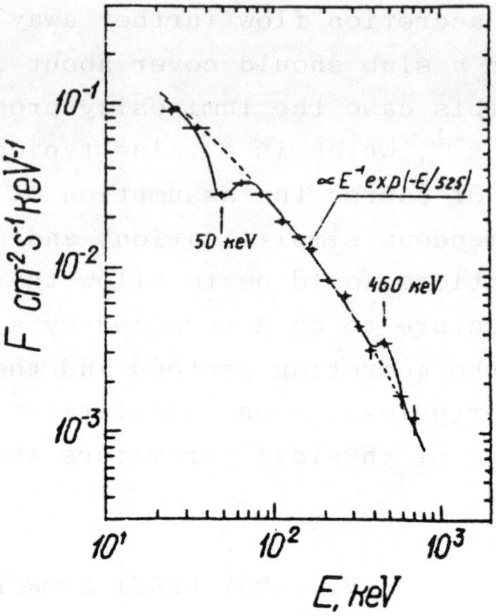

(b)

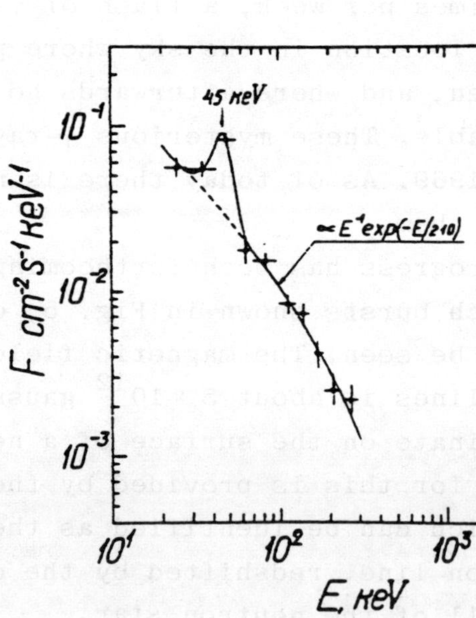

(c)

Fig. 5 A compilation of γ-ray burst spectra measured with the Konus experiment aboard the Venera 11 and 12 spacecraft[6]. a) 1979 March 7th burst b) 1979 May 26th burst c) 1979 June 22nd burst.

from the theoretical point of view. Largely unknown are the density of such objects in the galaxy, the minimum impact parameter which they must have to be captured and accreted, and the nature of the flash which would be produced upon accretion. In between flashes there is no particular prediction about the behaviour of the star.

The second type of model, on the other hand, is to be preferred because of the number of predictions it makes which are accessible to observation[42,43]. In this picture the flashes occur when a certain critical amount of highly explosive helium has been accumulated on the stellar surface by slow, steady accretion. The accretion rate must lie roughly within the range

$$10^{-16} < \dot{M}\ (M_\odot\ yr^{-1}) < 10^{-12} . \tag{17}$$

Slower rates enable helium to burn away steadily, thus defusing the bomb. Faster rates would lead to too many relatively weak explosions.

The energy which should be released in steady accretion is, when the long interval between bursts is taken into account, about ten times the energy released in the bursts themselves. Unfortunately, the largest accretion rate permitted by Eq. (17) is only on the borderline of being observable. An accretion rate $\dot{M}$ onto a neutron star of mass M and radius R must release energy at the rate

$$L = \dot{M}GM_*/R_*$$
$$= 8\times10^{33}\ erg\ s^{-1} \tag{18}$$

for $\dot{M} = 10^{-12}\ M_\odot\ yr^{-1}$, $M_* = 1\ M_\odot$ and R_*=10km.

This energy must be released from a surface which cannot be much larger than that of the neutron star, and may be much smaller (the magnetic polar cap). The temperature of this surface must, therefore, be higher than that required to radiate the energy at the black body rate. Thus, a lower limit T on the temperature is given by

$$L < 4\pi R_*^2 \sigma T^4 \tag{19}$$

or

$$T > 1.9\times10^6 K \tag{20}$$

for the same parameters as above.

The best possibility of observing such a neutron star was given by the Einstein satellite[44]. There, the detector was sensitive to photons of energy between 0.2 and 3 keV. The energy in this band depends very sensitively on T, when T is in the vicinity of 10^6 K. A source with the same parameters as those used above would have been observable provided it was within 1 kpc of the Earth. No such sources were seen at the sites of γray bursts. However, these models cannot be ruled out by this non-observation, because a lower accretion rate and a slightly larger but less massive neutron star would not have been observable.

There is another way in which the continuous accretion of γ-ray burst sources might be observed - one which has as yet not been tested. Going back to the calculation of the braking length of a proton in the atmosphere of a neutron star, Fig. 6 shows not only the evolution of the total energy, but also of $E_\perp$. Starting at a low value equal to the plasma temperature, $E_\perp$ rises during deceleration to several MeV, before ultimately falling back again to the plasma temperature. This energy is associated with gyration, and, consequently, with the cyclotron radiation of the proton. It is easy to calculate the energy which should be radiated, by adding the contribution of each proton. Before deceleration, (in the free-fall zone) the protons have a low $E_\perp$, but, on the other hand, this zone is quite large. During deceleration, the protons have a very high $E_{\perp}$, but this zone is rather thin. The energy radiated is shown in Table 1. Such fluxes are below observational limits unless the neutron star is closeby.

Table 1. Incoherent ion-cyclotron radiation from a neutron star with an accretion rate $\sim 10^{-13}$ $M_{\odot}$ yr^{-1}

	Free-fall zone	Deceleration zone
Linear Dimension (cm)	10^5	2
Maximum Density (cm^{-3})	5×10^{17}	10^{18}
Effective Temperature	1keV	5MeV
Surface Area (cm^2)	10^{11}	10^9
Luminosity (erg s^{-1})	2×10^{27}	8×10^{28}

However, the value of $E_{\perp}$ in Fig. 6 is so high, that it is possible for the protons to radiate coherently. In other words, the protons are so far removed from thermal equilibrium that they become unstable to the emission of cyclotron waves. In this case, one must add the contribution of each proton to the electric field of the wave, which gives an energy radiated proportional to the square of the number of particles involved. More detailed calculations[45] show that the flux to be expected lies at the proton gyro frequency

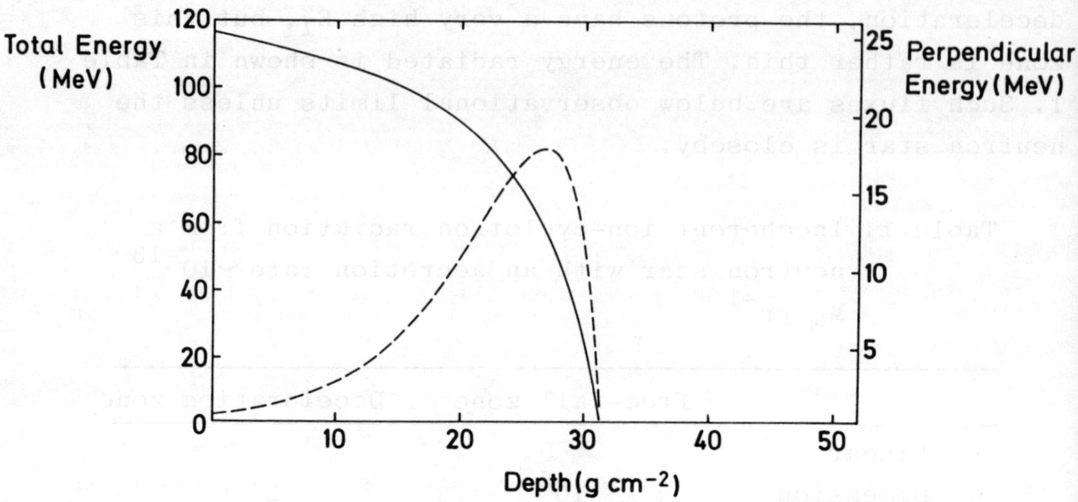

Fig. 6 Evolution of the total energy (solid line) and perpendicular energy (dashed line) for the same parameters as in Fig. 2 (b).

(in the far ultra-violet region), and is, in principle, strong enough to be observed.

At the present time, there is no instrument in operation which is suitable for such an observation. The major difficulty here lies in the uncertainty in the position of γ-ray burst sites. However, with the advent of imaging telescopes in the far ultra violet range it should be possible to obtain more much needed information about these objects.

REFERENCES

[1] Hawke, P.S., Burgess, T.J., Duerre, D. E., Huebel, J.G., Keeler, R.N., Klapper, H., Wallace, W.C., 1978, Phys. Rev. Lett., Vol. 41, 994.

[2] Ostriker, J.P., Gunn, J.E., 1969, Ap. J., Vol. 157, 1395.

[3] Trümper, J., Pietsch, W., Reppin, C., Voges, W., Staubert, R., Kendziorra, E., 1978, Ap. J., Vol. 219, L105.

[4] Wheaton et al., MIT/UCSD Collaboration, 1979, Nature, Vol. 282, 240.

[5] Pravdo, S.H., White, N.E., Boldt, E.A., Holt, S.S., Serlemitsos, P.J., Swand, J.H., Szymkowiak, A.E., Tuoky, I., Garmire, G., 1979, Ap. J., Vol. 231, 912.

[6] Mazets, E.P., Gotenetskii, S.V., Aptekor, R.L., Guryan, Y.A., Klyinskii, V.N., 1981, Nature, Vol. 290, 378.

[7] Schott, G.A., 1912, "Electromagnetic Radiation", Cambridge University Press, Cambridge, England.

[8] Schwinger, J., 1949, Phys. Rev., Vol. 75, 1212.

[9] Sokolov, A.A., Ternov, I.M., 1968, "Synchrotron Radiation", Pergamon Press, London.

[10] Melrose, D.B., 1980, "Plasma Astrophysics", Gordon and Breach, London.

[11] Rabi, I.I., 1928, Z. Physik, Vol. 49, 509.

[12] Landau, L. D., 1930, Z. Phys., Vol. 64, 629.

[13] Ventura, J., 1973, Phys. Rev., Vol. A8, 3021.

[14] Bussard, R.W., 1980, Ap. J., Vol. 237, 970.

[15] Spitzer, L., 1962, "The physics of fully ionized gases", Wiley and Sons, London.

[16] Kirk, J.G., Galloway, D.J., 1981, M.N.R.A.S., Vol. 195, 45P.

[17] Kirk, J.G., Galloway, D.J., 1982, "Plasma Physics" (in press).

[18] Langer, S.H., 1981, Phys. Rev., Vol. D23, 328.

[19] Nagel, W., Ventura, J., 1982, submitted to Astron. Astrophys.

[20] Herold, H., 1979, Phys. Rev., Vol. D19, 2868.

[21] Nagel, W., 1981, Ap. J., Vol. 251, 278.

[22] Nagel, W., 1981, Ap. J., Vol. 251, 288.

[23] Bonazzola, S., Heyvaerts, J., Puget, J.G., 1979, Astron. Astrophys., Vol. 78, 53.

[24] Wang, Y. M., Frank, J., 1981, Astron. Astrophys., Vol. 93, 255.

[25] Erber, T., 1966, Rev. Mod. Phys., Vol. 38, 626.

[26] Mészáros, P., Ventura, J., 1979, Phys. Rev., Vol. D 19, 3565.

[27] Pavlov, G.G., Shibanov, Y.A., Gnedin, Y.N., 1980, Sov. Phys. JETP Lett., Vol. 30, 125.

[28] Ventura, J., Nagel, W., Meszaros, P., 1979, Ap. J., Vol. 233, L125.

[29] Pavlov, G.G., Shibanov, Y.A., 1979, Zh. E.T.F., Vol. 76, 1457.

[30] Mészáros, P., 1981, to appear in "Proceedings of the 5th General Conference of the European Physical Society", Istanbul.

[31] Kirk, J.G., Trümper, J., 1982, to appear in "Accretion driven stellar X-ray sources", Lewin, van den Heuvel (eds.), Cambridge University Press, Cambridge, England.

[32] Basko, M.M., Sunyaev, R.A., 1976, M.N.R.A.S., Vol. 175, 395.

[33] Shapiro, S.L., Salpeter, E.E., 1975, Ap.J., Vol. 198,671.

[34] Langer, S.H., Rappaport, S., 1982, submitted to Ap. J.

[35] McKee, C.F., 1970, Phys. Rev. Lett., Vol. 24, L990.

[36] Alme, M.L., Wilson, J.R., 1973, Ap.J., Vol. 186, 1015.

[37] Basko, M.M., Sunyaev, R.A., 1975, Sov. Phys. JETP, Vol. 41, 52.

[38] Pavlov, G.G., Yakovlev, D.G., 1976, Sov. Phys. JETP, Vol. 43, 389.

[39] Chandrasekhar, S., 1942, "Principles of Stellar Dynamics", University of Chicago Press, Chicago.

[40] Colgate, S.A., Petschek, A.G., 1980, Ap.J., Vol. 236, L115.

[41] Harwit, M., Salpeter, E.E., 1973, Ap.J., Vol. 186, L37.

[42] Woosley, S.E., Wallace, R.K., 1982, submitted to Ap. J.

[43] Hameury, J.M., Bonazzola, S., Heyvaerts , J., Ventura, J., 1981, Submitted to Astron. Astrophys.

[44] Helfand, D.J., Chanan, G.A., Novick, R., 1980, Nature, Vol. 283, 337.

[45] Kirk, J.G., Stoneham, R.J., 1982, M.N.R.A.S. (in press).

ABNORMAL NEUTRON STARS WITH ABNORMAL PROTONS

QU Qinyue(曲钦岳), WANG Zhenru(汪珍如),
LU Tan(陆 埮), CHU Yi(初 一)

(Department of Astronomy, Nanjing University,
Nanjing)

On the basis of the abnormal nuclear state suggested by T. D. Lee et al.[1], a new kind of compact stars — abnormal neutron stars — was proposed and discussed in paper[2] where abnormal neutron stars consist solely of abnormal neutrons. However, because of the difference in electromagnetic property between abnormal neutrons and abnormal protons and the same value of ϕ-field in abnormal neutron stars, there should be mass difference between them. Hence the following process of weak decay could occur

$$N_a \longrightarrow P_a + e^- + \bar{\nu} \ , \tag{1}$$

where N_a and P_a represent abnormal neutron and abnormal proton, respectively. Let E_{Fe}, E_{Fn} and E_{Fp} be the Fermi energy of electrons, abnormal neutrons and abnormal protons, respectively. Only when the equation

$$E_{Fn} - E_{Fp} = E_{Fe} \tag{2}$$

is satisfied, the weak decay of abnormal neutrons will be forbidden, and abnormal neutrons become stable. From Eq.

(2) we can obtain that the number of abnormal protons can not be neglected in comparison with the number of abnormal neutrons.* So it is necessary to refine the previous work and to discuss the abnormal neutron stars with abnormal protons. The homogeneous models were discussed in paper[4], the main conclusions of which are as follows: a) The mass of abnormal neutron stars with abnormal protons is lower than that of abnormal neutron stars without abnormal protons. This result is expective because of the declination of Fermi energy of abnormal neutrons by the appearance of abnormal protons. b) The maximum mass of abnormal neutron stars with abnormal protons is about 3.5 $M_\odot$ and appears in higher density compared with those without abnormal protons. c) There are lots of abnormal protons in an abnormal neutron star, and n_P/n_N will increase with n. Its maximum is about 35% for stable stars.

Now, the inhomogeneous models are discussed.[5] They are based on the fundamental differential equations of stellar structure:

$$\frac{dP(r)}{dr} = -\frac{GM(r)\rho(r)}{r^2}\left[1+\frac{P(r)}{\rho(r)}\right]\left[1+\frac{4\pi r^3 p(r)}{M(r)}\right]\left[1-\frac{2GM(r)}{r}\right]^{-1} \tag{3}$$

$$\frac{dM(r)}{dr} = 4\pi r^2\rho(r); \tag{4}$$

and the equation of state

$$p = n\,\frac{d\rho}{dn} - \rho, \tag{5}$$

where the energy density

$$\rho = U_\phi + E_N + E_P + E_e$$

---

*Similar result has been obtained for relativistic case in normal neutron stars.[3]

$$= C + D\left(\frac{n_N^{4/3} + n_P^{4/3}}{1-\beta n^{1/3}} + n_P^{4/3}\right). \tag{6}$$

So we have

$$P = -C + \frac{Dn}{3\beta}\frac{(2-A)^2}{A-1}\left(\frac{1}{A^3+1}\right)^{1/3}\left[\frac{A}{2-A} + \frac{1}{A-1} + \frac{A^3}{A^3+1}\right], \tag{7}$$

where U_ϕ is the potential of ϕ-field, E_N, E_P and E_e are the energy density of abnormal neutrons, abnormal protons and electrons, respectively. n_N, n_P and n are the number densities of abnormal neutrons, abnormal protons and nucleons, respectively. Both C and D are constants,[5] $A = 2 - \beta n^{1/3}$, $\beta = (\frac{4\pi}{3})^{1/3}(0.8\alpha)$, where α is the range of repulsive and other quantities have their common meanings. 52 Models are calculated for two kinds of repulsive range (α= 0.60f, 0.62f) and 26 values of center number density. The main results are given in Figs. 1-4, and the following

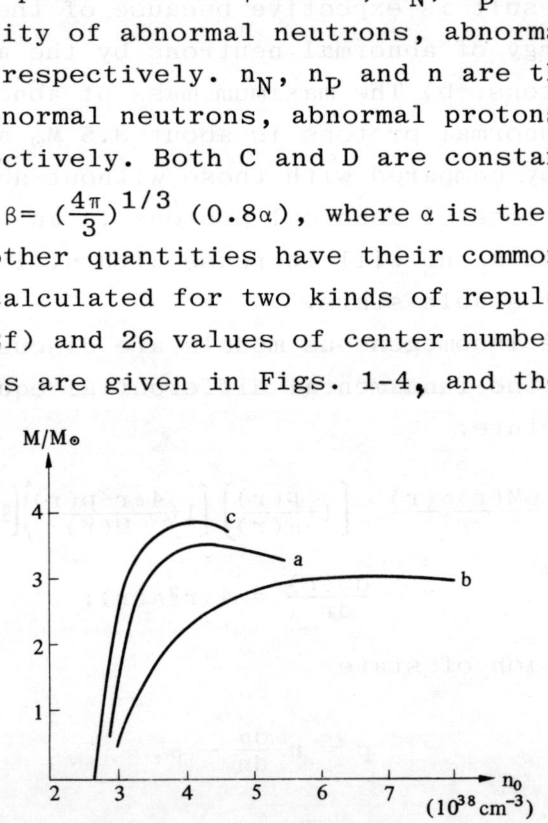

Fig. 1 M-n_0 relations of abnormal neutron stars (α= 0.60f)

a—homogeneous models[4]

b—inhomogeneous models

c—homogeneous models without abnormal protons[2]

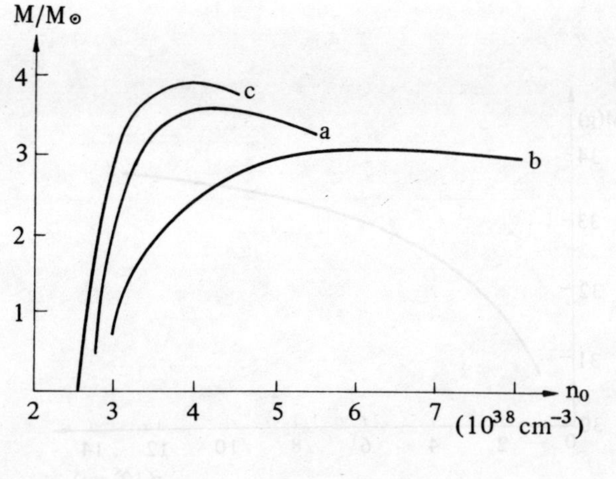

Fig. 2 M—n_0 relation of abnormal neutron stars (α = 0.62f)
a — homogeneous models[4]
b — inhomogeneous models
c — homogeneous models without abnormal prontons[2]

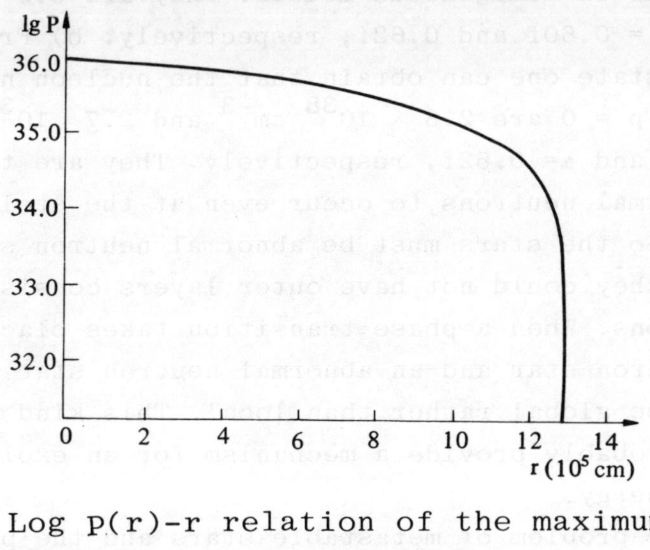

Fig. 3 Log p(r)-r relation of the maximum mass model for α = 0.62f
$n_0 = 6.20\times10^{38}$ cm^{-3} , M_{max} = 3.09 $M_\odot$

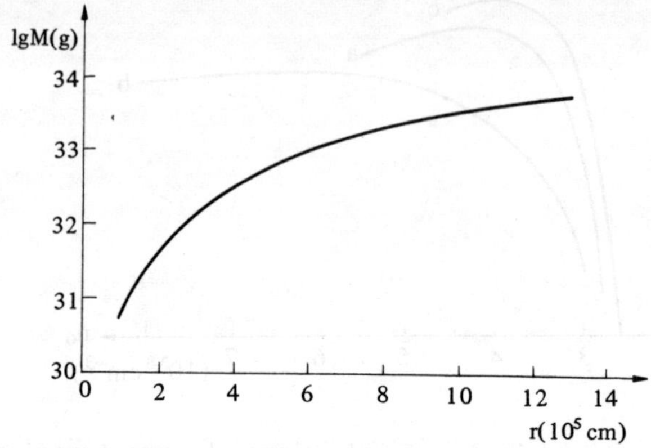

Fig. 4 Log M(r)-r relation of the maximum mass model for $\alpha = 0.62f$ $n_0 = 6.20 \times 10^{38}\ cm^{-3}$, $M_{max} = 3.09\ M_\odot$

conclusions can be drawn: a) The maximum mass of the inhomogeneous models of abnormal neutron stars is somewhat less than that of homogeneous models. They are 3.0 $M_\odot$ and 3.1 $M_\odot$ for $\alpha = 0.60f$ and $0.62f$, respectively. b) From the equation of state one can obtain that the nucleon number densities at $p = 0$ are $2.8 \times 10^{38}\ cm^{-3}$ and $2.7 \times 10^{38}\ cm^{-3}$ for $\alpha = 0.60f$ and $\alpha = 0.62f$, respectively. They are too large to permit normal neutrons to occur even at the stellar surface.[2] So the stars must be abnormal neutron stars as a whole and they could not have outer layers consisting of normal neutrons. When a phase transition takes place between a normal neutron star and an abnormal neutron star, the change must be global rather than local. This kind of processes may probably provide a mechanism for an explosive release of energy.

As to the problem of metastable stars and the phase transition between metastable and stable stars, we come to the following conclusions by use of homogeneous models[6]. Let M_n and M_a be the mass of a normal neutron star and an abnormal neutron star respectively. We obtain $M_n > M_a$ for the same value of the total nucleon number in the range of stellar mass from 0.2 $M_{\odot}$ to the maximum mass of normal neutron stars (Fig. 5). This means that abnormal neutron

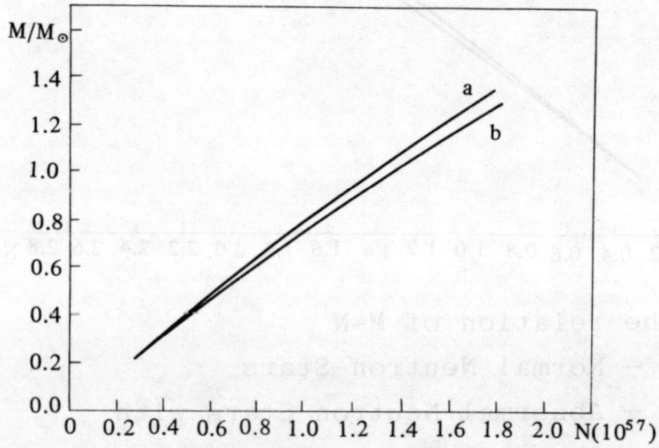

Fig. 5a The relation of M-N
a -- Normal Neutron Stars
b — Abnormal Neutron Stars with Abnormal Protons

stars are stable and normal neutron stars are metastable. An amount of energy of $10^{52} - 10^{53}$ erg would be released by a phase transition from a normal neutron star to an abnormal neutron star. The phase transition would take place through a strong interaction and a weak decay. In the strong interaction, all normal neutrons change into abnormal neutrons and the energy released is mainly kinetic energy (thermal energy). While in the weak decay, the abnormal neutrons change into abnormal protons by Eq. (1) until Eq. (2) is satisfied and the energy released is mainly in the form of neutrino emission accompanied by thermal energy. When a normal neutron star changes into

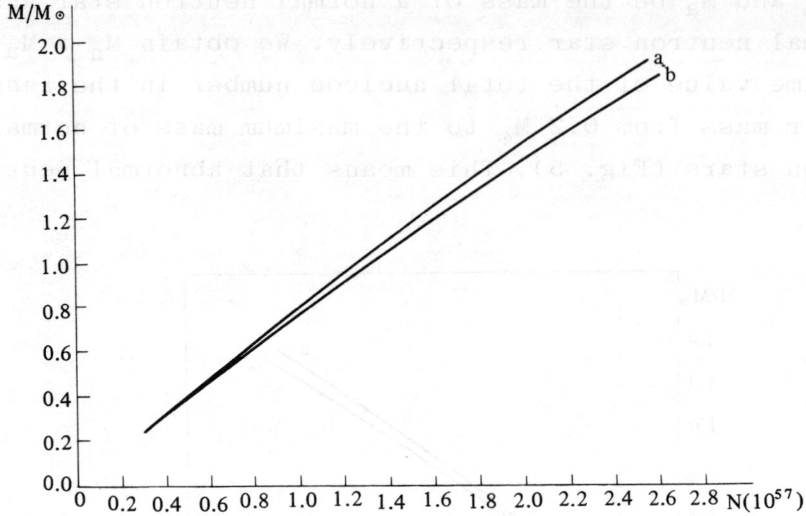

Fig. 5b The relation of M-N
a — Normal Neutron Stars
b — Abnormal Neutron Stars with Abnormal Protons

an abnormal neutron star, there would be a tremendous emission of neutrinos and electromagnetic radiation.

REFERENCES

[1] Lee, T.D., 1975, Rev. Mod. Phys., Vol. 47, 267.
[2] Fang Lizhi, Qu Qinyue, Wang Zhenru, Lu Tan, Luo Liaofu, 1978, Scientia Sinica, No. 4, 406 (in Chinese); 1979, Scientia Sinica, Vol. 22, No. 2, 187 (in English).
[3] Zeldovich, Y.B., Novikov, I.D., 1971, "Relativistic Astrophysics", Vol. 1, University of Chicago press, 189.
[4] Qu Qinyue, Wang Zhenru, Chu Yi, Lu Tan, 1981, Kexue Tongbao, Vol. 26, No. 23, 1436 (in Chinese).

[5] Wang Zhenru, Qu Qinyue, Lu Tan, Chu Yi, 1982, <u>Commun. of Theor. Phys</u>.
[6] Qu Qinyue, Wang Zhenru, 1982, <u>Commun. of Theor. Phys</u>. (to be published).

THE QUARK-CLUSTER PHASE OF SUPERDENSE MATTER AND A POSSIBLE MECHANISM OF STRONG FIELDS IN NEUTRON STARS

YANG Guochen (杨国琛)

(Hebei Institute of Technology)

LUO Liaofu (罗辽复)

(Neimenggu University)

The microscopic mechanism of the strong fields in neutron stars is not yet clear. We shall suggest a possible mechanism tentatively in this paper[1]. From the phenomenological QCD (quantum colordynamics) approach the interaction between quarks is expressed by[2]

$$V_{ij} = -c\lambda_i\sigma_i\cdot\lambda_j\sigma_j + e\lambda_i\cdot\lambda_j. \tag{1}$$

If the density of superdense matter is not too high to produce the free-quark phase transition[3] and the contribution of the strange quark is neglected, then one may assume that (u d d) in the matter forms a 3-quark cluster. It is assumed that the quark cluster has spin $\sigma_3=\pm 1$ and colour $\lambda_3=0$, $\lambda_8=0$. Based on these assumptions the following configurations of the cluster may be constructed. For $\sigma_3=+1$ one has

$$A_+ = \frac{1}{6}\{[(2d_1^\alpha d_1^\beta u_2^\gamma - u_1^\alpha d_1^\beta d_2^\gamma - d_1^\alpha u_1^\beta d_2^\gamma) - (2d_1^\beta d_1^\alpha u_2^\gamma - u_1^\beta d_1^\alpha d_2^\gamma - d_1^\beta u_1^\alpha d_2^\gamma)] + \text{permutations}\}$$

$$B_+ = \frac{1}{6}\{[(2d_1^\gamma d_1^\alpha u_2^\beta - u_1^\gamma d_1^\alpha d_2^\beta - d_1^\gamma u_1^\alpha d_2^\beta) - (2d_1^\alpha d_1^\gamma u_2^\beta - u_1^\alpha d_1^\gamma d_2^\beta - d_1^\alpha u_1^\gamma d_2^\beta)] + \text{permutations}\}$$

$$C_+ = \frac{1}{6}\{[(2d_1^\beta d_1^\gamma u_2^\alpha - u_1^\beta d_1^\gamma d_2^\alpha - d_1^\beta u_1^\gamma d_2^\alpha) - (2d_1^\gamma d_1^\beta u_2^\alpha - u_1^\gamma d_1^\beta d_2^\alpha - d_1^\gamma u_1^\beta d_2^\alpha)] + \text{permutations}\}$$

$$A'_+ = \frac{1}{\sqrt{12}}\{[(u_1^\alpha d_1^\beta d_2^\gamma - d_1^\alpha u_1^\beta d_2^\gamma) + (u_1^\beta d_1^\alpha d_2^\gamma - d_1^\beta u_1^\alpha d_2^\gamma)] + \text{permutations}\}$$

$$B'_+ = \frac{1}{\sqrt{12}}\{[(u_1^\gamma d_1^\alpha d_2^\beta - d_1^\gamma u_1^\alpha d_2^\beta) + (u_1^\alpha d_1^\gamma d_2^\beta - d_1^\alpha u_1^\gamma d_2^\beta)] + \text{permutations}\} \quad (2)$$

$$C'_+ = \frac{1}{\sqrt{12}}\{[(u_1^\beta d_1^\gamma d_2^\alpha - d_1^\beta u_1^\gamma d_2^\alpha) + (u_1^\gamma d_1^\beta d_2^\alpha - d_1^\gamma u_1^\beta d_2^\alpha)] + \text{permutations}\},$$

where the subscripts 1,2 refer to spin and α, β, γ to colour. The $\sigma_3 = -1$ clusters $A_-, B_-, C_-, A'_-, B'_-$ and C'_- are obtained by interchanging the subscripts 1 and 2. In constructing these wave functions the conservation of isospin has been used since the interaction (1) does not violate the symmetries of isospin.

The quantum numbers of these configurations are listed in Table 1:

Table 1. Quantum numbers of quark clusters

	λ_3	λ_8	σ_3	$\lambda_3\sigma_3$	$\sqrt{3}\lambda_8\sigma_3$
$A_\pm$	0	0	±1	0	±4
$B_\pm$	0	0	±1	±2	∓2
$C_\pm$	0	0	±1	∓2	∓2
$A'_\pm$	0	0	±1	0	±4
$B'_\pm$	0	0	±1	±2	∓2
$C'_\pm$	0	0	±1	∓2	∓2

From Eqs. (1) and (2) one finds the interaction energies in a cluster

$$U = \begin{cases} -\frac{8}{3}c - 4e & (\text{for } A_{\pm}, B_{\pm}, C_{\pm}) \\ \frac{4}{3}c & (\text{for } A'_{\pm}, B'_{\pm}, C'_{\pm}) \end{cases} \tag{3}$$

which are about 230 MeV or 430 MeV higher than the corresponding values (-8c-8e) for neutrons if

$$c \simeq 20 \text{ MeV}; \; e \simeq 30 \text{ MeV} \tag{4}$$

are used[2]. Therefore the cluster configurations are not energy-favourable. However, if the density of neutron matter is high enough (for example, 0.3 baryons fm^{-3}) and the distance between two neighbouring clusters is near the values between two quarks in the interior of a neutron, the colour magnetic interactions between clusters should not be neglected and one may use Eqs. (1) and (4) to estimate them. For normal neutrons of colour singlet the direct colour interaction between them does not exist though the two-gluon-exchange would cause the Van-der-Waals energy due to the colour polarization of neighbouring nucleons[4]. However the configurations (2) are not colour singlets and direct colour interaction between them is important. It the energy due to the quark exachange between two clusters is crudely equal to the nuclear force in normal states[5], then the additional energy between clusters in a new phase comes totally from the direct colour interaction. The calculated results are listed in Table. 2. Table 2 shows that the colour electric interaction contributes nothing because of $\lambda_3=\lambda_8=0$. The interaction energy for configurations $A'_+ A'_-$ etc. is equal to that of $A_+ A_-$ etc. Since the configurations $A'_{\pm}$ $B'_{\pm}$ and $C'_{\pm}$ are about 200 MeV higher than $A_{\pm}$ $B_{\pm}$ and $C_{\pm}$ the contribution of the former to the new phase can always be neglected unless the temperature T>200 MeV k^{-1}. Using the method of discussing the Keesom force in polar moleculars one may take the average for the interaction energy between clusters and obtain

Table 2. Interaction energy between clusters

Neighbouring configurations	Interaction energy
$A_+A_-, B_+B_-, C_+C_-, A_-A_+, B_-B_+, C_-C_+$	16/3 c
$A_+A_+, B_+B_+, C_+C_+, A_-A_-, B_-B_-, C_-C_-$	-16/3 c
$A_+B_+, B_+C_+, C_+A_+, A_-B_-, B_-C_-, C_-A_-$	8/3 c
$A_+B_-, B_+C_-, C_+A_-, A_-B_+, B_-C_+, C_-A_+$	-8/3 c

$$V_p = \left\{-\frac{16}{3}c\cdot\exp\left(\frac{\frac{16}{3}c}{kT}\right)+\frac{8}{3}c\cdot 2\exp\left(\frac{-\frac{8}{3}c}{kT}\right)\right\}\Big/\left\{\exp\left(\frac{\frac{16}{3}c}{kT}\right)+2\exp\left(\frac{-\frac{8}{3}c}{kT}\right)\right\}\simeq-\frac{16}{3}c \tag{5}$$

for parallel spins and

$$V_A = \left\{\frac{16}{3}c.\exp\left(\frac{-\frac{16}{3}c}{kT}\right)-\frac{8}{3}c\cdot 2\exp\left(\frac{\frac{8}{3}c}{kT}\right)\right\}\Big/\left\{\exp\left(\frac{-\frac{16}{3}c}{kT}\right)+2\exp\left(\frac{\frac{8}{3}c}{kT}\right)\right\}\simeq-\frac{8}{3}c \tag{6}$$

for opposite spins. Therefore the energy of quark-cluster phase is much lowered due to the colour magnetic interaction. For the allocation number 12, the total energy per baryon in quark cluster phase with parallel (antiparallel) spins is about 410 MeV (90 MeV). Lower than that in the normal phase of neutron. If neutron matter is indeed in quark cluster phase then the state with parallel spins is more energy-favourable. The Curie temperature

$$T_c = -2(V_p - V_A)/k \simeq 10^{12}K, \tag{7}$$

below which the long distance order in spins of quark clusters would occur. Of course, the above discussion is very crude. Some other factors should be considered seriously, such as the quark exchange effect due to colour nonsinglet and the contributions of colour confinement's force between clusters. However, the proposal on the possibility of the existence of a new phase between normal nucleon and free quark is intresting. The preliminary result, especially with regard to the long distance order in spins in superdense matter, is encouraging. In terms of quark moments, from Eq. (2) one can prove the magnetic moment of a cluster

$$\mu = \frac{1}{3} (4\mu_d - \mu_u), \tag{8}$$

which equals to a neutron's moment. If the number density n is taken to be 10^{39} cm^{-3}, then the field corresponding to the saturate magnetization is

$$B = 4\pi n\mu \simeq 10^{16} \text{ gauss.} \tag{9}$$

This explains the strong field in the center of the star. In fact, a ferromagnetic-like domain structure would appear and the average of the field in quark cluster phase is somewhat lower than the value given by Eq. (9). If the radius of superdense core in a neutron star is taken to be 1 km and the field in the core is taken to be 10^{15} gauss, then to the surface of the star the field strength would decrease by three orders of magnitude, i.e. $B \simeq 10^{12}$ gauss in agreement with observations.

The authors are indebted to Prof. Lu Tan for helpful discussions.

REFERENCES

[1] Yang Guochen, Luo Liaofu, 1982, Acta Scientiarum Naturalium Universitatis Intramongolicae, Vol. 13 (1), 59 (in Chinese); Yang Guochen, Luo Liaofu, Lu Tan, 1980, Report on the National Symposium on Cosmic Rays and High Energy Astrophysics, Qindao, China.

[2] DeGrand, T. et al., 1975, Phys. Rev., Vol. D12, 2060.

[3] Aerts, A.Th. M. et al., 1978, Phys. Rev., Vol. D17, 260.

[4] Freedman, B.A, McLerran, L.D., 1978, Phys. Rev., Vol. D17, 1109.

[5] Alvarez, E. et al., 1981, Astron. Astrophys., Vol. 98, 390.

[6] Fishbane, P.M., Grisaru, M.T., 1978, Phys. Lett., Vol. B74, 98.

[7] Fulli, Y., Mimo, K., 1978, Phys. Lett., Vol. 79B, 138.

[8] De Tar, C., 1978, Phys. Rev., Vol. D17, 323; 1979, Phys. Rev., Vol. D19, 1028.

[9] Harvey, M., 1981, Nucl. Phys., Vol. A352, 326.

[10] Vento, V. et al., 1981, Phys. Lett., Vol. 103B, 285.

NEUTRINO CYCLOTRON RADIATION FROM SUPERFLUID VORTEXES IN NEUTRON STARS: A NEW MECHANISM FOR PULSAR SPIN DOWN

PENG Qiuhe(彭秋和), HUANG Keliang(黄克谅), HUANG Jiehao(黄介浩)

(Department of Astronomy, Nanjing University, Nanjing)

In this paper we discuss a new type of neutrino emission mechanism — neutrino cyclotron radiation, and its effects.

According to the Weinberg-Salam theory on elementary particles, when neutrons move in a circle, neutrino pairs $(\nu,\bar{\nu})$ can be emitted owing to the weak interaction of the neutral current. This is a cyclotron-synchrotron emission mechanism which is different from those well-known neutrino processes. Even at absolute zero this radiation still exists.

The power spectrum for one neutron in circulation motion is[1]

$$p\ (\omega,\ r) = 4.7\times10^{-159}\ \omega^8 r^2, \tag{1}$$

where ω angular velocity, r a distance from axis. The frequency of emitted neutrino pairs is equal to the angular velocity of the neutron, ω.

The total power of neutrino cyclotron radiation from the whole neutron star is

$$W_\nu = 1.0\times10^{57}\ \overline{(\Delta^8/\rho^{2/3})}\ \frac{\overline{n^7}}{\overline{n}}\ R^3\Omega \equiv BG(n)\Omega, \tag{2}$$

where Δ the energy gap, ρ the mass density, n the circulation quantum number of vortex line, R the radius of neutron star, and Ω the angular velocity of neutron star, the bar is used to represent an average of some physical quantities for all the vortex lines.

In deriving the above formula, a few hypotheses have been taken. They are: the classical approximation and non-equilibrium state hypothesis. Because there is a process of "local circulation" thru which neutrino cyclotron radiation can be emitted without a change in the quantum number. This process is very much analogous to Ekman pumping[2], so we use the phenomenological method to explore the radiation problem in this paper.

On the other hand, we would expect that the interior of relative young neutron stars are not already in thermodynamic equilibrium state, but are still far from thermodynamic equilibrium. It would be expected to nucleate the superfluid vortexes in $n > 1$ state, so we could expect to have the superfluid vortexes in $n > 1$ state for the older pulsars also if we accept the Ekman mechanism.

Thinking only about magnetic decay model and neutrino cyclotron radiation, then we have the spindown rate of pulsars

$$\dot{P} = \frac{8\pi^2}{3c^3}\frac{\mu_0^2}{I}e^{-\xi t}P^{-1} + \frac{B}{2\pi I}G(n)P^2, \tag{3}$$

where I the moment of inertia, μ_0 the magnetic moment, ξ the characteristic time of magnetic decay, P the period of pulsars.

Obviously, there is a transition period at which one energy dissipation mechanism is replaced by another one, then the relation between $\dot{P}$ and P will be totally different.

The theoretical prediction of transition period is about 1 sec. The statistical study of 269 pulsars with measured P and $\dot{P}$ shows that for $P < 0.5$ sec and $P > 1.25$ sec the relation between log $\dot{P}$ and log P is significantly different at the 1% level which is just what is predicted from our model. This relation changes from an inverse proportionality to a direct one. The transition periods are $0.5 \text{ sec} < P < 1.25$ sec. The Fig. 1 is a diagram of log $\dot{P}$ vs log P for pulsars with

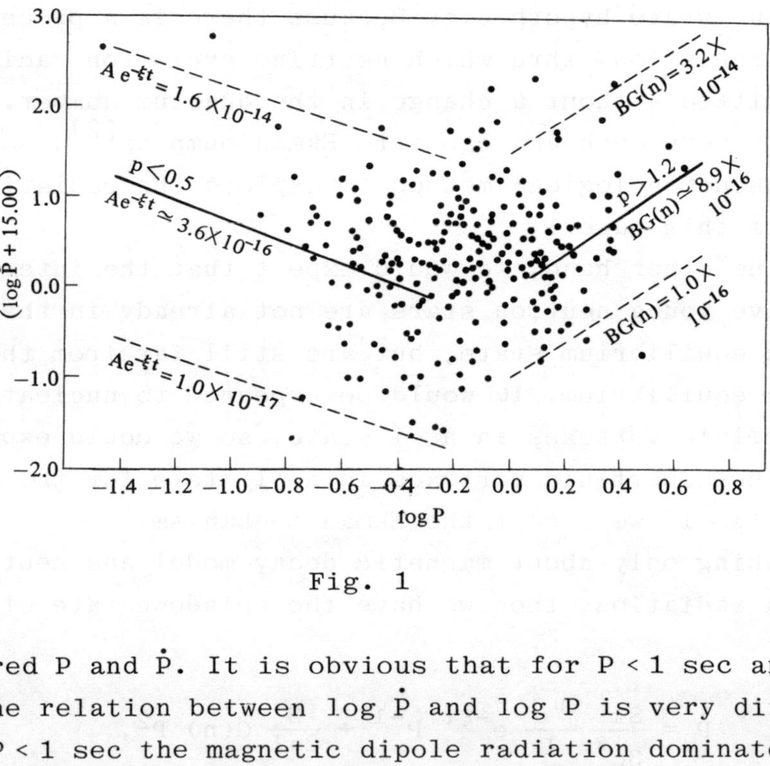

Fig. 1

measured P and $\dot{P}$. It is obvious that for $P < 1$ sec and $P > 1$ sec the relation between log $\dot{P}$ and log P is very different. When $P < 1$ sec the magnetic dipole radiation dominates; when $P > 1$ sec the neutrino radiation dominates.

REFERENCES

[1] Luo Liaofu, Lu Tan, Fang Lizhi, 1978, Nature Journal (in Chinese), Vol. 1, 84.

[2] Anderson, P.W., Pines, D., Ruderman, M., Shaham, J., 1978, J. Low Temperature Phys., Vol. 30, 839.

III. SUPERNOVAE

Xi
(1054)
The guest star had pointed rays on all sides ......
Chapter 25 of Sung-hui-yao
Oh! It agrees with my model excellently!
Hillebrandt

SUPERNOVA THEORY

Wolfgang Hillebrandt

(Max-Planck-Institut für Physik und Astrophysik, Institut für Astrophysik, München, Federal Republic of Germany)

SUMMARY

The present status of core-collapse models for type II supernovae is reviewed. It is demonstrated that stars of about 10 $M_\odot$ on the main sequence can eject some matter in a supernova explosion, leaving behind a neutron star of around 1.5 $M_\odot$. For these models the explosion mechanism is simply a reflected hydrodynamical shock wave from the central neutron star. This simple and straightforward mechanism probably does not work for more massive stars ($M \gtrsim 15\ M_\odot$). In fact, recent numerical studies indicate, that those stars are more likely to form black holes if they do not rotate sufficiently fast or possess strong magnetic fields, or both.

I. SOME OBSERVATIONAL FACTS

The modern era of supernova research began about 100 years ago on August 31, 1885, when Hartwig discovered a "nova" near the center of the Andromeda Galaxy (M31), which became invisible about 18 months later. In 1919 Lundmark estimated the distance of M31 to be about 7×10^5 lyr, and by that time it became obvious, that Hartwig's nova had been several 1000 times brighter than a normal nova. A similar event was observed in 1895 in NGC 5253 ("nova" Z Centauri), which appeared to be approximately 5 times brighter than the entire galaxy. Systematic searches for supernovae, performed predominantly by Zwicky, lead to the discovery of 54 supernovae in the years up to 1956, and, due to improved observational techniques, 82 further supernovae in the years 1958 to 1963, all of course in external galaxies. Recent estimates obtained from historical supernovae in our own galaxy, radio observations of old supernova remnants and supernovae in external galaxies give an average rate of about 1 supernova per (70±30) years in giant spirals like our own.

Table 1. Some observed properties of supernovae

Energetics	Light curves and spectra	type I	type II
Maximum luminosity $\simeq 10^{44}$ erg s^{-1}	Near maximum light for	$\simeq$30 days	$\lesssim$10 days
Energy in visual light $\simeq 10^{48}$–10^{50} erg	Plateau	no	yes, 100 days
Total energy output $\simeq 10^{51}$–10^{52} erg	Duration	$\simeq$2 years	~1 year
Expansion velocity $\simeq 10^4$ km s^{-1}	H-lines	no	strong
Temperature near maximum light $\gtrsim$15000 K	Abundances	Co, Fe?	solar

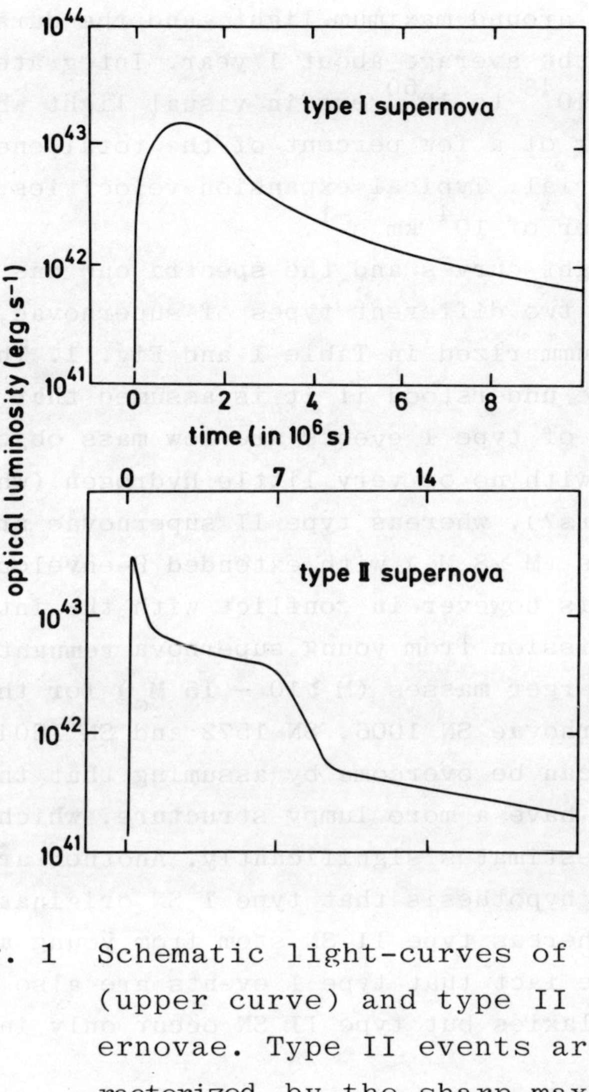

Fig. 1 Schematic light-curves of type I (upper curve) and type II supernovae. Type II events are characterized by the sharp maximum and by the pronounced plateau after maximum light, which indicates the presence of an extended stellar envelope.

For a typical supernova outburst the optical luminosity

is about 10^{10} $L_\odot$ around maximum light, and the duration of the event is on the average about 1 year. Integrated energies range from 10^{48} to 10^{50} erg in visual light which is only of the order of a few percent of the total energy of the ejected material. Typical expansion velocities are found to be of the order of 10^4 km s^{-1}.

From the light curves and the spectra one has to distinguish at least two different types of supernovae, the properties are summarized in Table 1 and Fig. 1. Those properties can be understood if it is assumed that the progenitor stars of type I events are low mass objects of about 1 to 2 $M_\odot$ with no or very little hydrogen (white dwarfs or He-stars?), whereas type II supernovae are exploding massive stars ($M \gtrsim 8$ $M_\odot$) with extended H-envelopes. This explanation is however in conflict with the interpretation of X-ray emission from young supernova remnants, which indicates much larger masses ($M \gtrsim 10 - 15$ $M_\odot$) for the historical type I supernovae SN 1006, SN 1572 and SN 1604. But this difficulty can be overcome by assuming that the X-ray emitting regions have a more lumpy structure, which would reduce the mass estimates significantly. Another argument in favour of the hypothesis that type I SN originate from low mass stars whereas type II SN stem from young and massive stars is the fact that type I events are also observed in elliptical galaxies but type II SN occur only in spiral arms.

In order to obtain further restrictions for theoretical models one would like to know the element abundances in specific objects. Unfortunately the observations are again not very decisive. In the Crab nebula only an overabundance of He is observed, consistent with about 1 $M_\odot$ of He being admixed to the nebula (we will return to this point later). In Cas A overabundances of N, S, Ar and maybe Ca are observed, indicating that the exploding star had undergone at least O-burning. In Pupis A overabundances of O and Ne are found which are products of C-burning, but the situation is not very clear.

Finally one can hope to learn something about the explosion mechanism from a correlation between supernovae and neutron stars, because it is likely that if a compact object is formed in a supernova, the explosion energy will ultimately come from the increase in gravitational binding. But again the observations do not give a clear correlation, as some young supernova remnants have a neutron star (Crab, Vela, probably RCW 103) and others do not (Cas A, SN 1006, probably also SN 1572, RCW 86, and others). In comparing the birthrates of pulsars and the supernova rates and the total mass of some low mass X-ray binaries one must also consider the possibility that some neutron stars are born without a supernova explosion.

So in conclusion we have to confess that supernova observations do not place strong restrictions on our theoretical models. Moreover it is very likely that supernovae are a much more heterogeneous class of objects than it is indicated by light curves and spectra only. In particular it may well be that we drastically oversimplify the models if we assume that all supernovae can be explained by a one parameter family, the only parameter being the stellar mass.

II. MODELS FOR TYPE II SUPERNOVAE

In order to explain the observed total energy of a supernova outburst of about 10^{51} erg one can think of two different sources which are in principle able to supply sufficient amounts of energy: thermonuclear burning and gravitational binding. If thermonuclear reactions are considered the requirement is that at least about 0.5 to 1 $M_\odot$ of He, C or O are fused to iron group elements. Consequently either white dwarfs or degenerate carbon cores of stars with masses of around 6 to 8 $M_\odot$ can meet this requirement. The first possibility will lead, if it should turn out to work, to type I supernovae because of the missing hydrogen envelope and therefore will not be discussed any further. The second scenario suffers from the fact that the energy available from nuclear burning is just sufficient and the question has to be answered whether or not really a detonation front forms which then can disrupt the star. This question has so far not been answered, mainly because of severe numerical difficulties. If finally a positive answer should be given, such a model would describe type II events without the formation of a neutron star.

In this article we will restrict ourselves to investigations of those models, which utilize gravitational energy to trigger the supernova explosion. In this case the situation is drastically different as far as the energy budget is concerned. The binding energy of a neutron star of 1 $M_\odot$ is about 2×10^{53} erg and therefore only about 1% (or even less) of the energy available is needed to explain the observations. But one may turn the argument around and say that if indeed gravitational energy is the source of the supernova event we have to search for very inefficient mechanisms with numerical codes. Such a situation is the nightmare of a theorist and it is not very surprising that until now there are discrepancies between different numerical studies.

(1) Evolution of Massive Stars—An Overview

As was mentioned before we are mainly interested here in those stars that can undergo gravitational collapse at the end of hydrostatic thermonuclear evolution. This requirement restricts the mass range of possible candidates to stars more massive than about 10 $M_\odot$. For a 15 $M_\odot$ star this evolution is summarized in Table 2, and similar evolutionary histories are obtained for stars of larger mass. The main feature all those models have in common is the formation of an iron-nickel-core of about 1.5 $M_\odot$, surrounded by "onionshells" in which nuclear burning is still going on. This fact is usually called "core-convergence" and reflects a general property of stellar evolution, namely that in order to use the "ashes" of a nuclear burning stage as "fuel" for the next one the star has to contract. Therefore the core mass M_{core} has to be larger than the Chandrasekhar mass M_{Ch},

$$M_{Ch} \simeq 5.7\, Y_e^2, \tag{1}$$

where Y_e is the electron concentration. After core-oxygen burning the electron concentration is still close to 0.5, which means that $M_{Ch} \simeq 1.4\, M_\odot$. There are, of course, uncertainties in the exact mass of the Fe-Ni core after Si-burning, which depends on both, the stellar mass (see Table 3) and the numerical treatment of convection, but also on the efficiency of neutrino cooling.

After core-Si burning Y_e has dropped to about 0.45, $M_{Ch} \simeq 1.2\, M_\odot$, and therefore the Fe-Ni cores will contract further and this time no other nuclear fuel can be ignited, because the nuclear binding energy per nucleon is maximum for nuclei of the iron group. The core becomes even dynamically unstable. At a central density $\rho_c \simeq 10^{10}$ g cm^{-3}, temperature $T_c \simeq 10^{10}$ K, and $Y_e^c \simeq 0.44$, the total pressure of the gas is about $p_c \simeq 10^{28}$ erg cm^{-3}, the energy density is about $\varepsilon_c \simeq 3 \times 10^{28}$ erg cm^{-3} and the entropy per baryon $S/N \simeq 1\, k_B$, where k_B is Boltzmann's constant. The pressure of the ions

$$p_{ion} \simeq n_{ion}\, k_B\, T \simeq 10^{26} \text{ erg cm}^{-3} \tag{2}$$

(mostly Cr, Fe, Ni) is only about 1% of the pressure of the electrons, which on the other hand, are highly relativistic

$$\mu_e \simeq 10 \text{ MeV} >> m_e c^2 \simeq 0.5 \text{ MeV} \qquad (3)$$

Table 2. Evolutionary stages of a 15 $M_\odot$ Pop. I star (from [1]). Given are central temperature, density and the radius of the star at the onset of each nuclear burning stage

	Temperature (K)	Density ($g\ cm^{-3}$)	Duration	Radius (km)
H-burning ($H \to He$)	3.4×10^7	5.9	1.2×10^7 yr	3.2×10^6
He-burning ($He \to C,O$)	1.6×10^8	1.3×10^3	1.4×10^6 yr	2.2×10^7
C-burning ($C \to Mg,Ne,\ldots$)	6.2×10^8	1.7×10^5	8.8×10^4 yr	3.7×10^8
Ne-burning ($Ne \to Mg,Si,\ldots$)	1.3×10^9	1.6×10^6	7 yr	3.9×10^8
O-burning ($O \to Si,S,\ldots$)	1.9×10^9	9.7×10^6	1.7 yr	3.9×10^8
Si-burning ($Si \to Fe,Ni,\ldots$)	3.1×10^9	2.3×10^8	4 d	3.9×10^8
Collapse	8.3×10^9	6.0×10^9	0.35 s	3.9×10^8

Table 3. Dependence of the mass of the iron-nickel core on the main sequence mass of the stellar models and He-core mass after hydrogen burning

$M_{tot}/M_\odot$	10	~13(*)	15	25
$M_{He}/M_\odot$	2.7	4	4.5	9.5
$M_{Fe-Ni}/M_\odot$	1.36	1.45	1.55	1.62

(*) Approximately the main sequence mass of a 4 $M_\odot$ He-core evolved by Arnett (1977)[2] to the stage of core collapse and consequently the adiabatic index is

$$\gamma = \left(\frac{\partial \ln P}{\partial \ln \rho}\right)_S \simeq \frac{4}{3}, \tag{4}$$

close to the critical one.

At such high electron chemical potentials e^--captures on free protons and heavy nuclei become energetically favorable:

$$e^- + \left\{\begin{matrix} p \\ A(Z,N) \end{matrix}\right\} \to \left\{\begin{matrix} n \\ A(Z-1,N+1) \end{matrix}\right\} + \nu_e, \tag{5}$$

therefore the electron concentration is reduced and γ becomes smaller than 4/3. In addition also photo-disintegration of heavy nuclei

$$\gamma + A(N,Z) \longrightarrow \left\{\begin{matrix} A(N-1,Z) + n \\ A(N,Z-1) + p \\ A(N-2,Z-2) + \alpha \end{matrix}\right. \tag{6}$$

reduce the internal energy and cause the adiabatic index to drop below 4/3.

(2) The Physics of Stellar Collapse

As we have demonstrated the cores of massive stars become dynamically unstable once the central density exceeds about 10^{10} g cm^{-3}. The further evolution therefore has to be followed by hydrodynamical rather than hydrostatic evolutionary techniques. For this purpose the following input data are needed.

1. Initial models. Stellar models available at present have 10, 15 and 25 $M_\odot$ on the main sequence[1,3]. In addition Arnett (see e.g.[2]) has evolved 4, 8, 16 and 32 $M_\odot$ He-cores

to the onset of core-collapse. The first two of these latter models correspond roughly to the 15 and 25 $M_\odot$ models of Weaver et al. (1978)[1]

2. Hydrodynamics. In almost all numerical studies the partial differential equations are replaced by finite difference equations and the solution at time t^{n+1} is obtained from t^n by simple forward extrapolation (explicit methods). These methods then require that the timestep $t = t^{n+1} - t^n$ has to be smaller than min $(\Delta r_i/c_s)$, where Δr_i is the zone spacing and c_s the local velocity of sound, and similar conditions hold for other transported quantities such as neutrino number and energy density. For non-rotating models one can use without difficulties the general relativistic equations, but for rotating models only Newtonian physics has been used so far.

3. Equation of state and rate equations. At the high temperatures and densities under consideration, matter is in equilibrium with respect to all strong and electromagnetic interactions. We then can compute the nuclear composition at fixed electron (lepton) concentration by minimizing the free energy of the system, which for each nuclear species i reads

$$f_i = - n_i k_B T \left\{1+\ln\left[\frac{1}{n_i}\left(\frac{m_i k_B T}{2\pi\hbar^2}\right)^{3/2} \times\omega_i(T)\right]\right\} + m_i c^2 n_i + (\text{interaction terms}), \quad (7)$$

where we have assumed that the ions are approximately a (free) Boltzmann gas. Here n_i is the number density of species i, m_i its rest mass, and $\omega_i(T)$ the nuclear partition function. If we define an "effective" chemical potential μ_i' by the Boltzmann equation

$$e^{-\mu_i'/k_BT} = \frac{1}{n_i}\left(\frac{m_i kT}{2\pi\hbar^2}\right)^{3/2} \omega_i(T), \tag{8}$$

we obtain for the real chemical potentials

$$\mu_i = \left(\frac{\partial f_i}{\partial n_i}\right)_{T,Y_e} = \mu_i' + m_i c^2 + \text{(interaction terms)}. \tag{9}$$

With $\beta = (k_BT)^{-1}$ we obtain the equilibrium condition

$$\beta\mu_i = (A_i - Z_i)\eta_n + Z_i\eta_p + \beta B_i + \text{(interaction terms)}, \tag{10}$$

where $\eta_{n,p} = \beta\cdot\mu_{n,p}$ ("degeneracy parameter") and $B_i = (A_i-Z_i)m_nc^2 + zm_pc^2 - m_i$ The indices n and p denote neutrons and protons, respectively. Finally the number densities n_i are given in equilibrium by

$$n_i = \omega_i(\beta)\,(2\pi m_i^{(0)}/\hbar^2\beta)^{3/2} e^{\beta B_i^{(0)}} \times e^{\left\{(A_i-Z_i)\eta_n + Z_i\eta_p\right\}}, \tag{11}$$

where the index (0) indicates nuclear ground state properties and the partition function is given by

$$\begin{aligned} \omega_i(\beta) &= \sum_J (2I_J+1)e^{-\beta(B_i^{(0)}-B_i^{(z)})} \\ &= \sum_J (2I_J+1)e^{-\beta E_J} \end{aligned} \tag{12}$$

Here the sum goes over all excited states J with spin I_J and excitation energy E_J.

Unkown quantities that have to be determined from nuclear physics are ground state binding energy and partition function for each species i under consideration. Moreover the unknown degeracy parameters η_n and η_p can be determined by requiring baryon number conservation and charge neutrality.

In general it is found (see e.g.[4]) that nuclear interaction corrections are small if the density is below nuclear matter density and that corrections due to the finite size of the nuclei and Coulomb effects are only important for the nuclear composition but not for the equation of state. So one can obtain good approximations by simply neglecting them.

Weak interactions, such as

$$e^- + \begin{Bmatrix} p \\ A(Z,N) \end{Bmatrix} \rightarrow \begin{Bmatrix} n+\nu_e \\ A(Z-1,N+1)+\nu_e \end{Bmatrix} ,$$

$$\nu_e + n \rightarrow p + e^- , \tag{13}$$

$$\nu_e + e^- \rightarrow \nu_e + e^- ,$$

are normally not in equilibrium. Therefore one has to solve explicitly time-dependent equations in order to compute the lepton concentrations. These equations are of the type

$$\dot{Y}_e = \frac{dY_e}{df} = \lambda_{\nu n} Y_n - \lambda_{ep} Y_p - \Sigma \lambda_{ei} Y_i , \tag{14}$$

$$\dot{Y} = - \dot{Y}_e - q_\nu ,$$

where λ denotes (weak) reaction rates and q_ν is the net gain or loss of neutrinos due to reactions and/or transport. Similar equations hold for positrons and anti-neutrinos, but these are normally not very important because the positron chemical potential is

$$\mu_{e+} = -\mu_{e-} - 2m_e c^2 < -(20 - 100) \text{ MeV} \tag{15}$$

and therefore their concentration $n_{e+} \simeq \exp\{\mu_{e+}/k_B T\}$ is very small. In a realistic collapse calculation most of the e^--captures occur at densities between 10^{10} and several times 10^{11} g cm^{-3}. For typical entropies of around 1 to 2 k_B/N

this means that the free proton concentration is always sufficiently large so that e^--captures on heavy nuclei do not dominate the rates. Electron captures on free protons can be computed numerically exactly (see e.g. [5]) and therefore Eq. (14) can be solved without huge difficulties, provided a numerical scheme for the treatment of neutrino transport is available.

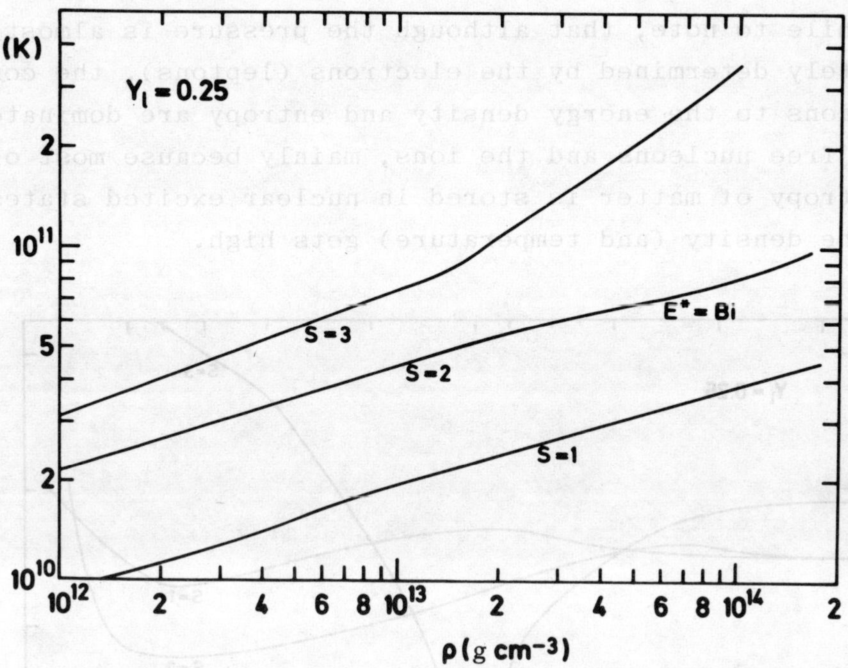

Fig. 2 Curves of constant entropy (from [4]). A total lepton concentration Y_e = 0.25 was chosen. The curves are labeled with the entropy in units of Boltzmann's constant per baryon. The slopes of S = 1 and 2 are close to those of a relativistic electron gas.

At very high densities ($\rho \gtrsim 5 \times 10^{12}$ g cm^{-3}) also weak interaction rates are sufficiently large to justify the assumption of a complete equilibrium, provided neutrinos are trapped, which is indeed the case, as we will demonstrate in the next section. Eq. (14) can then be replaced by the condition

$$\mu_{e^-} + \mu_p = \mu_n + \mu_\nu . \tag{16}$$

The results of a typical example for such a computation of the equation of state are shown in Figs. 2 and 3. Here it is worthwhile to note, that although the pressure is almost completely determined by the electrons (leptons), the contributions to the energy density and entropy are dominated by the free nucleons and the ions, mainly because most of the entropy of matter is stored in nuclear excited states once the density (and temperature) gets high.

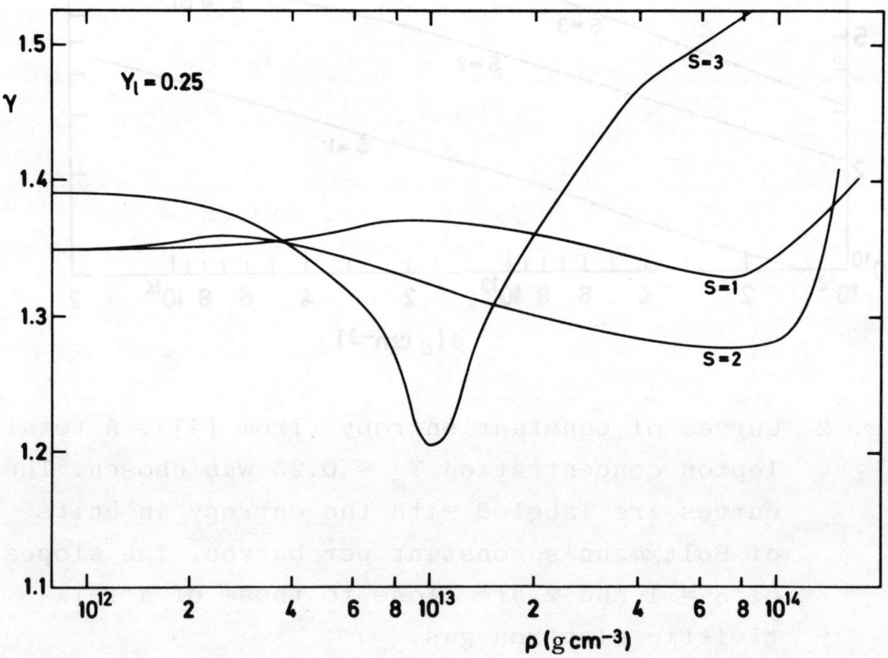

Fig. 3 Adiabatic index γ as a function of density for different values of S. One can see that up to nuclear matter density and for $S \lesssim 2$ γ stays very close the value 4/3 of a relativistic electron gas.

4. Neutrino transport. We do not want to discuss here in detail numerical schemes that have been applied to the supernova problem (see e.g.[6,7]) but only describe some of the basic physics. As we have seen in the preceding section, neutrinos are mostly produced by e^--captures on free protons. Therefore the average neutrino energy will be around $E_\nu \simeq \frac{2}{3} \varepsilon_f^e \simeq 10 - 100$ MeV, where ε_f^e is the electron Fermi energy. Contributions to the ν-opacities come from elastic scattering off nuclei,

$$\sigma \simeq \sigma_0 A^{-2} \left(\frac{E_\nu}{\text{MeV}}\right)^2 ; \ \sigma_0 \simeq 10^{-44} \ \text{cm}^2 , \qquad (17)$$

absorption by free neutrons $\nu_e + n \rightarrow p + e^-$, and electron scattering $\nu + e^- \rightarrow \nu + e^-$. At the densities where most of the neutrinos are produced the average mass number of heavy nuclei is $\bar{A} \simeq 80 - 100$, which gives a ν-mean free path of

$$\lambda_\nu \simeq \frac{1}{n_A \sigma} \simeq 10^7 \ \text{cm} \left(\frac{\rho}{10^{12}}\right)^{-1} \bar{A}^{-1} \left(\frac{E_\nu}{10 \ \text{MeV}}\right)^{-2} , \qquad (18)$$

and consequently matter becomes opaque for neutrinos if $\rho \gtrsim 5 \times 10^{11}$ g cm^{-3}. Assuming a typical length scale of 10^7 cm we find a diffusion time scale

$$\tau_{\text{diff}} \simeq \frac{l^2}{\frac{1}{3} c \lambda_\nu} \simeq 2\text{s} , \qquad (19)$$

which is much larger than the collapse timescale

$$E_{\text{coll}} \simeq \frac{400}{\sqrt{\rho}} \simeq 10^{-3} \ \text{s} , \qquad (20)$$

which means that ν's are trapped at such densities. Neutrino-electron scattering, on the other hand, tends to rapidly thermalize the neutrino distribution. So only for densities $5 \times 10^{10} \lesssim \rho(\text{g cm}^{-3}) \lesssim 5 \times 10^{11}$ one has to worry about neutrino transport.

There are two important consequences:

a) stellar collapse is almost adiabatic because only non-equilibrium weak interactions change the entropy.

b) Nuclei exist up to nuclear matter density, because the entropy remains low.

One expects, therefore, that the gravitational collapse will not be stopped before nuclear matter density ($\rho \gtrsim 2\times10^{14}$ g cm^{-3}), where the nuclei dissolve and the pressure from free nucleons becomes dominant.

(3) Core-collapse Supernova Models

1. Adiabatic models. Motivated by the fact, that during collapse the entropy of the core increases by less than about 0.5 k_B/N, one can think of simplifying the problem further by assuming that the entropy is strictly conserved unless in shock fronts kinetic energy is dissipated into heat. These models are called "adiabatic" and we will discuss them in some detail, because most properties of gravitational collapse are quite well described by this assumption. The energy conservation equation becomes simply

$$dE = -pdv + dQ_v \tag{21}$$

or $Tds = dQ_v$, where dQ_v gives the entropy increase due to viscous heating. Note, however, that in such an approach also the electron concentration stays constant ($Y_e \equiv Y_e^0 \simeq 0.44$; $\gamma \simeq \frac{4}{3}$). One can show quite general, that for $\gamma = \frac{4}{3}$ approximately the Chandrasekhar mass collapses homologously, because the infall velocity remains subsonic. With $Y_e \simeq 0.44$ one finds $M_{Ch} \simeq 1.2\ M_\odot$, which gives roughly the position of the sonic point ($v(r) = v_s$). Once the density exceeds nuclear matter density γ gets larger than 5/3 (equal to $\frac{5}{3}$

for a non-relativistic gas of free nucleons) and the central parts of the core decelerate, stop and bounce back, since, because of its inertia, matter overshoots its equilibrium density. A shock wave forms at the sonic point (see Fig. 4) near M_{Ch}. From the kinetic energy at the "last good homology" and the work done up to maximum compression one can estimate the energy originally put into the shock wave and finds $E_{tot} \simeq 10^{52}$ erg.

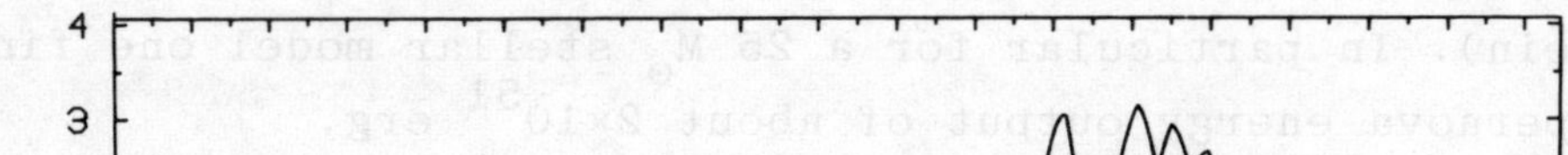

Fig. 4 Velocity versus mass for an adiabatically collapsing 25 $M_\odot$ star (only the innermost 1.8 $M_\odot$ are shown). Because electron captures were suppressed, the shock forms at 1.1 $M_\odot$ (curve on the left). About 10 ms later the shock has reached the edge of the iron nickel core (curve on the right). At that time, (positive) outward velocities are already close to escape velocity.

On its way out through the outer parts of the stellar core the shock is heavily damped by nuclear photodisintegrations of heavy nuclei, which cost about 8 MeV per nucleon. From the shock energy of about 10^{52} erg one estimates that this is sufficient to photodisintegrate about 0.5 $M_\odot$ and additional energy is added to the shock from the kinetic energy of the infalling matter. One therefore expects that the shock propagates out to at least 1.7 $M_\odot$, which is more than the core mass and consequently a supernova explosion should result. These general considerations are indeed confirmed by numerical studies (see e.g. [8], and references therein). In particular for a 25 $M_\odot$ stellar model one finds a supernova energy output of about 2×10^{51} erg.

There is, however, one necessary modification that has to be made. As we have discussed earlier, electron captures have to be included. The main effect is to decrease Y_e from 0.44 to about 0.35 at core bounce. Consequently the mass of the homologous core will be reduced to about 0.7 $M_\odot$. The problem is then, that much more matter has to be disintegrated. In fact numerical studies[9] show that adiabatic models with electron capture included do not give the desired explosion: The shock forms at 0.65 $M_\odot$ and changes into a standing accretion shock at about 1.35 $M_\odot$, again in agreement with our qualitative considerations.

2. Non-adiabatic models. Recently several core-collapse computations have been performed in which the assumption of adiabatic hydrodynamics has been dropped and neutrinos were included in a more or less sophisticated way (see e.g. the reviews by Hillebrandt (1982)[8] or Trimble (1982))[10]. These computations dealt with stars in the mass-range from 15 $M_\odot$ to about 50 $M_\odot$ but the outcome was always the same: None of the models exploded, the reason being mostly the damping of the shock due to disintegrations of heavy nuclei. One could have hoped that neutrino diffusion behind the shock through matter of low neutrino opacity ($\bar{A}\simeq1$ in matter of high entropy, $S\gtrsim5\ k_B/N$, behind the shock) would supply

enough energy to the shock in order to bring it to the edge of the core, but in all numerical studies this energy supply turned out to be too small, because neutrino diffusion times for the unshocked core were too long and in the shocked matter not enough energy was stored in neutrinos. Only models based on a 4 $M_\odot$ He-core, evolved by Arnett (1977)[2,7] to the stage of collapse were found to be close to an explosion. A typical result of such a computation is shown in Fig. 5.

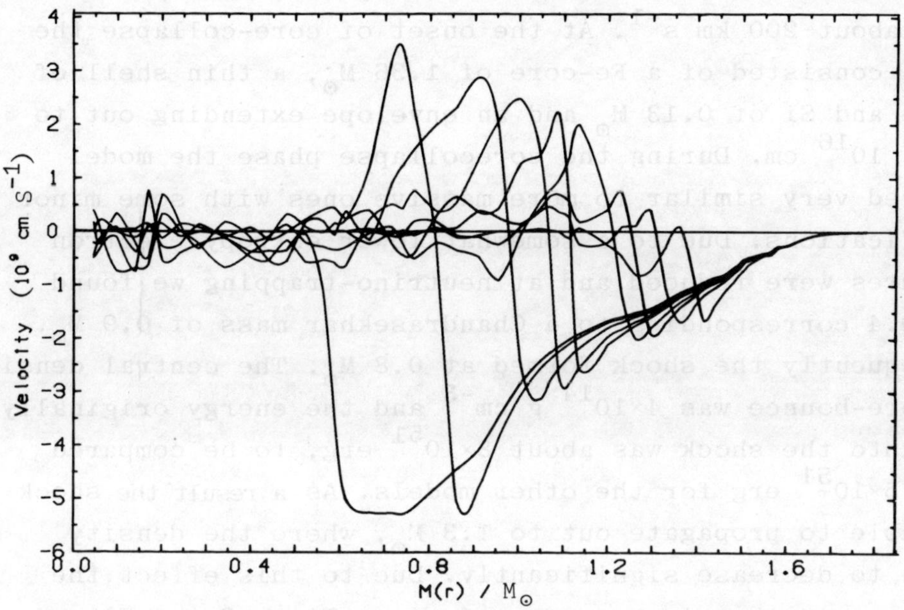

Fig. 5 Same as Fig. 4, but for a non-adiabatically collapsing 4 $M_\odot$ He-star. It can be seen that now the shock forms at 0.7 $M_\odot$ and changes into a standing accretion shock at about 1.3 $M_\odot$. In this case we did not find an explosion.

It became clear, however, that the core structure of

the model prior to collapse may play an important role in deciding which stellar models may lead to supernova explosions. Hillebrandt (1982)[11] therefore investigated a 10 $M_\odot$ model, which had been evolved from the main sequence by Weaver and Woosley (1980)[3]. This particular model differed from others that had been studied so far in several important aspects. After H-burning the He-core had 2.7 $M_\odot$ only. After core-carbon burning Ne and O burnt off-center, and finally the central Ne ignited under semi-degenerate conditions, driving several Ne-flashes and a nuclear deflagration front, which in the end ejected the He-H mantle with about 200 km s^{-1}. At the onset of core-collapse the model consisted of a Fe-core of 1.36 $M_\odot$, a thin shell of O, Ne and Si of 0.13 $M_\odot$ and an envelope extending out to about 10^{16} cm. During the corecollapse phase the model evolved very similar to more massive ones with some minor modifications. Due to a somewhat lower entropy electron captures were reduced and at neutrino-trapping we found $Y_e \simeq 0.4$ corresponding to a Chandrasekhar mass of 0.9 $M_\odot$. Consequently the shock formed at 0.8 $M_\odot$. The central density at core-bounce was 4×10^{14} g cm^{-3} and the energy originally put into the shock was about 8×10^{51} erg, to be compared with 5×10^{51} erg for the other models. As a result the shock was able to propagate out to 1.3 $M_\odot$, where the density began to decrease significantly. Due to this effect the shock speed then increased again (see Figs. 6 and 7) and finally ejected about 0.06 $M_\odot$ from the core with velocities of about 25000 km s^{-1} and a total energy of about 5×10^{50} erg, leaving behind a neutron star of 1.44 $M_\odot$.

This first successful core-collapse supernova model shows several similarities with SN 1054, the Crab. It explains the fact that SN 1054 was a very bright event, that the Crab nebula is only enriched in He but not in heavy elements, the existence of a neutron star and the observed hydrogen halo around the Crab of about (8±2) $M_\odot$. But it is probably not characteristic for the average type II event.

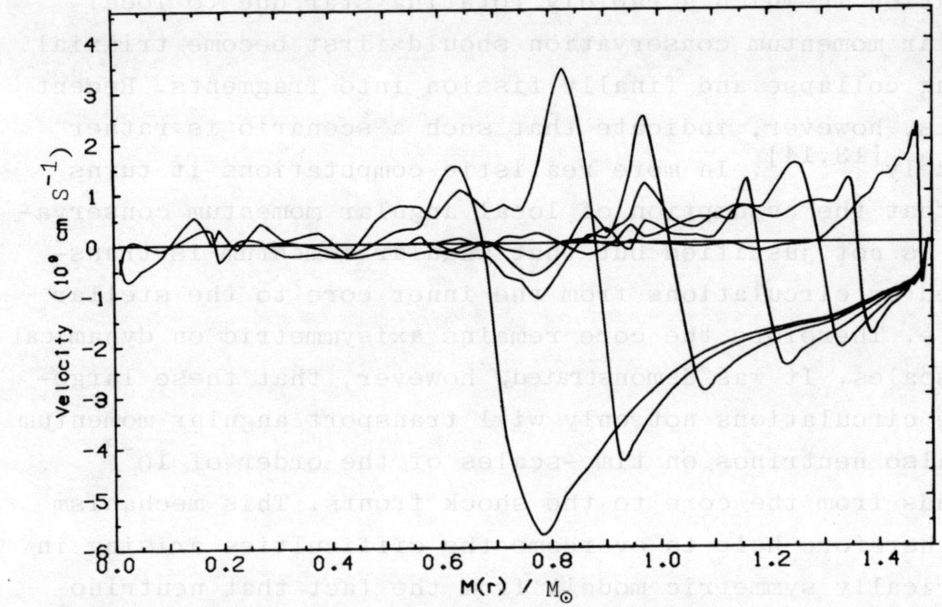

Fig. 6 Same as Fig. 4, but for a non-adiabatically collapsing core of a 10 $M_\odot$ star. Due to the different core-structure of this model the shock speeds up near the edge of the core ($\simeq$1.36 $M_\odot$) and finally ejects 0.06 $M_\odot$ from the core. The total energy in the ejecta was about 5×10^{50} erg.

(4) Rotating Supernova Models

Motivated by the apparent difficulties of spherically symmetric models to give the desired supernova explosion, rotating stellar cores have recently been investigated in some detail. But due to the lack of stellar models, that have been evolved from the main sequence to core collapse with rotation included, only parameter studies have been performed so far.

The idea, that rotation may cause supernova explosions,

was first discussed by Hoyle (1946)[12], who envisioned a situation in which a rapidly rotating star due to local angular momentum conservation should first become triaxial during collapse and finally fission into fragments. Recent models, however, indicate that such a scenario is rather unlikely[13,14]. In more realistic computations it turns out that the assumption of local angular momentum conservation is not justified but that angular momentum is transported by circulations from the inner core to the stellar mantle. Therefore the core remains axisymmetric on dynamical timescales. It was demonstrated, however, that these large-scale circulations not only will transport angular momentum but also neutrinos on time-scales of the order of 10^{-3} seconds from the core to the shock fronts. This mechanism may therefore help to overcome the difficulties arising in spherically symmetric models from the fact that neutrino diffusion time-scales are too long.

A second possibility indicated by the computations of Müller and Hillebrandt (1981)[14] is, that the inner cores of rapidly rotating stars may become unstable to triaxial deformations on secular time-scales. If this should indeed be the case one can imagine a situation in which the triaxial core dissipates a significant fraction of its rotational energy (several 10^{51} erg) into heat at the boundary to the still axisymmetrically infalling envelope, which then could give rise to a supernova explosion.

Finally we want to discuss briefly the effects of magnetic fields in rotating collapsing stellar cores. From the magnetic fields observed in stars, where the ratio of magnetic energy to gravitational energy is found to be less than 10^{-10} even in the most extreme cases, it is obvious that effects from magnetic fields can only become important if the fields are largely amplified locally by differential rotation. At present it cannot be ruled out that differential rotation in some parts of the star may increase the magnetic field strength sufficiently so that the magnetic pressure becomes comparable to the thermal pressure[15,16]. A buoyancy instability will then set in which may or may not eject matter. Realistic computations which require a 2-dimensional magneto-hydrodynamical code, however, are still missing. The only complete calculation[17] has been performed with rather unrealistic stellar models.

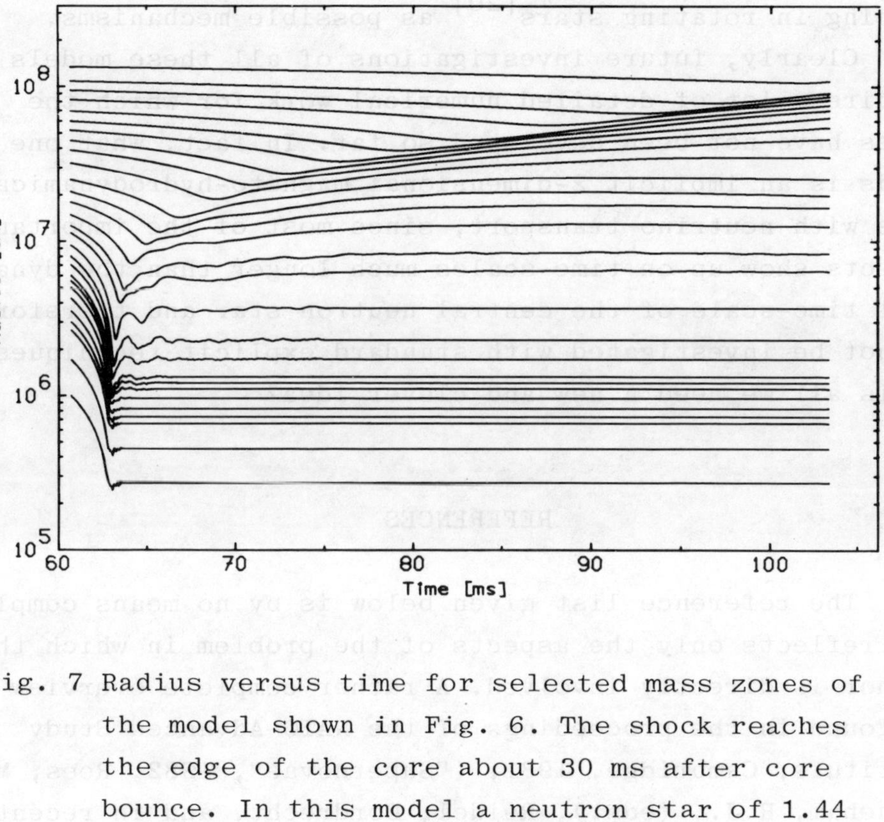

Fig. 7 Radius versus time for selected mass zones of the model shown in Fig. 6. The shock reaches the edge of the core about 30 ms after core bounce. In this model a neutron star of 1.44 $M_\odot$ as well as a supernova would result.

III. A FEW CONCLUDING REMARKS

It has been demonstrated that most of the spherically symmetric models of collapsing stars do not lead to the desired (type II) supernova explosion. If the stellar models used in these computations are not completely wrong, one has to search for other probably more complicated scenarios, because at least some massive stars should explode in order to explain the heavy element content of the galaxy. Preliminary investigations of this type include rotation as the main trigger but all those models are so far not conclusive. Here we should also mention lepto-convection driven by rotation[18,19] or explosive oxygen burning in rotating stars[20] as possible mechanisms.

Clearly, future investigations of all these models reguire a lot of detailed numerical work for which the codes have not been developed so far. In fact, what one needs is an implicit 2-dimensional magneto-hydrodynamical code with neutrino transport, since most of the important effects show up on time-scales much longer than the dynamical time-scale of the central neutron star and therefore cannot be investigated with standard explicit techniques. Or is all we need a new and clever idea?

REFERENCES

The reference list given below is by no means complete and reflects only the aspects of the problem in which the author is directly involved. A rather complete overview can be found in the proceedings of the NATO-Advanced Study Institute, Cambridge, 1981, ("Supernovae", 1982, Rees, M.J., Stoneham, R.J., (eds.), Reidel, Dordrecht, and in recent review articles by Lattimer, J.M., 1981, Ann. Rev. Nucl. Particle Science, Vol. 31, 337, Hillebrandt, 1982 and Trimble, 1982 (l.c.).

[1] Weaver, T. A., Zimmerman, G. B., Woosley, S.E., 1978, Ap. J., Vol. 225, 1021.

[2] Arnett, W.D., 1977, Ap. J. Suppl., Vol. 35, 145.

[3] Weaver, T. A., Woosley, S. E., 1980, B.A.A.S., Vol. 4, 724.

[4] El Eid, M. F., Hillebrandt, W., 1980, Astron. Astrophys. Suppl., Vol. 42, 215.

[5] Takahashi, K., El Eid, M. F., Hillebrandt, W., 1978, Astron. Astrophys., Vol. 67. 185.

[6] Wilson, J. R., 1978, Proc. Int. Sch. Phys. "Enrico Fermi", Vol. 65, 644.

[7] Arnett, W. D., 1977, Ap. J., Vol. 218, 815.

[8] Hillebrandt, W., 1982, in "Supernovae", Rees, M. J., Stoneham, R. J., (eds.), Reidel, Dordrecht. 123.

[9] Hillebrandt, W., Müller, E., 1981, Astron. Astrophys., Vol. 103, 147.

[10] Trimble,V., 1982, Rev. Mod. Phys.(in press).

[11] Hillebrandt, W., 1982, Astron. Astrophys., Vol. 110, L3.

[12] Hoyle, F., 1946, M.N.R.A.S., Vol. 106, 343.

[13] Müller, E., Rozcyczka, M., Hillebrandt, W., 1980, Astron. Astrophys., Vol. 81, 288.

[14] Müller, E., Hillebrandt, W., 1981, Astron. Astrophys., Vol. 103, 358.

[15] Meier, D. L., Epstein, R. I., Arnett, W. D., Schramm, D. N., 1976, Ap. J., Vol. 204, 869.

[16] Müller, E., Hillebrandt, W., 1979, Astron. Astrophys., Vol. 80, 147.

[17] Le Blanc, J. M., Wilson, J. R., 1970, Ap. J., Vol. 161, 541.

[18] Epstein, R. J., 1979, M.N.R.A.S., Vol. 188, 305.

[19] Smarr, L. L., Wilson, J. R., Barton, R., Bowers, R., 1981, Ap. J., Vol. 246, 515.

[20 Bodenheimer, P., Woosley, S. E., 1980, B.A.A.S., Vol. 12, 833.

THE APPLICATION OF HISTORICAL RECORDS TO ASTROPHYSICAL PROBLEMS

XI Zezong（席泽宗）

(Institute for the History of Natural Science, Academia Sinica, Beijing)

I. STELLAR EVOLUTION

V.F. Weisskopf, a former president of American Council of Atomic Energy, has said that in the history of mankind there are two "4 July" which should be written down in annals. One is in 1776 when the United States of America was founded, another is July 1054 when Chinese and Japanese astronomers recorded the supernova explosion in Taurus[1]. In 1921 the Swedish astronomer K. Lundmark first noted that the position of supernova recorded in 1054 was very close to that of the Crab Nebula, and suspected that they were related[2]. Soon after that, the calculation from the expansion speed of the Crab Nebula by E. Hubble showed that the explosion has taken place 900 years before—it is consistent with the year of explosion of the supernova of 1054[3]. In 1942 J.H. Oort and J.J. Duyvendark confirmed that the Crab Nebula was a remmant of the explosion in 1054 and the explosion was one of supernova rather than of nova[4–6].

The identification of 1054 supernova with the Crab Nebula is an important evidence of the value of historical records to astrophysics and the Crab Nebula is considered the Rosetta stone of astrophysics. Most of what we know about the origin of cosmic rays, synchrotron radiation and heavy elements derives from our knowledge of the Crab Nebula, and much of what we can deduce from observations of the Crab Nebula is aided by the records of supernova explosion in 1054. In addition to this, the historical records of galactic supernovae can give some information on the frequency of such outbursts and on the development of their remnants. For example, from Sedov's equation[7]

$$D = 4.3 \times 10^{-11}(E_0/n)^{1/5} t^{2/5}, \qquad (1)$$

where D (pc) is the diameter of the shock wave preceding the expanding shell of swept-up interstellar material, t(yr) is the time elapsed since the explosion, E_0 (erg) is the energy released in the outburst, and n(cm^{-3}) is the number of H atoms in the interstellar medium, we can obtain E_0/N, if D is known by observation and t from historical record. In 1976 Clark and Caswell concluded that the best estimate was $E_0/n \simeq 5 \times 10^{51}$ erg cm^3 [8].

Most early observations of supernovae were made in China, Japan and Korea. In ancient and mediaeval Europe and the Arab Lands there seems to have been little interest in such phenomena. In "The Historical Supernovae" written by D.H. Clark and F.R. Stephenson in 1977, they listed seven supernovae:

	year	Constellation	Magnitude	Duration	Records
1	AD 185	Centaurus	-8	20 months	C
2	393	Scorpius	-1	8 months	C
3	1006	Lupus	-9.5	several years	A,C,J,E
4	1054	Taurus	-5	22 months	C,J
5	1181	Cassiopeia	0	6 months	C,J
6	1572	Cassiopeia	-4	18 months	C,K,E
7	1604	Ophiuchus	-2.5	12 months	C,K,E

C: Chinese, A: Arabic, J: Japanese,

E: European, K: Korean.

Among them the supernova of 1006 is the only one known to be recorded in both European and Arabian literatures before the Renaissance. It was also carefully observed by astronomers in China and Japan. It was the most brilliant supernova ever recorded and was observed for several years. There are various opinions about its remnants: NGC 5882[9] PKS 1459—42.[10]. G 327.6 + 14.5[11] or PKS 1527—42[12]. However, perhaps most startlingly, re-explosion of the supernova probably took place on May 16, 1016 in the light of Chinese literature[13]. This would support the theory of abnormal neutron, put forward by Fang Lizhi and Qu Qinyue, if it were true. According to the theory, as a remnant of supernova, a neutron star may transfer its phase from normal to abnormal in some conditions by releasing vast quantities of energy, on the order of 10^{52} erg (equivalent to supernova outburst)[14].

In 1955 the author compiled "A new catalogue of ancient novae"[15] and in 1965 with his colleague Bo Shuren revised and republished it as an appendix of "Nova and supernova records of Chinese. Japanese and korean annals and its signficance in radio astronomy"[16]. Both catalogues attracted much attention of astronomers over the world and have been translated into English and Russian. In 1980 G.G.C. Palumbo, G.K. Miley and P. Schiavo Campo selected seven objects from the latter catalogue, and observed the field 1°.5×1°.5 of each object with Westerbork radio telescope in Holland to discover non-thermal radio sources. Even though no diffuse radio emission was detected they suggested that the investigation be continued[17].

In 1978 Li Qibin found one entry in our catalogue, managed no less than nine references to this event, which refers to an event occured on September 10,1408. He considered it a progenitor of Cyg X-1, a prime black hole candidate[18]. In addition to the Chinese sightings, Imaeda and T. Kiang pointed out that the guest star was also seen in Japan. The first Japanese account is dated at July 14, 1408, which precedes the first Chinese record by 58 days. Combining the Chinese and Japanese dates, we can conclude that the event of 1408 was visible for 102 days. The long duration indicates that the event was a supernova[19]. However, the position deduced for the supernova of 1408 places it near two astronomical curiosities, Cyg X-1 and CTB 80. There is some confusion over the interpretation of the original Chinese as to whether the location should be "to the southeast of" or "in the southeast of" Niandao (17 Cyg). The latter interpretation would favor Cyg X-1 and the former—CTB 80[20]. The final solution of these arguments should be interesting, no matter which object is ultimately favored, because the association of the event with either would constrain physical models of the source evolution.

Cassiopeia A, the strongest radio source in the sky, was considered as a supernova remnant as early as 1950s before any record of it had been found[21]. In 1968 Korean Scholar Chu Sunil suggested that it be a remnant of supernova recorded to the west of the first star of Wang Liang (located on the west side of β Cas) from December 6, 1592 to March 5, 1593 in the Korean book "Real Records of Lee Dynasty" and "Chungbo MunhunBigo" (Supplement of historical records for reference) for their position seems to be very close[22]. But S.F. Gull pointed out that the preferred age of the remnant is much less than this[23]. Based on the study of proper motion of filamentary structure in this optical remnant, Sidney Van den Bergh and W.W. Dodd placed its occurence in the year 1667±8[24]. In 1980 a new discovery made by W.B. Ashworth revealed that the first British Astronomer Royal, John Flamsteed, had recorded the event after all, fixing the actual year of the supernova explosion as 1680[25]. K. Brecher immediatly wrote a paper on the finding by combining the actual date of the event with the extrapolated date to determine the mass ejected from the supernova, which turned out to be ten times greater than the Sun[26]. However. R.R. Broughton and K.W. Kamper quickly denied Ashworth's discovery[27–28]. So up to now the identification of Cas A with supernova remnant has not yet been solved.

Chinese annals recorded not only the sudden change of the process of stellar evolution, such as supernova explosion, but also the gradual change. In the astronomical chapter of "Shiji" ("Historical Records") by Sima Qian about the first century B.C., he said:"The yellow is like Betelgeuse (α Ori)," but Ptolemy included it in the list of red stars in his <u>Almagest</u> which was written about 150 A. D. Normal stellar evolution, even of a supergiant, from yellow to red in 250 years seems unlikely. However, a shell of gas and dust of about 30" radius (about 5500 AU at the 190 pc distance to the star) appears to be moving away from α Ori with a present speed of about 10 km s^{-1}. Brecher has explored the possibility that the ejection of the shell occurred about 2700 years ago giving rise temporarily to a smaller, hotter and whiter photosphere which then readjusted nearly to its original equilibrium state between the times of the writing of the "Shiji" and the "Almagest"[29–30].

II. SOLAR ACTIVITY

Another use of Ancient astronomical records in astrophysics in recent years is to discuss the periodicity of solar activity. In 1976 J.A. Eddy again strengthened the reality of the Maunder Minimum[31] that the sunspot cycle essentially vanished between 1645 and 1715 from a reanalysis of the contemporary literature detailing sunspot counts, auroral records, and observations of the Sun at eclipse, plus indirect evidence from C^{14} in tree rings. Eddy even claimed that there was insufficient evidence to establish whether the 11-year cycle existed before the onset of the Maunder Minimum and after the introduction of the telescope[32]. Eddy's claim caused a sensation. If it were true, solar physics would need rewritting. The only way to solve this problem is to study the historical sunspot records of China, Japan and Korea, because the Orient was free from the stranglehold of Aristotelian doctrine of celestial perfection, and sunspots were recorded from very early times.

To accept Eddy's challenge Yunnan Observatory sorted out more than 100 items of sunspots from Chinese annals from 28 B.C. to 1683 A.D., and through self-corelation analysis showed that the 11 year cycle existed in the past 2000 years[33–34].

At the same time Zou Yixin (Beijing) estimated that the weighted mean cycle of solar activity was 10.42±0.19 years, using Chinese historical sunspot records from 165 B. C. to 1884 A.D. and auroral records as well[35]

In 1979 Xu Zhentao and his wife Jiang Yaotiao (Nanjing) published a paper entitled "The solar activity of the 17th century viewed in the light of sunspots (1637.5, 1647.4, 1656.2, 1673.4, 1684.2, 1709.6) and their weighted mean of solar activity is 10.54±0.62 years[36]. The study of Xu and Jiang shows that the 11 year cycle still exists within the Maunder Minimum.

It is necessary to point out that in the paper "Sunspot records in China, Korea and Japan" by S. Kanda (1932) there is no one sunspot from 1639 to 1720[37]. Kanda's paper is one of Eddy's bases. Now Xu and Jiang replenish six items of sunspots to the Maunder Minimum, and consequently make the first basis of Eddy's idea to loss efficiency. Therefore their paper attracts considerably interest.

In 1980 Dai Nianzu and Chen Meidong (Beijing) published "The aurora borealis and solar activity in history" as well as "Chronological table of the aurora borealis in Chinese, Korean and Japanese history from the legendary period to A.D. 1747"[38]. Using 929 items of aurorae in an all-round way they negated the second basis of Eddy's idea. They considered that from 217B.C. to A.D. 1749 there were 180 peak years of solar activity, and their mean cycle was equal to 11 years. The auroral records also indicated that there were several minima of solar activity in history, and the Maunder Minimum was one of them, but in minimum periods the mean cycle was also 11 years. This is consistent with the result of Xu and Jiang.

Tree-ring widths have failed to show convincing evidence of past solar cyclic activity, because local, rather than global effects dominate their patterns[39]. As to C^{14}, it is also impossible to detect the 11-year cycle due to the appreciable delay (10—50 years) between variations in C^{14} production and resultant changes in the biospheric abundance, although Eddy has used the C^{14} history to recognize six major excursion in solar behaviour in the past 2000 years with possibly a total of 12 in the past 5000 years[40]. Thus it is unsatisfied to study solar cycle using natural "unwritten" records, and the solar cycle as well as its origin still remain a problem for astrophysics.

III. CONSTANT G

The relationship between Ephemeris Time (ET) and Universal Time (UT) is

$$ET = UT + \Delta T, \tag{2}$$

where ΔT is determined by observation. If the steady increase in the length of the day is one millisecond (ms) per century^2, the mean length of day over the past 20 centuries is 10 ms shorter than that of the present day, and the sum of ΔT is

$$10\times10^{-3}\times365.25\times100\times20 = 7305^s = 2^h01^m45^s.$$

Thus the difference between the time (UT) recorded by observation of solar eclipses in the first century and the time (ET) calculated according to gravitional theory may reach two hours. The earlier the solar eclipse takes place, the greater is the difference of time. Moreover, the difference of time may be reflected by the difference between observed place and calculated path of eclipse. For example, in the seventh year of Luhou (The Empress of Han Dynasty) Chinese astronomers reported a total eclipse of the Sun at Changan on a date corresponding to March 4,181 B.C. The calculation of the path of this totality by Stephenson and Clark on the assumption of no acceleration of the Earth's spin shows that the track comes out well to the west of Changan. Arotation of the Earth of 75 degrees would be necessary to make the eclipse total at Changan. The inference is that in the intervening time between 181 B.C. and the present day, the Earth has lost five hours relative to an ideal clock, i.e. the increase rate of day length is more than 2 ms century^{-2}[41].

The application of historical records of solar eclipses to the study of the acceleration of the Earth's rotation can obtain the same precise result as current observations do, if the exact date and place of the observation as well as the magnitude of eclipse have been described in these records. The more magnitude, the better. It would be better if we could get the historical records of the occurence of total solar eclipses. In 1980 Wu Shouxian checked Chinese eclipse records, used by four western astronomers (D.R. Curott, 1966; R.R. Newton, 1970; P.M. Mullar and F.R. Stephenson, 1975). He found out that apart from repetition of each other, the total number of records they used was 30 items, one-third of which are incorrect[42]. Since it is so, their results should be questionable.

Another new time system. Atomic Time (AT), was proposed in 1967, for the intrinsic accuracy of Ephemeris Time is not easy to reach. It utilizes the natural resonance frequency of CS^{133} to make extremely precise measurements of time intervals, and is not related to gravitational force. If the universal constant of gravitation, G, varies with time as Paul Dirac expected, there is difference between the two results of historical solar eclipese reckoned by Ephemeris Time and by Atomic Time.

In 1975 Van Flandern made an interesting observational test based on modern and historical date. His analysis of lunar occultations over the period 1955 to 1974 gave a result for G:

$$\dot{G}/G = (-8\pm5) \times 10^{-11} \ yr^{-1}, \qquad (3)$$

i.e. G is on decrease[43]. But this problem has recenly been analysed again using the lunar laser ranging measurements and the eclipes data from ancient China. The close agreement between both results is evidence in favour of the constancy of G[44]. However, definite evidence against (or for) the non-constancy of G must await a further study.

* * *

In addition to the fields discussed above, there are many others where the historical records may have applications, such as in cometary investigations, planetary orbital perturbation studies, solar-terrestrial relationship researches,...etc. We believe that combining the current technology with information recovered from the records of our ancestors will certainly bring astrophysics to the fore in terms of increasing our understanding of the universe, so we have collected and arranged all historical records of celestial phenomena in China, and are now preparing to publish it.

REFERENCES

[1] A talk with Wu Youxun at Beijing in 1972.

[2] Lundmark, K., 1921, P.A.S.P., Vol. 33, 225.

[3] E. Hubble, 1928, Astr. Soc. Pacific Leaflet, No. 14.

[4] Duyvendark, J.J., Oort, J.H., 1972, T'oung Pao, Vol. 36, 174.

[5] Duyvendark, J.J., 1942, P.A.S.P., Vol. 54, 91.

[6] Mayall, N.U., Oort, J.H., 1942, ibid, Vol. 54, 95.

[7] Sedov, L.I., 1959, "Similarity and Dimensional Methods in Mechanics", New York, Academic Press.

[8] Clark, D.H., Caswell, J.L., 1976, M.N.R.A.S., Vol. 174, 267.

[9] Goldstein, B.R., Ho Peng-yoke, 1965, A.J., Vol. 70, 748.

[10] Gardner, F.F., Milime D.K., 1965, A.J., Vol. 70, 754.

[11] Stephenson, F.R. et al., 1977, M.N.R.A.S., Vol. 180, 567.

[12] Bo Shuren et al., 1978, "collected Essaies of History of Science and Technology-Special Issue for History of Astronomy (No. 1)", 79.

[13] Wang Jianmin, 1979, Publications of the Beijing Astr. Obs., No. 1, 69.

[14] Fang Lizhi et al., 1978, Scientia Sinica, No. 4, 406.

[15] Hsi Tze-Tsung, 1955, Acta Astronomica Sinica, Vol. 3, 183.

[16] Xi Zezong, Bo Shuren, 1965, Acta Astronomica Sinica, Vol. 13, 1.

[17] Palumbo G.G.C. et al., 1980, Acta Astronomica Sinica, Vol. 21, 334.

[18] Li Qibin, 1978, Acta Astronomica Sinica, Vol. 19, 210.

[19] Imaeda, K., Kiang, T., 1980, Journal for the History of Astronomy, Vol. 11, 77.

[20] Baade, W., Minkowski, R., 1954, Ap. J., Vol. 119, 206.

[21] Angerhofer, P.E., 1981, Archaeoastronomy, Vol. 4, 22.

[22] Chu Sun-Il, 1968, Journal of the Korean Astr. Soc., Vol. 1, 29.

[23] Gull, S.F., 1973, M.N.R.A.S., Vol. 161, 47.

[24] Sidney Van den Bergh, Dodd, W.W., 1970, Ap. J., Vol. 162, 485.

[25] Ashworth, W.B., 1980 Journal for the History of Astronomy, Vol. 11, Part. 1.

[26] Brecher K., Wasserman, I., 1980, Ap. J., Vol. 240, 105.

[27] Broughton R.P., 1980, J.R.A.S.C., Vol. 173, 381.

[28] Kamper K.K., 1980, Observatory, Vol. 100, 3.

[29] Bo Shuren et al.,1978,"Collected Essaies of History of Science and Technology-Special Issue for History of Astronomy (No.1)", 75.

[30] Brecher K., 1980, B.A.A.S.

[31] Maunder W., 1890, M.N.R.A.S., Vol. 50, 251.

[32] Eddy J.A., 1976, Science, Vol. 192, 1189; 1976, Proceeding of the Inter. Syn. on solar-Terr. Phys., Vol. 11, 958; 1976, Sky and Telescope, Vol. 52, 394.

[33] Ding Youji et al., 1978, Kexue Tongbao, Vol. 23, 107.

[34] Lo Paoyung, Li Weipao, 1978, Kexue Tongbao, Vol. 23, 362.

[35] Zou Yixin, 1978, Publications of the Beijing Astr. Obs., No. 12, 87.

[36] Xu Zhentao et al., 1979, Journal of Nanjing University (Natural Science), No. 2, 31.

[37] Kanda S., Tokyo Astron. Obs., 1932, Rep. 1, No. 1.

[38] Dai Nianzu et al., 1980, "Collected Essaies of History of Science and Technology-Special Issue for History of Astronomy (No.2)", 69.

[39] Marche La, Fritts H.C., 1972, Tree Bulletin, Vol. 32, 21.

[40] Eddy J.A., 1977, "The Solar Output and Its Variation", Colorado Associated Univ. Press, Boulder.

[41] Stephenson F.R., Clark D.H., 1977, Q.J.R. Astr. Soc., Vol. 18, 340.

[42] Wu Shouxian, 1980, Publications of the Shanxi Astr. Obs., No. 2, 23.

[43] Van Flandern T.C., 1955, M.N.R.A.S., Vol. 170, 333.

[44] Stephenson F.R., 1979, New Scientist., Vol. 81, 560.

[45] Stephenson F.R., Clark D.H. (eds.), 1978, "Applications of Early Astronomical Records", Adam Hilger Ltd, Bristol.

IV. BINARY SYSTEMS

Börner
PSR 1913 + 16
He(?)
Do you believe Einstein ?
No! It is because HEINE has said "only an IDIOT expects to answer whether q_0 is larger than 1/2
Hu

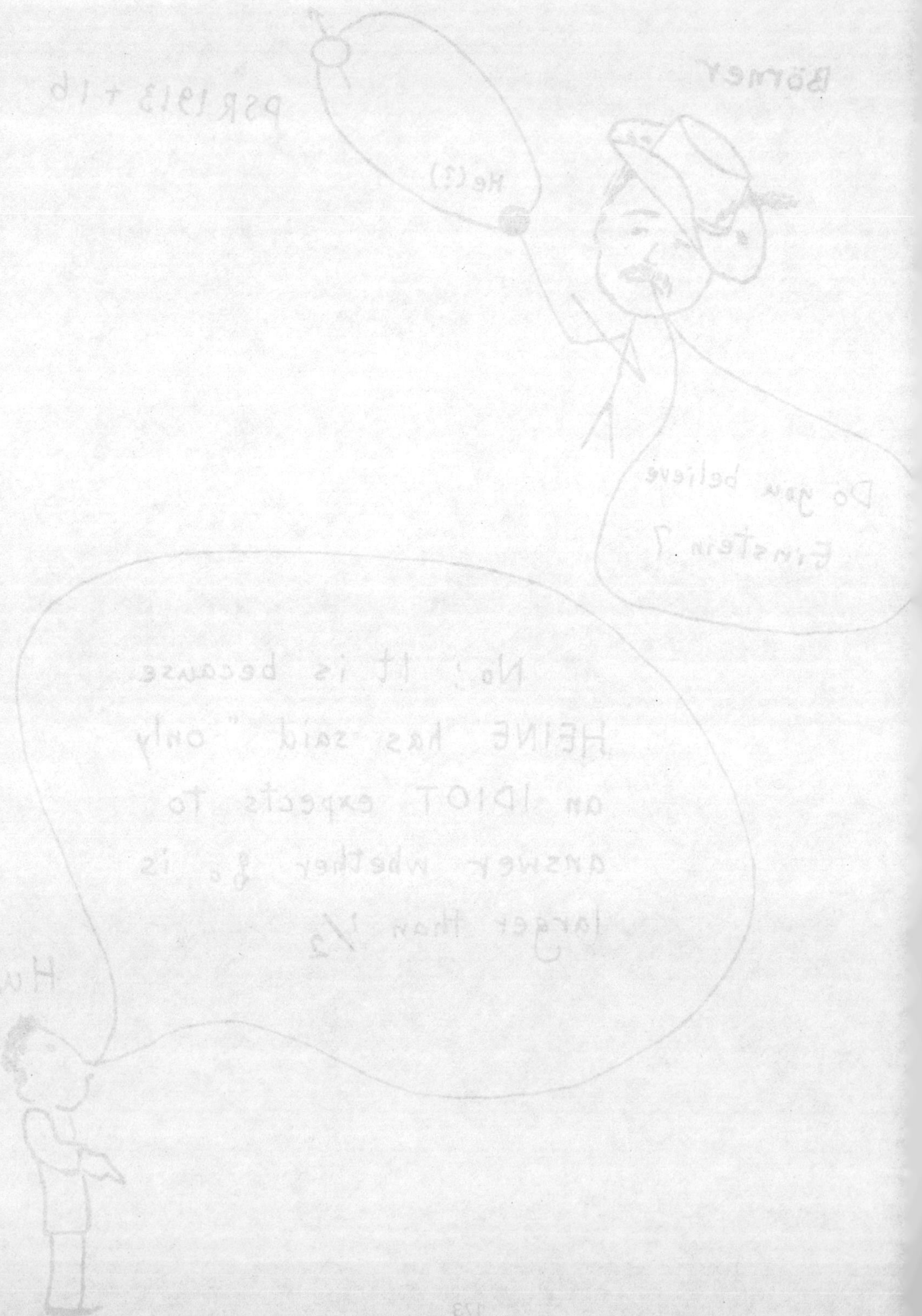

THE BINARY PULSAR-A TEST FOR GENERAL RELATIVITY?*

Gerhard Börner
(Max-Planck-Institut für Astrophysik,
München, Federal Republic of Germany)

ABSTRACT

The binary system of PSR1913+16 contains the pulsar and an, as yet unknown, companion. If this star is a compact object too, then the data can be interpreted in terms of general relativistic effects. This leads to the conclusion that the decay of the orbit must be due to the emission of gravitational waves. The nature of the unseen companion is discussed in detail.

I. INTRODUCTION

It may well be that the binary pulsar PSR 1913+16, discovered in September 1974[1], is a celestial laboratory for measuring relativistic effects:

The pulsar, with its period of 59 msec, is a precise clock moving with large velocities ($\frac{v}{c} \simeq 10^{-3}$) in a high eccentricity orbit (e = 0.617) through a relatively strong

---

*This paper is a modified version of a talk given at EPS Symposium on "Neutron Stars". (Istanbul, Sept. 1981)

gravitational field ($\frac{GM}{c^2r} \simeq 10^{-6}$). Several special and general relativistic effects have become observable in the meantime, such as the periastron precession, the variable time delay due to a combination of the transverse Doppler shift and the gravitational redshift, and the rate of change of the orbital period - possibly due to the emission of gravitational waves. These measurements yield enough information to determine all parameters of the system, and to allow in addition one test of the gravitational theory - Einstein's theory of general relativity.

All this is true provided the pulsar and its companion can both be treated as point-masses in the Newtonian sense, i.e. as masses with negligible quadrupole moment, and negligible tidal interaction. The companion star has not been observed yet, and there is the possibility that it may not be a compact object, but a helium star. As always in life a small uncertainty remains, but it seems reasonable to discuss in detail the consequences of PSR 1913+16 being a "clean" relativistic system.

In this talk I shall concentrate on this binary system and omit the other two recently observed binary pulsars[2] (because of their long orbital periods of 24 hours and 4 years respectively; they are not as interesting from the point of view of relativity).

II. ARRIVAL TIME ANALYSIS FOR THE PULSES

The first observations[1] showed a slow periodic change in the pulsar period of 0.059 sec which immediately led to a curve for the radial velocities as shown in Fig. 1. The orbital period turned out to be 7.75 hours, i.e. the system had to be rather narrow (see Table 1 for the best values of the parameters, taken from[3]).

The whole binary system could fit inside our sun. A

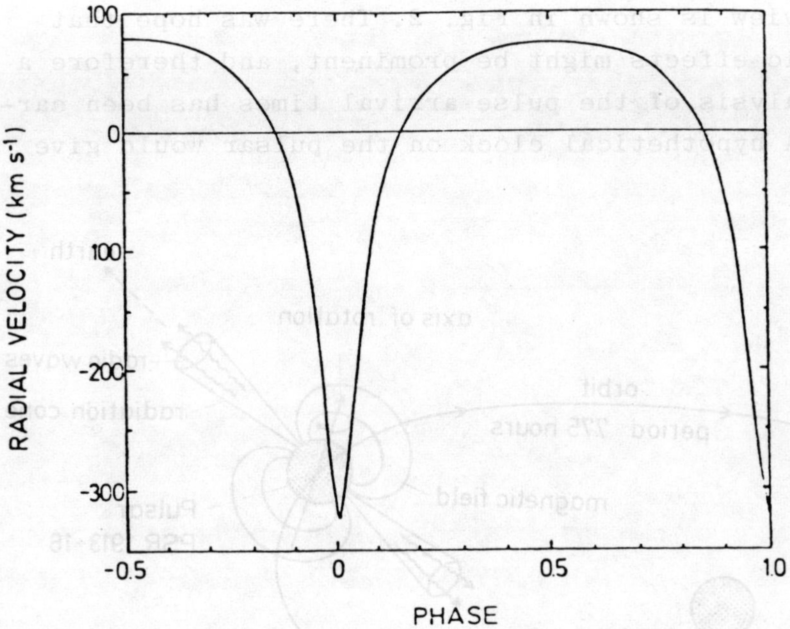

Fig. 1 Radial velocity

Table 1. Parameter estimates obtained from pulse arrival times

$$\text{Right ascension (1950.0)}: 19^{h}13^{m}\ 12\overset{s}{.}47$$
$$\text{declination (1950.0)}: 16°01'\ 08''\!.2$$
$$\text{period } P = 0.0590299952709\ (\pm 2\times 10^{-12})\ \text{sec}$$
$$\dot{P} = 8.628\times 10^{-18}\ \text{sec sec}^{-1}$$
$$\ddot{P} = (-58\pm 1200)\ 10^{-30}\ \text{sec}^{-1}$$
$$a_p \sin i = 2.34186\ \text{lightsec}$$
$$e = 0.61714$$
$$P_b = 27906.9816\ \text{sec}$$
$$\dot{\omega} = 4.226\ \text{deg yr}^{-1}$$
$$\gamma = 0.00438\ (\pm 0.00024)\ \text{sec}$$
$$\dot{P}_b = (-2.30\pm .22)\times 10^{-12}\ \text{sec sec}^{-1}$$

schematic view is shown in Fig. 2. There was hope that relativistic effects might be prominent, and therefore a careful analysis of the pulse arrival times has been carried out. A hypothetical clock on the pulsar would give

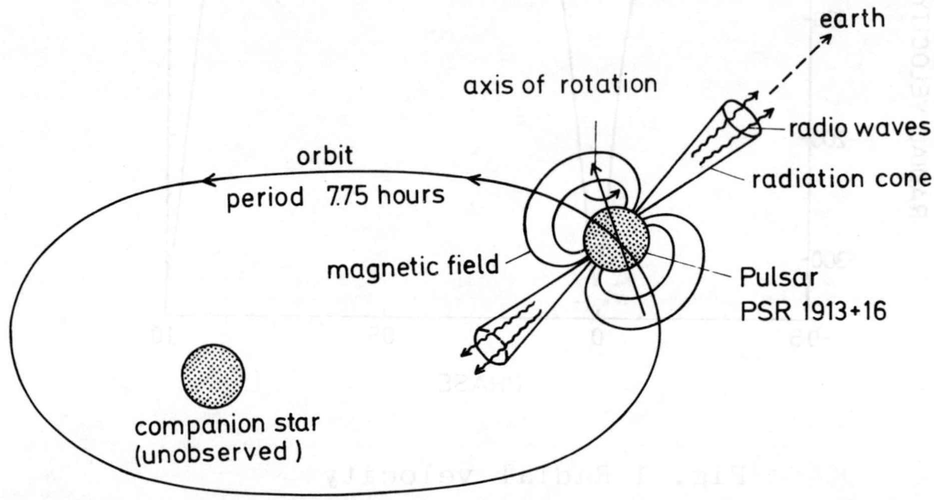

Fig. 2 Schematic view of the binary system

the time of emission of the Nth pulse as T_p (proper time), and

$$N = N_0 + \Omega T_p + \frac{1}{2}\dot{\Omega}T_p^2 + \frac{1}{6}\ddot{\Omega}T_p^3 \qquad (1)$$

(cf. Ref. [3] and [4]).

Here N_0 is an arbitrary constant, while $\Omega, \dot{\Omega}, \ddot{\Omega}$ are the rotation frequency of the pulsar and its first and second time derivatives. Experience with other pulsars has shown that no higher derivatives in Eq. (1) are needed, although discontinuous frequency jumps (glitches) and noise may occur. During its first 7 years of observation PSR 1913+16 has been extremely well-behaved: no glitches and no noise have been found. The time T_p of Eq. (1) has then to be converted to the time of arrival of the Nth pulse at the earth as measured by an observer on earth. This conversion was performed in several steps[3]: First the dispersion delay caused by the interstellar plasma was removed. The dispersion of the signals had, in fact, to be compensated right from the beginning: The rather large dispersion measure (DM = 168.77 cm^{-3} pc) would lead to differential time-delays of 70 ms across a typical bandwidth of 4 MHz at 430 MHz. This is more than the pulsar period, and without dispersion-compensating hardware, the pulsar would be unobservable. The large dispersion measure indicates that it is also far away, at a distance of 5kpc. The flux density at 430 MHz is 5 mJy which makes PSR 1913+16 one of the weakest known pulsars. Therefore individual pulses cannot be observed, rather about 5 minutes of data have been superimposed synchronously to yield an average pulse profile (cf. Fig. 3, from [3]). The time resolution at 430 MHz has been

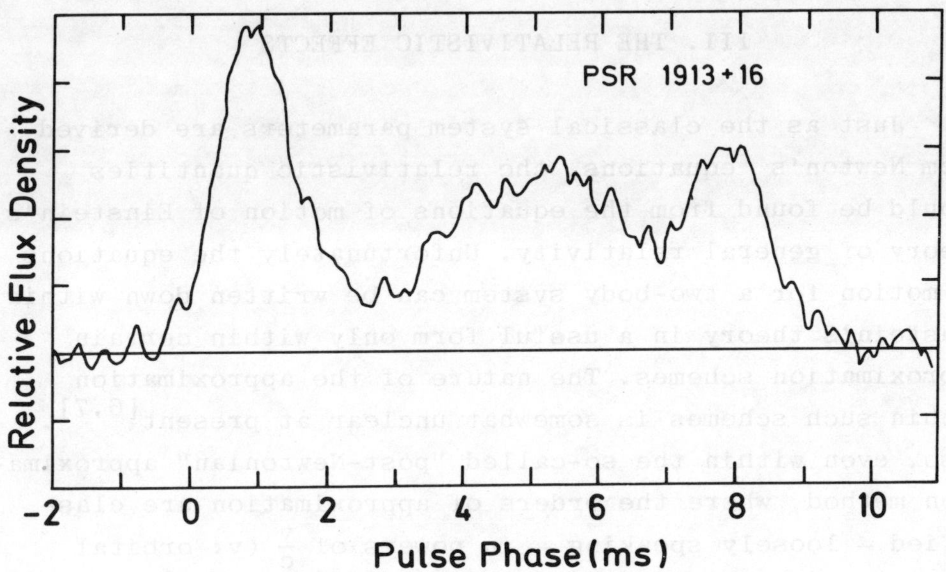

Fig. 3 Average pulse shape

improved from 5000 μs (1974) to presently 43 μs.

Then the observed arrival times were corrected for solar system motions (a special relativistic correction for the annual variation in the rate of terrestrial clocks due to velocity variations of the earth was included).

These arrival times[3] were then corrected to the pulsar frame by taking into account pulsar variables such as period P, orbital period P_b, etc., as well as various relativistic effects to be described below.

The classical quantities are: right ascension α, declination δ, initial pulse phase, ϕ_0, period P, first and second derivatives $\dot{P}$, $\ddot{P}$; eccentricity e, projected semimajor axis of the pulsar orbit $a_p \sin i$, orbital period P_b, longitude of periastron ω_0, and time of periastron passage T_0. Table 1 contains the most exact values to date of those quantities. It may be interesting to note that the value for the time derivative of the pulsar period is rather low $\dot{P} = 8.8\times10^{-18}$, leading – via the usual dipole radiation argument[5] – to a low value of the magnetic field of $B = 3\times10^{10}$ gauss.

III. THE RELATIVISTIC EFFECTS

Just as the classical system parameters are derived from Newton's equations, the relativistic quantities should be found from the equations of motion of Einstein's theory of general relativity. Unfortunately the equations of motion for a two-body system can be written down within Einstein's theory in a useful form only within certain approximation schemes. The nature of the approximation within such schemes is somewhat unclear at present[6,7]. Also, even within the so-called "post-Newtonian" approximation method, where the orders of approximation are classified – loosely speaking – in powers of $\frac{v}{c}$ (v: orbital velocity), the various effects have been derived by successively determining the $\frac{v}{c}$ (Newtonian), $\frac{v^2}{c^2}$ (periastron precession, transverse Doppler + gravitational redshift), $\frac{v^3}{c^3}$ (post-Newtonian deviation of orbit from ellipse, light travel time), $\frac{v^4}{c^4}$ (?), $\frac{v^5}{c^5}$ (gravitational radiation) corrections. It might be aesthetically more satisfying to write down the equations of motion in the "slow motion" approximation up to $(\frac{v}{c})^5$ terms, integrate them, and collect the various terms corresponding to specific effects. This remains to be done.

(1) Periastron Advance

The $(\frac{v^2}{c^2})$ terms involve the periastron precession. The theory of Einstein predicts an advance of the periastron of the elliptical orbit (a_p, e, P_b: orbital period) of two point masses m_p and m_c of

$$\dot{\omega} = 3G^{2/3}c^{-2}(P_b/2\pi)^{-5/3}(1-e^2)^{-1}(m_p+m_c)^{2/3}\deg \mathrm{yr}^{-1} \quad (2)$$

(2) Transverse Doppler Shift+Variable Part of Gravitational Redshift

These $(\frac{v}{c})^2$ terms include the transverse Doppler shift of signals (the pulsar motion perpendicular to the line of sight) and the variable part of the gravitational redshift. Both terms have the same dependence on the classical quantities, and their sum gives therefore only one measurable term γ:

$$\gamma = G^{2/3}c^{-2}e(P_b/2\pi)^{1/3}m_c(m_p+2m_c)(m_p+m_c)^{-4/3}\ \mathrm{sec}. \quad (3)$$

(3) Other Effects

The $(\frac{v}{c})^3$ terms include the variation of the relativistic time delay and deviations from the elliptic orbit. The relativistic time delay across the orbit has an amplitude of typically $\simeq$ 20 μsec[4], unobservable at present, where the arrival time measurement has an accuracy of $\simeq$ 40 μsec. Post-Newtonian corrections to the elliptic orbit are also of this magnitude. These effects can be used to define either m_p, m_c or measure sin i.

(4) Results

The observed periastron advance is (cf. Table 1) $\dot{\omega}$ = 4.226 deg yr^{-1}, about 600 times the relativistic effect of the planet Mercury. γ is measured as γ = 0.00436 (±.00024) sec. If there is no classical contribution to these quantities, then together with the five measurable classical orbital elements a_p sin i, e, P_b, T_0, ω_0 (Table 1) we have seven quantities — enough to determine completely the orbital parameters and the masses of the two stars. The results are listed in Table 2. m_p = (1.42±0.06) $M_\odot$ would

Table 2. Derived Quantities

total mass	$m = 2.83\ M_\odot$
pulsar mass	$m_p = (1.42\pm.06)\ M_\odot$
companion mass	$m_c = (1.41\pm0.06)\ M_\odot$
inclination	$\sin i = 0.72\pm.03$
pulsar semimajor axis	$a_p = 0.324\pm0.13$ light sec

be the first precise value for the mass of a radio pulsar. It is slightly above the Chandrasekhar mass limit for a white dwarf. The value of sin i = 0.72±0.03 corresponds to an inclination angle of about 45°.

(5) Spin Precession

In a close binary system the intrinsic angular momentum ("spin") $\underline{S}$ of, say, the first body precesses at a rate $\frac{dS}{dt}$ = $\underline{\Omega}_S \times \underline{S}$, which is determined by three contributions:

a) the motion of the first body around the center of mass of the system (de Sitter - Fokker precession);

b) the spin - spin interaction between the two bodies;

c) the dragging of the inertial frame at the first body due to the translational motion of the second body.

Averaging over one orbital period, neglecting the spin - spin interaction, one finds

$$\langle\Omega_S\rangle = \frac{1}{4}\left(2 + \frac{3}{2}\nu\right) \frac{\nu}{(1+\nu)^2}\, \dot{\omega}\, \underline{n}, \tag{4}$$

where $\nu = \frac{m_c}{m_p}$; $\underline{n}$ is the unit vector normal to the orbital plane, and $\dot{\omega}$ the periastron advance.[8-10] The effect could show up as a slow change in the pulse shape, or as a polarization change with an expected rate of about 1 deg yr^{-1}. No such effects have been found in 2 years of observation indicating either $\underline{S}$ parallel to $\underline{n}$, or a present location near the axis of the radiation cone.

It should be realized that a proper physical characterization of the frame in which $\underline{\Omega}_S$ can be measured is non-trivial. Just like the orientation of a gyroscope, the axis of the comoving coordinate system should be fixed relative to the light from distant stars after averaging over the periodic motions of the system. Then the precession of $\underline{S}$ would be measured relative to the directions of distant stars over times large compared to one period. This interpretation can be generalized to the case of a non-periodic motion (periastron shift). It still does not show, however, how $\underline{\Omega}$ could be measured by a distant observer.

If $\underline{S}$ is parallel to $\underline{n}$, and since the orbital plane is inclined by about 45° to the line of sight, then the observed radio emission originates — if one believes in a narrow radiation cone — at pulsar latitudes around 45°. This is compatible with polar cap emission models (a 45°

inclined dipole axis), but does exclude radiation mechanisms that involve relativistic beaming at the velocity-of-light cylinder.

IV. GRAVITATIONAL RADIATION

This effect is of order $(\frac{v}{c})^5$, and its calculation in the post-Newtonian formalism requires quite an effort (see e.g.[11]). The time variation of the quadrupole moment in the binary system leads within Einstein's theory necessarily to the emission of gravitational waves. The gravitational radiation from point masses in a Keplerian orbit has been written down in a useful form in [12]. (It was subsequently rediscovered in [13] and [14]). The radiation reaction results in a secular decrease of the orbital period:

$$\dot{P}_b = -\frac{192\pi}{5}\frac{G^{5/2}m_c m_p(m_c+m_p)^{1/2}}{c^5 a^{5/2}(1-e^2)^{7/2}}(1+\frac{73}{24}e^2+\frac{37}{96}e^4). \quad (5)$$

The derivation of this formula has not been achieved so far in a complete and well-defined way in [6], see also[7]. It is used, nevertheless, and gives the prediction – with relevant quantities from Table 1 and 2 –

$$\dot{P}_b = -(2.403\pm0.005)\times10^{-12}. \quad (6)$$

The pulse arrival-time analysis yields excellent agreement: $\dot{P}_b = (-2.30\pm0.22)\times10^{-12}$, thus giving for the first time direct evidence of the emission of gravitational waves, and supporting the old Einstein "quadrupole formula"; a result of unrivalled importance, unless some nasty Newtonian properties of the companion star screw up the clean relativistic interpretation of the binary system.

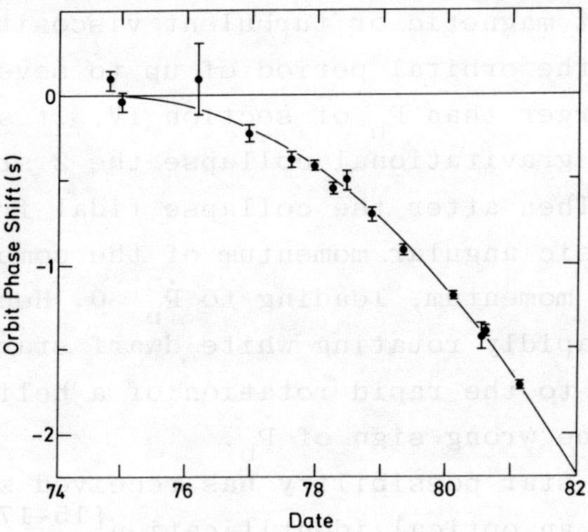

Fig. 4 Orbit phase residuals, i.e. departures from integral multiples of P_b of the periastron passage times, are plotted. The drown-out line is the prediction from general relativity theory.

V. THE (UNSEEN) COMPANION STAR

The use of PSR 1913+16 as a laboratory for general relativity rests on the assumption that classical effects, such as tidal or rotational distortion of the companion and viscous dissipation of orbital energy, are unimportant. These effects are completely negligible for a neutron star, black hole, or slowly rotating white dwarf companion.

A main sequence star of normal composition is difficult to fit into the system and leads to a classical periastron advance much larger than the observed value (cf. III.(4)). The only remaining candidates are either a rapidly rotating white dwarf or a helium star.

The dissipative action of tidal forces is difficult to estimate, but magnetic or turbulent viscosity might lead to a change of the orbital period of up to several orders of magnitude larger than $\dot{P}_b$ of section IV. It seems probable that before the gravitational collapse the 2 stars rotate synchronously. Then after the collapse tidal friction will transfer intrinsic angular momentum of the components to orbital angular momentum, leading to $\dot{P}_b > 0$. Hence it seems likely that a rapidly rotating white dwarf star (the same arguments apply to the rapid rotation of a helium star) would lead to the wrong sign of $\dot{P}_b$.

The helium star possibility has received some support by the claim of an optical identification[15-17] of the pulsar companion. A star with $m_V = 22.5$, $m_R = 20.9$ has been found close to the position of the pulsar (.13" ± .21" east, .36"±.22" north [3]). This may well be a chance coincidence, although the small discrepancy in declination is not enough to reject the association. Recently the interesting suggestion has been made that the optical light is due to the pulsar itself.[18]

There are several points making a helium star companion unlikely: a) The measurement of the transverse Doppler shift and gravitational redshift term γ limits the possible masses regardless of the nature of the companion. The interval $0.00412 \leq \gamma \leq 0.00460$ sec translates into an interval $0.8 \leq m_c/m_\odot \leq 1.9$ for the mass of the companion, if $0 \leq m_p/m_\odot \leq 3$ for the pulsar mass. The periastron advance for a helium star companion is given by [19]:

$$\dot{\omega}_{tot} = 13.6 \left(\frac{m_c}{m_\odot}\right)^{4.59} \frac{m_p}{m_c} \left[(m_p + m_c)/m_\odot\right]^{-5/3} + 2.11 \left[(m_p + m_c)/m_\odot\right]^{2/3}. \quad (7)$$

Here the first term describes the classical, tidal effect and the second term is the relativistic contribution. The structure function $k_2(R_c/R_\odot)^5 \simeq 4.02\times10^{-6}(m_c/m_\odot)^{4.59}$ has been inserted[19], as well as the classical orbital parameters. The restrictions imposed by the value of γ and the requirement that $\dot{\omega}_{tot} \leq \dot{\omega}_{obs}$ lead to the conclusion that $m_c \leq 1.0 m_\odot$ and $m_p \leq 0.3 m_\odot$ for a helium star companion. It seems reasonable to require the lightest neutron stars to be more tightly bound than presupernova cores (i.e. ^{12}C or ^{56}Fe), i.e. $m_p \geq 0.2\ m_\odot$. There seems to be an error in the usual derivation of the classical apsidal motion formula[20]. The correct formula gives $2\ \dot{\omega}_{ct}$ in Eq. (7), and leads to the limits $m_c \leq 0.8\ m_\odot$, $m_p \leq 0.2\ m_\odot$. The allowed values for the pulsar are thus uncomfortably close to the minimum value, and not at all in agreement with the mass determinations in binary X-ray sources (where the neutron stars have masses $\geq 0.6\ m_\odot$).

b) The large eccentricity speaks for a compact companion, since tidal interactions would tend to circularize the orbit on a short time scale.

c) Helium stars have strong, radiation-driven winds. Searches for variations of the pulsar's dispersion measure with respect to orbital phase have given an upper limit of 0.002 pc cm^{-3}, i.e. the mean electron density inside the pulsar orbit is less than $10^5\ cm^{-3}$[3]. This restricts the stellar wind parameters to $\frac{\dot{M}}{V} < 10^{-17} M_\odot\ yr^{-1}\ km^{-1}\ sec$ ($\simeq 10^9\ g\ sec^{-1}\ v^{-1}$).

There is no conclusive evidence yet, but observations and theoretical considerations strongly suggest a compact object as the companion star of PSR 1913+16.

Various evolutionary paths have been suggested [21,22] which could lead to the presently observed configuration. The most natural way seems to start out with a pair of massive main sequence stars (20 to 30 $M_\odot$), orbiting each other with a period of several days. The more massive component evolved more rapidly, transferred all its envelope to the smaller one, retaining a helium core, then — bang! — the supernova explosion creates PSR 1913+16. After a X-ray binary stage the secondary expanded, engulfed the neutron star, and made it spiral inward. The frictionally produced luminosity blew away the envelope, then — bang! — explosion of the secondary, and two neutron stars orbit each other in an eccentric binary system.

In this scheme PSR 1913+16 is the older neutron star which has undergone a spin-up during its life as an accreting X-ray source. The low value of $\dot{P}$, and correspondingly $B \simeq 3\times10^{10}$ gauss (derived from the usual radiating dipole model) ties in with an old pulsar. The main difficulty is to explain the present tight orbit with its rather large eccentricity. Another difficulty is the timescale of the spin-up phase. For typical parameters of such a system (Helium core of 15 $M_{\odot}$ supergiant and 1.4 $M_{\odot}$ neutron star) the common envelope phase last about 1500 years[23]. This time is too short by a factor 100 to achieve the short period of 59 msec, for a pulsar accreting at the Eddington limit, and changing its angular momentum accordingly (F. Meyer, personal communication).

ACKNOWLEDGEMENTS

I have to thank J.H. Taylor and J.L. Weisberg for making available to me the most recent representation of the data prior to publication. Thanks are also due to H. Ritter for discussions and to U. Anzer for a critical reading of the manuscript.

REFERENCES

[1] Hulse, R.A., Taylor, J.H., 1975, Ap. J., Vol. 195,L51.

[2] Börner, G., 1981, Abstract of EPS-Neutron Star Symposium talk.
[3] Taylor, J.H., Weisberg, J.L., 1981, preprint (submitted to _Ap. J_.).
[4] Blandford, R.D., Teukolsky, S.A., 1976, _Ap. J._, Vol. 205, 580.
[5] Ruderman, M., 1972, _Ann. Rev. Astron. Astrophys_., Vol. 10, 427.
[6] Ehlers, J., Rosenblum, A., Goldberg, J.N., Havas, P., 1976, _Ap. J_., Vol. 208, L77.
[7] Walker, M., Will, C.M., 1980, _Ap. J._, Vol. 242, L129.
[8] Börner, G., Ehlers, J., Rudolph, E., 1975, _Astron_. _Astrophys_., Vol. 44, 417.
[9] Hari Dass, N.D., Radhakrishnan, V., 1975, _Ap. Lett_. Vol. 16, 135.
[10] Barker, B.M., O'Connel, R.F., 1975, _Ap. J._,Vol. 199, L25.
[11] Breuer, R., Rudolph, E., 1981, _GRG_, Vol. 13, 777.
[12] Peters, P.C., Mathews, J., 1963, _Phys. Rev_., Vol. 131, 435.
[13] Wagoner, R., 1975, _Ap. J._, Vol. 196. L63.
[14] Esposito, L.W., Harrison, E.R., 1975, _Ap. J._, Vol. 196, L1.
[15] Kristian, J., Westphal, J.A., 1978, _IAU Circ._,No.
[16] Crane, P., Nelson, J.E., Tyson, J.A., 1979, _Nature_, Vol. 280, 367.
[17] Elliott, K.H. et al., 1980, _M.N.R.A.S._, Vol. 192, 51p.
[18] Kundt, W., 1981, preprint (Bonn Univ.).
[19] Roberts, D.H., Masters, A.R., Arnett, W.D., 1976, _Ap. J_.,Vol. 203, 196.
[20] Straumann, N., unpublished (personal comm.).
[21] Smarr, L.L., Blandford, R., 1976, _Ap.J_.,Vol. 207, 574.
[22] Srinivasan, G., Van den Heuvel, E.P.J., 1982, _Astron_. _Astrophys_., Vol. 108, 143.
[23] Meyer, F., Meyer-Hofmeister, E., 1979, _Astron_. _Astrophys_., Vol. 78, 167.

EVOLUTION OF CLOSE BINARY SYSTEMS

Hans-Christoph Thomas

(Max-Planck-Institut für Physik und Astrophysik, München, Federal Republic of Germany)

I. INTRODUCTION

The really interesting observations are those, which seem to contradict our theoretical expectations. Since we believe strongly in our theories, we tend to think of such observations as being paradox. One such example is called the Algol paradox. About 1700 years ago Ji-Shi (the Chinese name of Algol) was recorded in Xing Jing(Astronomical Records) as a symbol for the dying of a population. This is probably the first indication of its variable brightness. The Arabs also must have known this property of Algol, because they named it Rhâs al ghûl (spirit of unrest). However, exchange of such knowledge was non-existent at those times and so 311 years ago the Italian astronomer Gemiani Montanari (1671)[1] rediscovered, that the second brightest star in the Perseus constellation sometimes became a rather dark star. 200 Years ago the English astronomer John Goodricke (1783)[2], again independently, detected the variability of Algol, determined a period and suggested the explanation, that the star is from time to time partially covered by another object. Another 100 years later this explanation was finally accepted as the correct one, when Vogel (1890)[3] obtained the first spectroscopic observations of Algol and so established its binary nature. It became clear, that the more massive component, Algol A, is a normal main sequence star of spectral class B8, while Algol B seemed to be a subgiant of class late G to early K.

But then, starting with Eddington, the theory of stellar structure developed, which postulates that, of two stars of the same age, the more massive one must be more evolved than its companion, in contradiction to what we see in the Algol system. Another well-known example, which violates the above postulate, is Sirius, a system, where the highly evolved star, the white dwarf Sirius B, has about half the mass of its main sequence companion. The Algol paradox was removed independently by Crawford (1955)[4] and Kopal (1955)[5], who pointed to a fundamental difference between binary star evolution and the evolution of single stars, because in a close binary each star has only a finite volume of space at its disposal. At some stage of its evolution the more massive star will expand above this limiting volume and matter will flow from its surface towards the companion star. This has been worked out in more detail by Morton (1960)[6] with the help of stellar evolution calculations by Kushwaha (1957)[7] and Schwarzschild and Härm (1958)[8]. He discussed the different time scales involved and showed, that mass transfer from one component to the other will reverse the mass ratio of the system within a thermal time scale, rendering the unevolved component to be the more massive one. Under this assumption, that all matter lost by one star is absorbed by the other, and that no orbital angular momentum is lost in the process, we have the following equation for the separation of the two stars

$$d = \frac{J^2}{GM^3} \; [\, q_1(\, 1 - q_1)]^{-2},$$

where d is the separation, J the total orbital angular momentum, G the gravitational constant, M the sum of the two stellar masses M_1 and M_2 and $q_1 = M_1/M$. The function in square brackets is a parabola, which reaches its maximum value 0.25 at $q_1 = 0.5$, so starting with $q_1 > 0.5$ a decrease in q_1 will result in a decrease of the separation ($d \log d/d \log M_1 \simeq 2$ for $q_1 = 2/3$). This means that the volume available for the evolving star shrinks. Up to this stage the evolution of the star usually proceeded on a nuclear time scale, close to thermal equilibrium. But in thermal equilibrium the change in mass corresponds to a change in radius which would not keep the star inside its maximum volume ($d \log R/d \log M \simeq 0.64$ for the main sequence). So the star cannot reach thermal equilibrium without loosing more mass to its companion. The mass transfer will proceed on a much smaller time scale than the evolution before and when we finally see the result in the sky, the mass ratio will have been reversed, because then the separation starts to increase and thermal equilibrium becomes possible again.

II. BASIC CONCEPTS AND ASSUMPTIONS

First I have to define a close binary. Following Paczýnski (1971)[9] I count all binary systems as such, if at some stage of their evolution mass transfer changes the system drastically. The frequency of early type binary systems as a function of their separation (after Van Albada, 1968)[10], and the period distribution of low mass binaries (after Abt and Levy, 1976)[11] are shown in Figs. 1 and 2 respectively. The maximum radius a star can reach during its evolution lies between 100 and 500 solar radii, depending on its mass. For low mass binaries a separation of 100 solar radii very roughly corresponds to a period of 100 days. So we see,

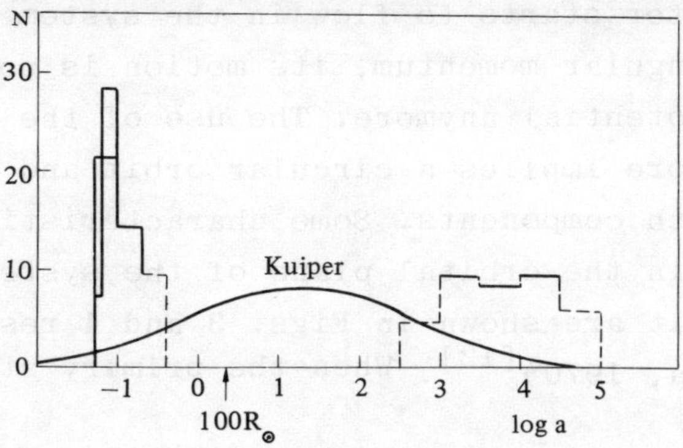

Fig. 1

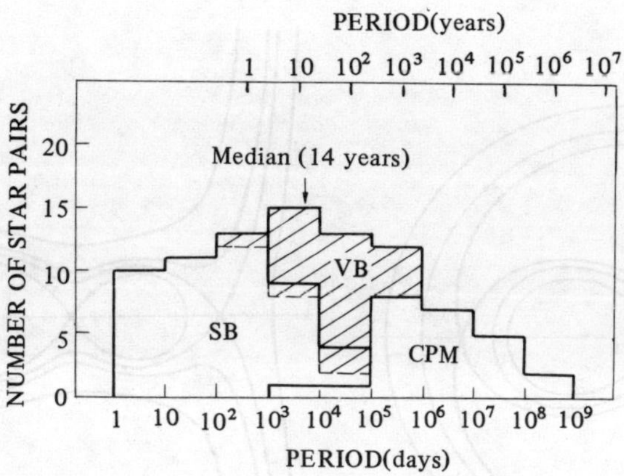

Fig. 2

that close binaries are quite frequent, they amount to about one quarter of all binaries. To calculate the maximum volume available for each component, the Roche potential has been of great importance. This is the sum of the gravitational potentials of two point masses and the potential of centrifugal force assuming the same angular velocity for each point in space. Since stars are centrally condensed and do not deviate greatly from spherical symmetry, the Roche potential is a good approximation for matter, which corotates with the binary system. But, and this is often ignored, as soon as matter starts to flow in the system without changing its angular momentum, its motion is not determined by the Roche potential anymore. The use of the Roche potential therefore implies a circular orbit and synchronous rotation of both components. Some characteristic equipotential lines in the orbital plane of the system and perpendicular to it are shown in Figs. 3 and 4 respectively (after Kitamura, 1970)[12]. When the primary star has ex-

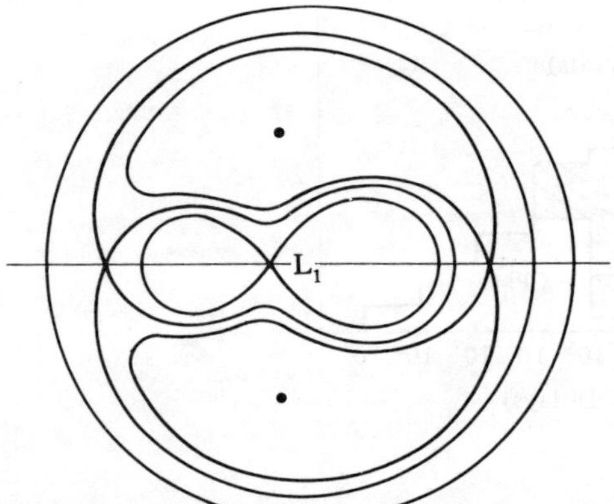

Fig. 3

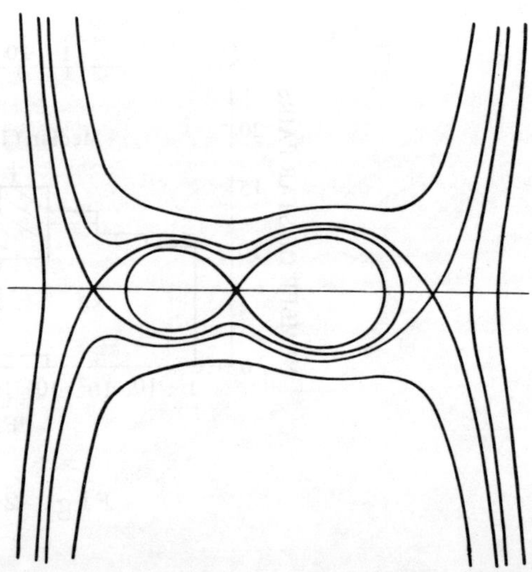

Fig. 4

panded above its limiting Roche volume given by the surface through the point L_1 (inner Lagrangian point), one may assume that all matter outside this surface flows towards the other component either directly or through forming a disk of negligible mass and that synchronous rotation is soon reestablished. This assumption has been named "conservative", because it conserves the total mass and angular momentum of the system. The assumption is not proven to be justified, some matter may leave the system or a significant part of the angular momentum might be stored in rotational motion. To follow the mass transfer process, two other simplifying assumptions may be used: the mass-loosing star remains in hydrostatic (but not in thermal) equilibrium, the mass-gaining star remains in hydrostatic and thermal equilibrium. Again these are not always justified, especially the second one will often be violated. Since deviations from spherical symmetry are confined to the outer layers of a star, one can compute the volume of a spherical star and compare it to the volume inside the limiting Roche surface, in order to obtain the mass transfer rate.

III. CONSERVATIVE MASS TRANSFER

Using the conservative assumption it is the evolutionary state of the primary component which determines the outcome of a phase of mass transfer. Following the notation introduced by Kippenhahn and Weigert(1967)[13] and Lauterborn (1970)[14] I will distinguish between cases A, B, and C, depending upon whether mass transfer starts, while the star is close to the main sequence (A), has not yet started helium burning (B), or has exhausted its helium content at the center (C). Fig. 5 shows as a function of primary mass, where these three cases are expected to occur (from Thomas, 1977)[15].

I will first discuss the properties of case A, following Kippenhahn and Weigert(1967)[13], who started with a system of 9 and 5 solar masses, having a period of 1.49 days. The 9 solar mass star filled the limiting Roche volume after 1.25×10^7 years, its central hydrogen

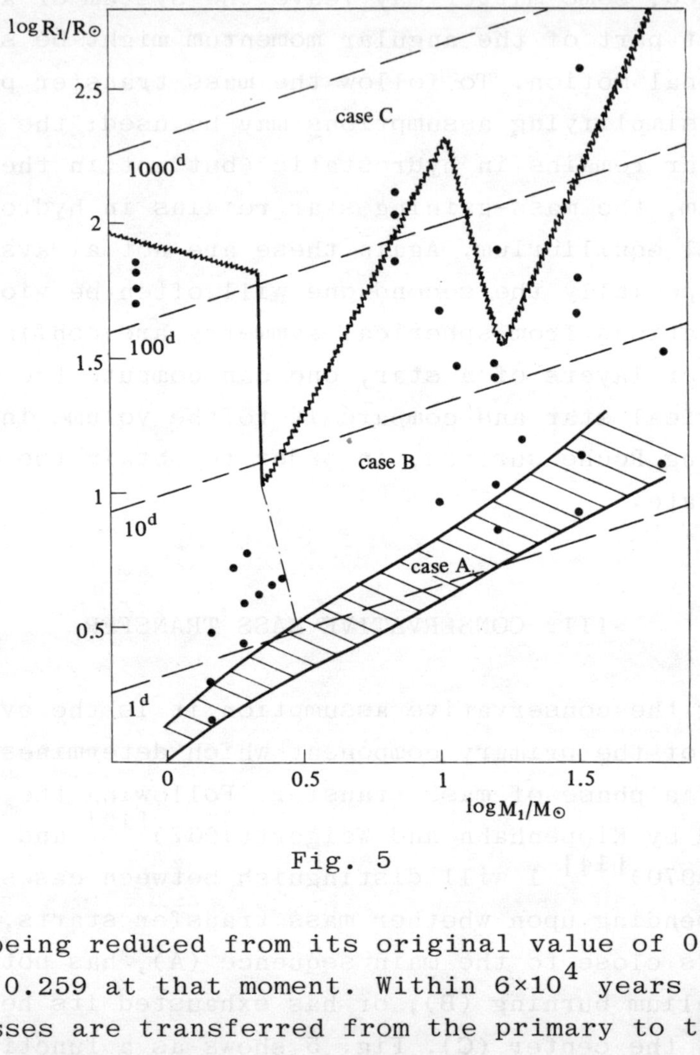

Fig. 5

content being reduced from its original value of 0.602 by mass to 0.259 at that moment. Within 6×10^4 years 5.27 solar masses are transferred from the primary to the secondary. The position of the primary in the Hertzsprung-Russell diagram is shown in Fig. 6 (evolution from point a to point b). The luminosity at point b is smaller than at point a, because the reduction in mass causes a reduction in the rate of nuclear burning. That the luminosity reaches lower values between these points, shows, that these models are out of thermal equilibrium: energy is

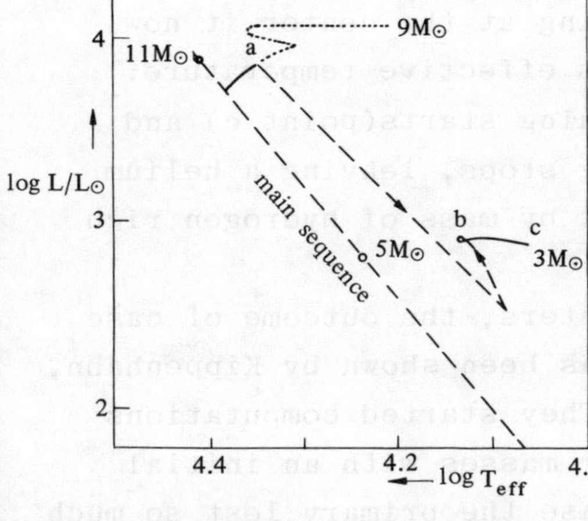

Fig. 6

needed to expand the outer layers during the mass transfer phase. After point b the primary expands on a nuclear time scale, transferring mass at a rate of 4×10^{-8} solar masses per year, until its hydrogen is exhausted at the center after another 1.8×10^{7} years. During this phase it is overluminous for its mass and has a lower mass than its still unevolved companion.

The other computation of Kippenhahn and Weigert (1967)[13], referred to as case B, started with a system of 9 and 3.13 solar masses and a period of 5.34 days. When the primary reached its limiting volume, it contained a pure helium core of 1.5 solar masses and energy was produced in a hydrogen burning shell above this core. Within 4×10^{4} years the star lost about 7 solar masses to its companion. One can see from Fig. 7, that at the end of mass transfer the luminosity was the same as before, because in

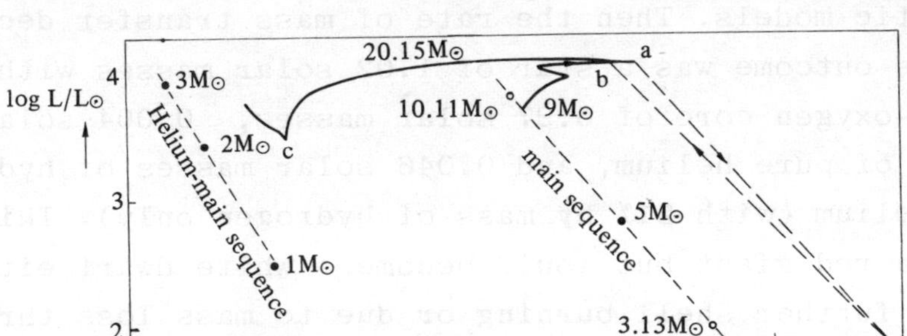

Fig. 7

shell burning models with dense cores it depends on the core mass and not the total mass of the star (Paczýnski, 1970)[16]. During mass transfer thermal equilibrium again cannot be reached. But at the end of mass transfer the star has a shell source rather close to the surface, and in order to ignite helium burning at the center it now has to contract, increasing its effective temperature. After 3×10^5 years helium burning starts(point c) and somewhat later hydrogen burning stops, leaving a helium burning star with a few percent by mass of hydrogen rich material on top.

Varying the initial parameters, the outcome of case B can be quite different, as has been shown by Kippenhahn, Kohl, and Weigert (1967)[17]. They started computations with a system of 2 plus 1 solar masses with an initial period of 1.14 days. In this case the primary lost so much mass, that it ended as a white dwarf of 0.264 solar masses.

The first example of case C mass transfer was presented by Lauterborn (1970)[14]. His initial parameters were 5 plus 2 solar masses and a period of 227 days. Mass transfer started, when the primary had a carbon-oxygen core of 0.5 solar masses, a helium burning shell with 0.45 solar masses of pure helium on top, and an envelope of 4.05 solar masses with the original chemical composition. Most of the envelope was lost during a first very rapid phase of mass transfer, which could not be calculated with hydrostatic models. Then the rate of mass transfer decreased and the outcome was a star of 1.02 solar masses with a carbon-oxygen core of 0.97 solar masses, 0.004 solar masses of pure helium, and 0.046 solar masses of hydrogen plus helium (with 21% by mass of hydrogen only). This was still a red giant but would become a white dwarf either due to further shell burning or due to mass loss through a stellar wind. The final period of the system was 2.8 years, the original secondary (with now 6 solar masses) still almost unevolved. During the rapid phase of mass transfer in case C the conservative assumption almost certainly brakes down, the loss of mass and angular momentum from the system reducing the mass of the original secondary and changing the final period. Also the mass-receiving component might not always be able to maintain thermal equilibrium (see for instance Kippenhahn and Meyer-Hofmeister, 1977[18]), its expansion then leads to a system where both stars reach their limiting volumes (contact configuration) and again mass and angular momentum can be lost from the system. Even if this is not the case, one cannot be sure, whether not some material finds its way out of the system and, if magnetic fields are present, carry a high specific angular momentum with it. To cope with this problem, one may look at observed systems and test, whether they can be explained with the conservative assumptions or whether a certain amount of mass and angular momentum loss during mass transfer is required.

IV. CONSEQUENCES OF MASS LOSS FROM THE BINARY SYSTEM

I introduce two parameters f and g, to prescribe non-conservative mass transfer. f is the amount of mass transferred to the companion star divided by the amount of mass leaving the primary component, and g =(d log J/(d log M) is the average specific angular momentum carried by the matter leaving the system (J: total angular momentum, M: total mass of the binary system). This means that we now have five free parameters, which need to be specified, if we want to model close binary evolution. Since f and g only hide our lack of physical understanding, one has to vary them within plausible ranges. This has the consequence, that one cannot simply trace back the evolution of an observed binary system to its origin. But then it is not worthwhile to do step by step integration for the evolution of the two components, but rather use crude results from single star evolution together with our knowledge of the mass transfer in cases A, B, and C. Here cases B and C can be treated quite simply, because as a result of the strong dependance of luminosity on core mass the mass transfer stops, when only a small amount of mass is left on top of the core. The remnant core will either become a white dwarf or explode as a supernova. Kippenhahn and Thomas (1980)[19] together with. Meyer-Hofmeister and Ritter used this simplified treatment to model the evolution of massive binaries. Figs. 8 and 9 present some of their results involving case B mass transfer and a supernova explosion. The original mass ratio was assumed as 1.2. Specifying the original separation A (vertical scale) and primary mass M_1 (upper horizontal scale) one can read off within certain limits (solid curves) the final period (hatched curves) of a system consisting of a normal star with mass M_{opt} (lower horizontal scale) and a neutron star left after the supernova explosion, assumed to have a mass of one solar mass. Fig. 8 used the conservative assumptions($f = 1$); Fig. 9 is an example for the other extreme ($f = o$, $g = 2$). Outside the solid curves a second phase of mass transfer occurred, the outcome of which could not be predicted. Comparing the results with observed systems, which consist of a neutron star and a massive companion, namely the x-ray binaries, we see, that under the conservative assumptions one does not end with the rather low periods observed (of the order of a few days) and in some cases the loss of angular momentum is probably higher than implied with $g = 2$.

A different approach to the problem was taken by de Grève and Vanbeveren (1980)[20], who fitted analytical formulae to 151 results of mass transfer computations for cases A and B. These formulae give the primary's mass after

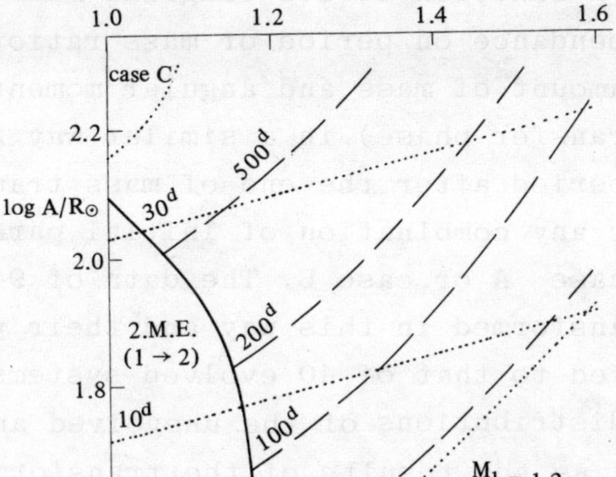

log $M_1/M_\odot$
1.0
1.2
1.4
1.6
case C
2.2
500d
log $A/R_\odot$
30d
2.0
200d
2.M.E.
(1 → 2)
1.8
10d
100d
$\frac{M_1}{M_2} = 1.2$
1.6
50d
case B
f = 1, g = 0
1.4
3d
case A
20.5
25.4
31.3
38.6
46
$M_{opt}/M_\odot$

Fig. 8

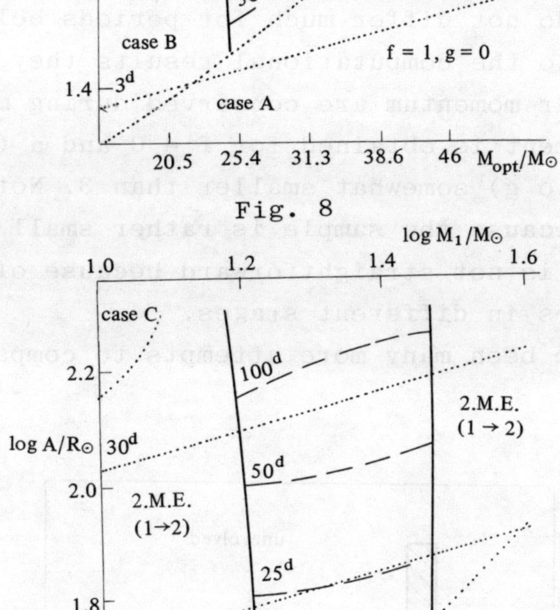

log $M_1/M_\odot$
1.0
1.2
1.4
1.6
case C
2.2
100d
2.M.E.
(1 → 2)
log $A/R_\odot$
30d
50d
2.0
2.M.E.
(1→2)
25d
1.8
10d
15d
$\frac{M_1}{M_2} = 1.2$
1.6
10d
case B
f = 0, g = 2
1.4
3d
case A
10.5
13.2
16.6
20.9
26.4
$M_{opt}/M_\odot$

Fig. 9

(mass transfer as a function of its original mass, neglecting, however, any dependance on period or mass ratio. After specifying the amount of mass and angular momentum (averaged over the transfer phase) in a similar way as above, the masses and period after the end of mass transfer can be predicted for any combination of initial parameters, which leads to case A or case E. The data of 94 unevolved systems were transformed in this way and their period distribution compared to that of 40 evolved systems. Figs. 10-13 show period distributions of the unevolved and evolved systems, as well as the results of the transformation in the conservative case and the case $f = 0$, $g \simeq 3$. The observed distributions do not differ much for periods below 60 days, but according to the computational results they should, if mass and angular momentum are conserved during mass transfer. A better agreement is obtained for $f = 0$ and α (which is roughly equal to g) somewhat smaller than 3. Not much more can be said, because the sample is rather small and the transformation is not straightforward because of the different lifetimes in different stages.

There have been many more attempts to compare observa-

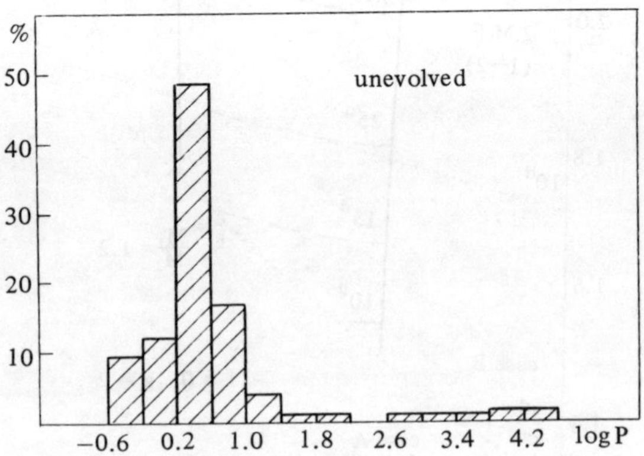

Fig. 10

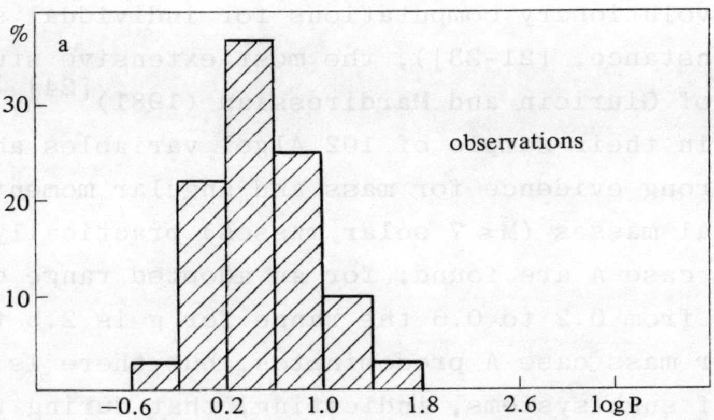

Fig. 11

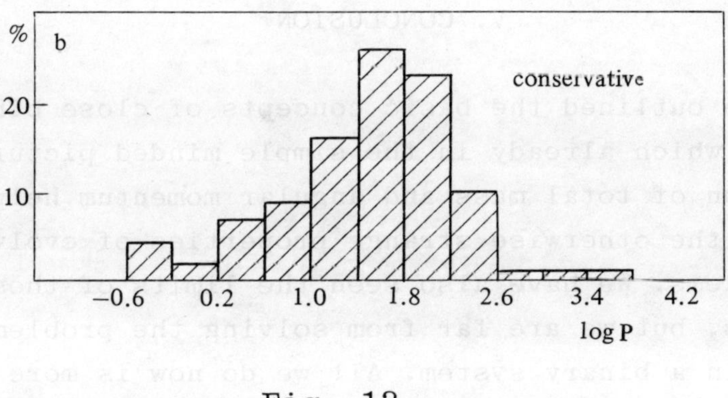

Fig. 12

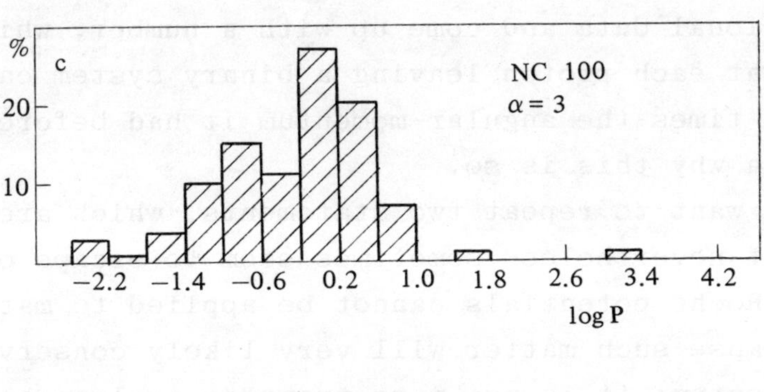

Fig. 13

tions and evolutionary computations for individual systems (see, for instance, [21-23]), the most extensive study being that of Giuricin and Mardirossian (1981)[24], who find, that in their sample of 102 Algol variables about 20% show strong evidence for mass and angular momentum loss. For low total masses ($M \leq 7$ solar masses) practically no remnants of case A are found; for an adopted range of the parameter f from 0.2 to 0.6 the range for g is 2.5 to 3. Among higher mass case A predominates, but there is still a deficit of such systems, indicating, that during the early stages of mass transfer contact configurations may occur, which lead to different final properties of the system.

V. CONCLUSION

I have outlined the basic concepts of close binary evolution, which already in the simple minded picture of conservation of total mass and angular momentum helped to understand the otherwise strange properties of evolved binary systems. We have also seen the limits of these assumptions, but we are far from solving the problem of mass flow in a binary system. All we do now is more or less parameter fitting, which is not very satisfactory. Statistics may give one a direction where to look, but never a reason. Although we have processed large amounts of observational data and come up with a number, which tells us that each proton leaving a binary system carries about three times the angular momentum it had before, we have no idea why this is so.

I also want to repeat two statements, which are quite obvious, but nevertheless sometimes seem to escape our attention: Roche potentials cannot be applied to matter in motion, because such matter will very likely conserve its angular momentum; it is not mass transfer or loss alone but transfer of angular momentum, which changes the distance of the two components. There exist a large number of binary systems, which I have not mentioned, which are discussed extensively in the literature, such as W UMa systems, RS CVn stars, cataclysmic variables, Wolf-Rayet binaries, X-ray bursters, and a wealth of computations aimed to explain the evolution into and out of such systems. Some of these will be discussed here, for the others an overview can be found in Sahade and Wood (1978)[25] or Kopal(1978)[26].

REFERENCES

[1] Montanari, G., 1671, Sopra la sparizione d'alcune stelle e altre novità celesti, in: Prose di Signori Academici Gelati di Bologna.

[2] Goodricke, J., 1783, Phil. Trans. Roy. Soc. London, Vol. 73, 474.

[3] Vogel, H.C., 1890, Astron. Nachr., Vol. 232, 1.

[4] Crawford, J.A., 1955, Ap. J., Vol. 121, 71.

[5] Kopal, Z., 1955, Ann. d'Astrophys., Vol. 18, 379.

[6] Morton, D.C., 1960, Ap. J., Vol. 132, 146.

[7] Kushwaha, R.S., 1957, Ap. J., Vol. 125, 242.

[8] Schwarzschild, M., Harm, R., 1958, Ap. J., Vol. 128, 348.

[9] Paczynski, B., 1971, Ann. Rev. Astron. Astrophys. Vol. 9, 183.

[10] Van Albada, T.S., 1968, B.A.N., Vol. 20, 47.

[11] Abt, H.A., Levy, S.G., 1976, Ap. J. Suppl., Vol. 30, 273.

[12] Kitamura, M., 1970, Astrophys. Space Sci., Vol. 7, 272.

[13] Kippenhahn, R., Weigert, A., 1967, Z. Astrophys., Vol. 65, 251.

[14] Lauterborn, D., 1970, Astron. Astrophys., Vol. 7, 150.

[15] Thomas, H. -C., 1977, Ann. Rev. Astron. Astrophys. Vol. 15, 127.

[16] Paczýnski, B., 1970, Acta Astron., Vol. 20, 47.

[17] Kippenhahn, R., Kohl, K., Weigert, A., 1967, Z. Astrophys., Vol. 66, 58.

[18] Kippenhahn, R., Meyer-Hofmeister, E., 1977, Astron. Astrophys., Vol. 54, 539.

[19] Kippenhahn, R., Thomas, H. -C., 1980, in: Ninth Texas Symposium on Relativistic Astrophysics, Ann. N.Y. Acad. Sci., Vol. 336, 579.

[20] De Grève, J.P., Vanbeveren, D., 1980, Astrophys. Space Sci., Vol. 68, 433.

[21] Kreiner, J.M., Ziólkowski, J., 1978, Acta Astron., Vol. 28, 497.

[22] Vanbeveren, D., de Loore, C., 1980, Astron. Astrophys., Vol. 86, 21.

[23] De Grève, J.P., 1980, Astrophys. Space Sci., Vol. 72, 411.

[24] Giuricin, G., Mardirossian, F., 1981, Ap. J. Suppl., Vol. 46, 1.

[25] Sahade, J., Wood, F.B., 1978, Interacting Binary Stars, Pergamon Press.

[26] Kopal, Z., 1978, Dynamics of close binary systems, Reidel, Dordrecht.

CATACLYSMIC BINARIES AND RELATED OBJECTS IN THE CONTEXT OF STELLAR EVOLUTION

Hans Ritter

(Max-Planck-Institut-für Physik und Astrophysik, Institut für Astrophysik, München, Federal Republic of Germany)

I. INTRODUCTION

The scope of this paper is to discuss cataclysmic binaries (hereafter CB's) and related objects in the context of stellar evolution. Before doing so we first define more precisely what types of objects we shall address here. In the next two sections we will then give a brief outline of what is known about these objects. The subsequent sections will be devoted to the evolutionary problems.

The objects we are dealing with in this paper share the following characteristics:

a) The orbital period, P_{orb}, is shorter than $\simeq 1$ day.

b) One of the components is a highly evolved star, i.e. normally a white dwarf or a neutron star.

c) The companion star has a low mass and a low luminosity, i.e. $M \lesssim 1M_{\odot}$, $L \lesssim 1L_{\odot}$. It may also be highly evolved.

The binaries which conform to these criteria are conveniently subdivided into different classes according to the type of the evolved component and depending on whether the binary is a detached system or not, as is shown in the following scheme:

Type of Evolved Component		Magn. Field	Detached Systems	Semi-detached Systems
subdwarfs sd O/B			Central Stars of Planetary Nebulae Objects like UX CVn AA Dor	not observed
White Dwarf		low	Systems like V471 Tau GK Vir HZ 9	Cataclysmic Binaries Old Novae Dwarf Novae Nova-like Objects AM CVn-Stars
White Dwarf		medium	not observed	DQ Her-Stars
White Dwarf		strong	not observed	AM Her-Stars
Neutron Star		weak	not observed and probably not observable	Low-mass X-ray Binaries Sco X-1 like obj. X-ray transients X-ray bursters
Neutron Star		strong	Binary Pulsar	X-ray pulsars

According to this scheme we can distinguish three main classes of binaries, i.e. the Cataclysmic Binaries, the Low-Mass X-ray Binaries and the Detached Binaries. Among these, the CB's are the most important group, not only with regard to the number of objects known but even more with regard to the amount of observational data available. Therefore we shall concentrate mainly on CB's and refer to the related objects where appropriate.

Since it is not the scope of this paper to give a detailed discussion of the observations and of the observational problems, we should mention a number of exellent reviews where readers interested in these questions will find the relevant information. The observational characteristics of cataclysmic variables (hereafter CV's) in general have been reviewed by Robinson (1976)[1] and Warner (1976)[2]. Observational and theoretical aspects of novae are discussed by Gallagher and Starrfield (1978)[3]. The properties of accreting degenerate dwarfs in close binaries have been reviewed by Cordova and Mason (1981)[4]. Observational and theoretical aspects of low-mass X-ray sources may be found in the reviews of Lewin and Joss (1981)[5] and of van Paradijs (1981)[6]. A complete list of all CB's and related objects known at present, together with references to the relevant literature is given by Ritter (1982)[7].

II. THE BINARY MODEL

The wealth of observational data on CV's, which has been accumulated during the past 20 years since Kraft's pioneering papers appeared (see e.g.[8]), is consistent with the following standard model of a CB (Fig. 1):

The binary consists of a white-dwarf primary and a low-mass secondary, presumably a main sequence star, which fills its critical Roche lobe. The secondary spills matter through the inner Lagrangian point L_1. The matter falling towards the white dwarf enters an accretion disk around the primary. At the point where the matter coming from L_1 intersects the disk, a shock front is formed which is usually referred to as the hot spot. The standard model

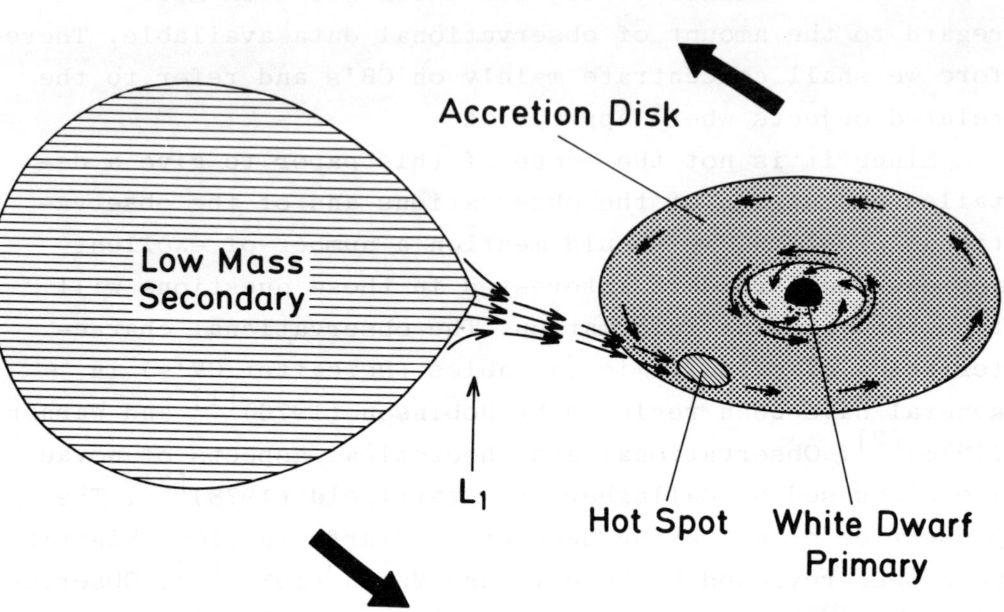

Fig. 1 The binary model of cataclysmic variables

applies to all types of CV's except to AM Her stars. In AM Her stars the strong magnetic field of the white dwarf prevents the formation of an accretion disk. Rather, the matter falling towards the synchronously rotating white dwarf is funnelled towards its magnetic poles.

In low-mass X-ray binaries the primary is a neutron star rather than a white dwarf. Otherwise the standard model of CB's holds also for these objects (see e.g.[5]).

The related detached binaries all contain a white dwarf or an object in pre-white-dwarf stage and a low mass companion which is detached from its Roche lobe. Therefore these objects lack all the properties of CB's which are due to the mass transfer, i.e. a disk, a hot spot, outbursts etc.

Fig. 2 shows the histogram of the orbital periods of CB's Most of the periods are in the range $80^{min} \lesssim P_{orb} \lesssim 8^{h}$.The

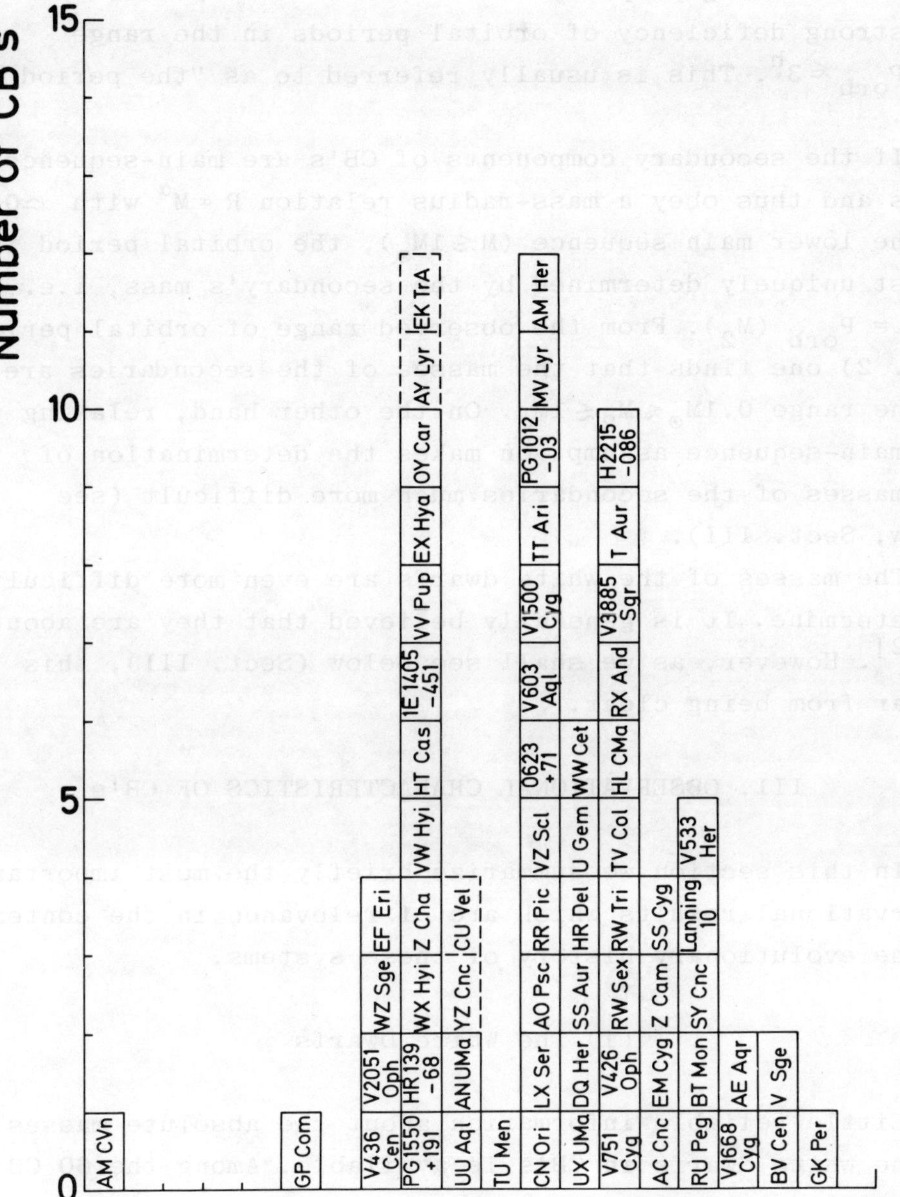

Fig. 2 Period histogram of known CB's

most outstanding property of the period distribution is the strong deficiency of orbital periods in the range $2^h \lesssim P_{orb} \lesssim 3^h$. This is usually referred to as "the period gap".

If the secondary components of CB's are main-sequence stars and thus obey a mass-radius relation $R \propto M^\alpha$ with $\alpha \simeq 0.8$ on the lower main sequence ($M \lesssim 1M_\odot$), the orbital period is almost uniquely determined by the secondary's mass, i.e. $P_{orb} = P_{orb}(M_2)$. From the observed range of orbital periods (Fig. 2) one finds that the masses of the secondaries are in the range $0.1M_\odot \lesssim M_2 \lesssim 1M_\odot$. On the other hand, relaxing the main-sequence assumption makes the determination of the masses of the secondaries much more difficult (see below, Sect. III).

The masses of the white dwarfs are even more difficult to determine. It is generally believed that they are about $1M_\odot$[2]. However, as we shall see below (Sect. III), this is far from being clear.

III. OBSERVATIONAL CHARACTERISTICS OF CB's

In this section we summarize briefly the most important observational results which are of relevance in the context of the evolutionary history of these systems.

(1) The White Dwarfs

Little reliable information about the absolute masses of the white dwarfs in CB's is available. Among the 60 CB's known at present only two are double-lined spectroscopic eclipsing binaries, namely EM Cyg and U Gem. Only these two systems fulfil the classical criteria for a mass determination. The corresponding white-dwarf masses are listed in Table 1a. In a few other systems which are double-lined spectroscopic, but not eclipsing, binaries, constraints on the inclination are possible which allows an estimate of

Table 1. Masses of white dwarfs in CB's

a) The probably most reliable masses

Object	DL/SL*	Ecl.	$M_{WD}/M_\odot$	References
EM Cyg	DL	+	0.56±0.05	Stover et al.(1981)[9]
U Gem	DL	+	1.18±0.15	Stover (1981)[10]
Z Cam	DL		1.20±0.20	Robinson(1976)[1], Warner(1976)[2]
AE Aqr	DL		0.94±0.10	Patterson (1979)[11]
BV Cen	DL		1.3±0.1	Vogt and Breysacher(1980)[12]

b) Two cases with very discrepant determinations

Object	DL/SL*	Ecl.	$M_{WD}/M_\odot$	References
SS Cyg	DL		0.80±0.10	Cowley et al.(1980)[13] Walker (1981)[14]
			1.33±0.10	Stover et al. (1980)[15]
DQ Her	SL	+	0.45±0.05	Young and Schneider(1980)[16]
			1.20±0.20	Hutchings et al. (1979)[17]

*DL = double-line spectroscopic binary
SL = single-line spectroscopic binary

the mass of the white dwarf. These masses are also listed in Table 1a. There are, however, also a number of well-observed CB's, even double-lined spectroscopic ones, for which depending on the authors, very different masses have been obtained. Two examples are given in Table 1b. The reason for these discrepancies may be traced back to different assumptions and to the intricacy of determining radial velocities of these objects. The examples of Table 1b should make us very cautious about the reliability of any white-dwarf mass published, even in the cases of EM

Cyg and U Gem. This is because these masses are also based on radial-velocity measurements whose reliability is very difficult to estimate.

Taking the masses of the white dwarfs listed in Table 1a at face value, one arrives at a surprisingly high mean for the white-dwarf mass of $\langle M_{WD}\rangle \simeq 1.04\ M_\odot$. The standard deviation, however, is considerable $\sigma_M \simeq 0.3\ M_\odot$. The above value of $\langle M_{WD}\rangle$ has to be compared with the corresponding value for single white dwarfs which is $M_{WD} = (0.60\pm0.10)\ M_\odot$ [18,19].

In addition to the systems mentioned so far, white dwarf masses have been derived for an additional 15 - 20 objects. Although these estimates are even less reliable one can at least try to derive a mass spectrum of the white dwarfs. The result of such an attempt is shown in Fig. 3. Ignoring strong fluctuations which result from the still rather small number of objects involved, the distribution is remarkably flat in the range $0.4\ M_\odot \lesssim M_{WD} \lesssim 1.3\ M_\odot$. The mean mass of this distribution is $\langle M_{WD}\rangle \simeq 0.8\ M_\odot$. When compared with the observed mass distribution of single white dwarfs[19] as is shown in Fig. 3, we note the following facts:

a) The mass spectrum of white dwarfs in CB's is much broader.

b) The relative number of massive white dwarfs, i.e. $M_{WD} \gtrsim 1 M_\odot$, is much higher in CB's than among single white dwarfs.

Again we have to be very suspicious about the significance of this result. It might be entirely due to systematic and statistical errors of the methods used to derive the

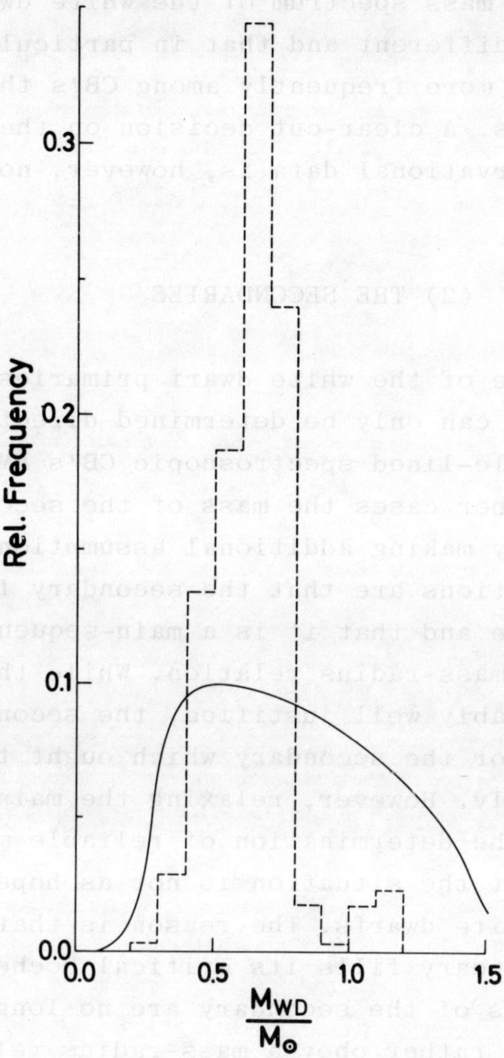

Fig. 3 "Observed" mass spectrum of the white dwarfs in CB's (full line) and observed mass spectrum of single white dwarfs (broken line). The single white dwarf data are taken from Koester and Weidemann (1980)[19]

individual masses. Nevertheless we should also be prepared to accept that the mass spectrum of the white dwarfs in CB's may be quite different and that in particular massive white dwarfs occur more frequently among CB's than among single white dwarfs. A clear-cut decision on the basis of the available observational data is, however, not possible at present.

(2) THE SECONDARIES

As in the case of the white dwarf primaries, the masses of the secondaries can only be determined directly for the two eclipsing double-lined spectroscopic CB's EM Cyg and U Gem. In all the other cases the mass of the secondary can only be inferred by making additional assumptions. The most widely used assumptions are that the secondary fills its critical Roche lobe and that it is a main-sequence star obeying a certain mass-radius relation. While the first assumption is probably well justified, the second one assumes a property for the secondary which ought to be confirmed independently. However, relaxing the main-sequence assumption makes the determination of reliable masses much more difficult. Yet the situation is not as hopeless as in the case of the white dwarfs. The reason is that due to the fact that the secondary fills its critical Roche lobe the mass and the radius of the secondary are no longer independent quantities but rather obey a mass-radius relation of the following form:

$$M_2 = K(P_{orb})R_2^3, \tag{1}$$

where the value of the constant K depends only on the orbital period of the system. (For details see e.g.[20,2,21]). In a few cases it is then possible to put constraints on either the radius R_2 or the mass M_2 of the secondary and thus in favorable cases to get an estimate of M_2 and R_2. Possible constraints are e.g. that the mass of the white dwarf should be less than the Chandrasekhar mass, that the system does or does not eclipse or that the spectral type of the secondary should be roughly consistent with its mass. Fig. 4a shows the (M_2, R_2) values of a number of systems which have been derived in the manner just described (cf.[22,23]).

By comparing the position of the secondaries in the mass-radius diagram with theoretical main-sequence mass-radius relations (Figs. 4b, 4c), a number of authors have concluded that at least some of the secondaries are slightly evolved. In particular this has been claimed for BV Cen, AE Aqr, DQ Her, U Gem and AM Her. At least in the last three cases, the mass of the secondary is so low that normal nuclear evolution cannot result in any significant evolutionary effect over the age of the universe. Therefore, if these secondaries are in fact evolved something unusual must have happened to them in their previous history. When discussing CB's in the context of stellar evolution, it is therefore of the highest importance to know whether the secondaries are evolved or not. From looking at Fig. 4c one gets the impression that at least some of them are. However, if we compare the stars from Fig. 4a with the mass-radius diagram of components of visual (astrometric) and spectroscopic but detached binaries shown in Fig. 4d, we find that the two sets of stars match perfectly within the uncertainties (Fig. 4e). Since the stars shown in Fig. 4d are components of detached, in fact very wide, binaries their mass has not changed since their formation. Furthermore, since they all have low masses they represent the observed zero-age main sequence of low-mass stars. The conclusion from Fig. 4e is thus that the secondaries of CB's are unevolved and that the mismatch observed in Fig. 4c is probably due to a systematic error in the theoretical

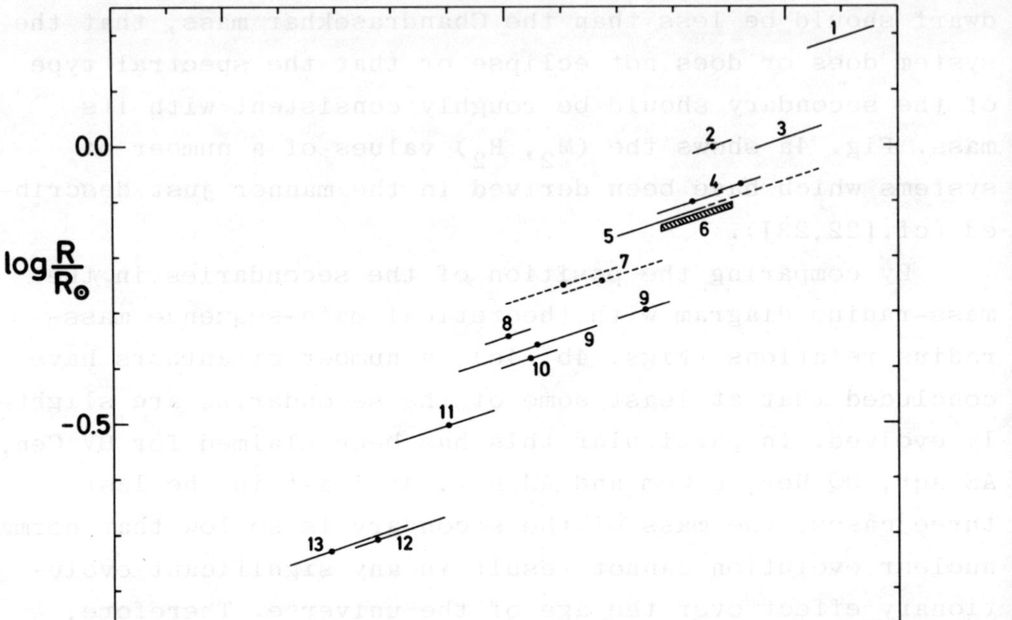

Fig. 4a Mass-radius diagram of the secondary components of selected CB's. The numbers refer to the following objects:
1 = BV Cen, 2 = AE Aqr, 3 = RU Peg,
4 = EM Cyg, 5 = Z Cam, 6 = SS Cyg,
7 = RW Tri, 8 = DQ Her, 9 = U Gem,
10= LX Ser, 11= AM Her, 12 = HT Cas,
13= Z Cha

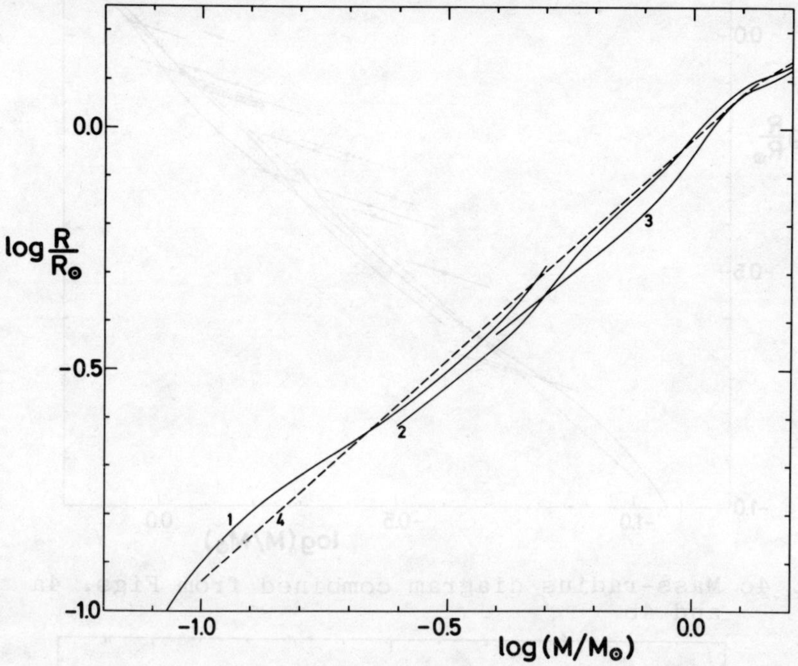

Fig. 4b Theoretical mass-radius diagram of the lower main sequence. The numbers refer to the following computations:

1 = Grossmann, Hays and Graboske (1974)[24], $X = 0.68$, $Z = 0.03$, $1/H_p = 1.0$;

2 = Eggleton (private communication), $X = 0.7$, $Z = 0.02$, $1/H_p = 1.5$;

3 = Copeland, Jensen and Jørgensen (1970)[25], $X = 0.7$, $Z = 0.02$, $1/H_p = 1.5$.

For comparison, the dashed line represents the empirical main sequence according to Lacy (1977)[26].

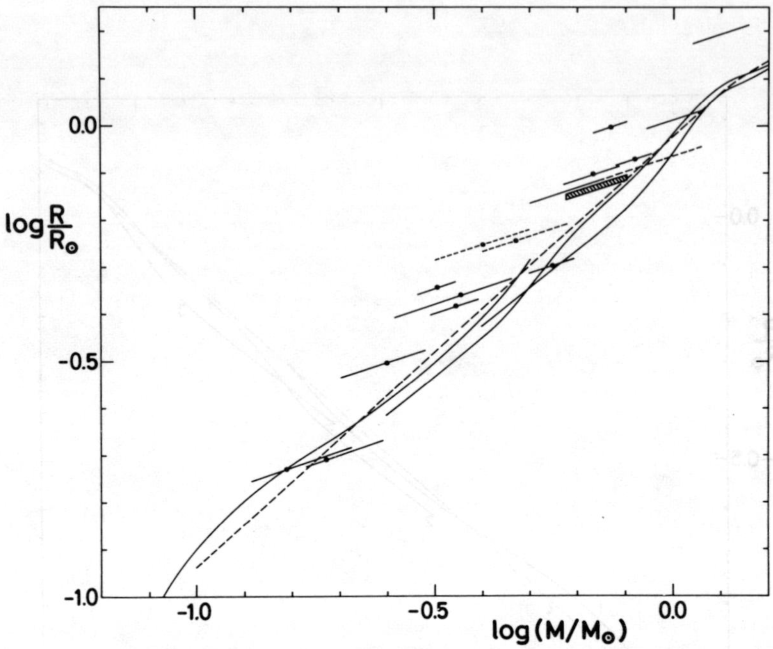

Fig. 4c Mass-radius diagram combined from Figs. 4a and 4b

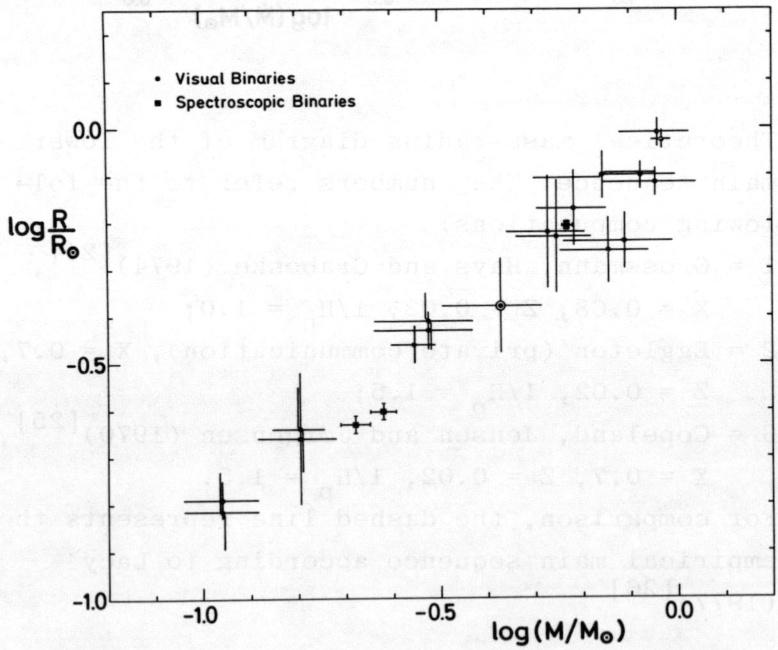

Fig. 4d Mass-radius diagram of low mass components of visual and detached spectroscopic binaries. The data are taken from Popper (1980)[27].

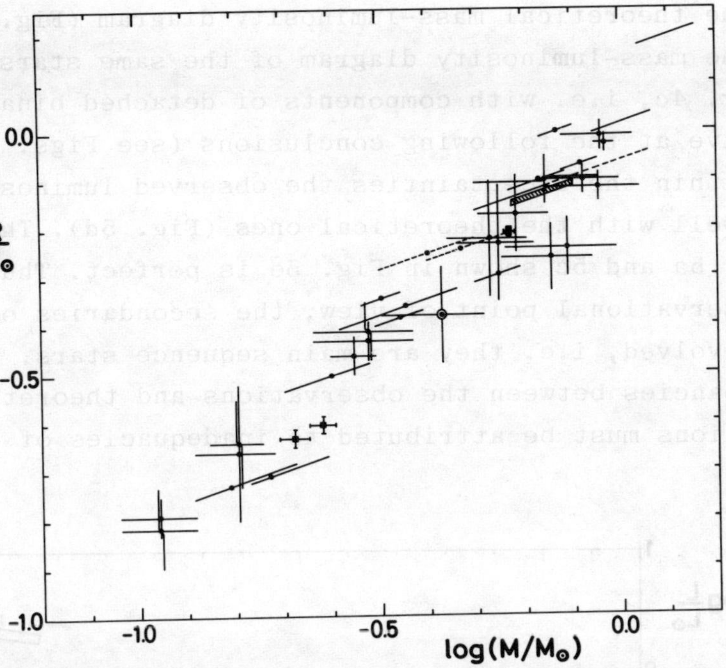

Fig. 4e Mass-radius diagram combined from Figs. 4a and 4d.

computations[22,23].

A much more sensitive probe of the evolutionary status of the secondaries is their position in the mass-luminosity diagram. The determination of absolute luminosities is notoriously difficult for many reasons and in the case of CB's seems to be beyond what can be done at present. There is, however, an elegant way out: In those cases where the position of the secondary in the mass-radius diagram (Fig. 4a) and at the same time its spectral type are known, its luminosity can be computed by using an observationally calibrated relation between the spectral type and the effective temperature T_{eff}. This has been done[23] by using a relation given by Popper (1980)[27]. The result is the mass-luminosity diagram shown in Fig. 5a. When compared with the theoretical mass-luminosity diagram (Fig. 5b) or with the mass-luminosity diagram of the same stars as used for Fig. 4c, i.e. with components of detached binaries[27] we arrive at the following conclusions (see Figs. 5c, 5d, 5e): Within the uncertainties the observed luminosities agree well with the theoretical ones (Fig. 5d). The match of Fig. 5a and 5c shown in Fig. 5e is perfect. Thus from the observational point of view, the secondaries of CB's are unevolved, i.e. they are main sequence stars. The discrepancies between the observations and theoretical predictions must be attributed to inadequacies of the theory.

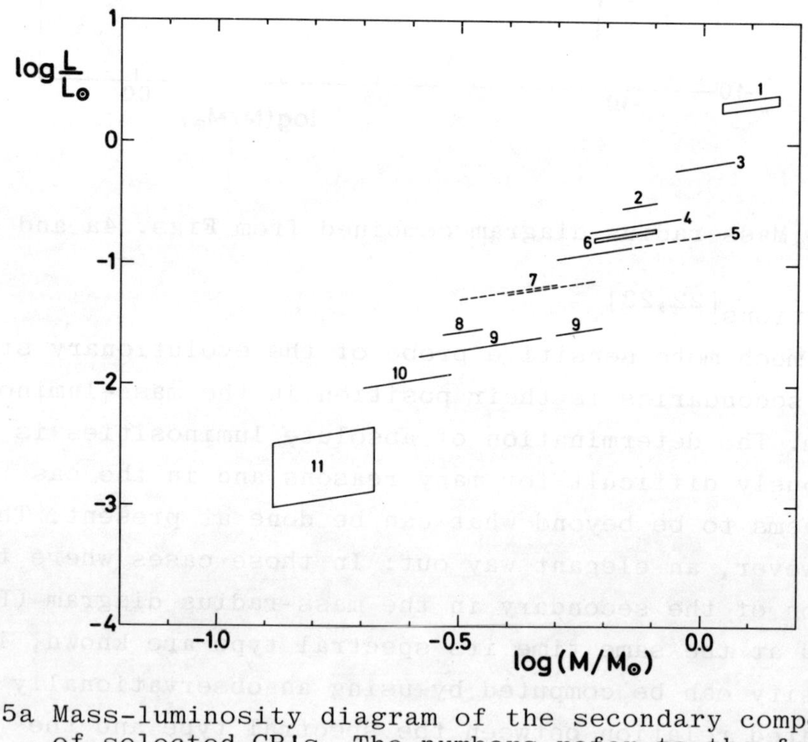

Fig. 5a Mass-luminosity diagram of the secondary components of selected CB's. The numbers refer to the following objects:
1 = BV Cen, 2 = AE Aqr, 3 = RU Peg, 4 = EM Cyg,
5 = Z Cam, 6 = SS Cyg, 7 = RW Tri, 8 = DQ Her,
9 = U Gem, 10 = AM Her, 11 = Z Cha

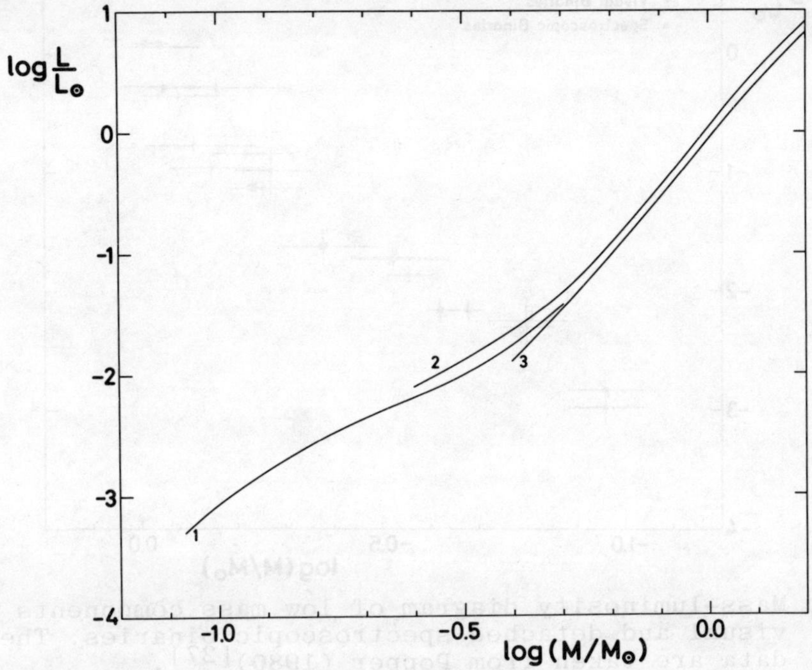

Fig. 5b Theoretical mass-luminosity diagram of the lower main sequence. The numbers refer to the following computations:

1 = Grossmann, Hays and Graboske (1974)[24], $X = 0.68$, $Z = 0.03$, $1/H_p = 1.0$;

2 = Eggleton (private communication), $X = 0.70$, $Z = 0.02$, $1/H_p = 1.5$;

3 = Copeland, Jensen, Jørgensen (1970)[25], $X = 0.70$, $Z = 0.02$, $1/H_p = 1.5$.

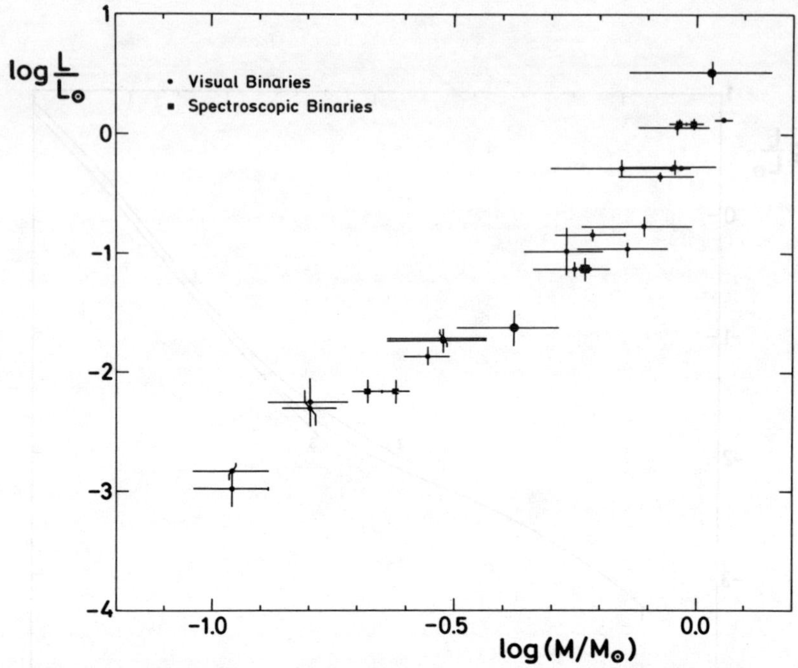

Fig. 5c Mass-luminosity diagram of low mass components of visual and detached spectroscopic binaries. The data are taken from Popper (1980)[27].

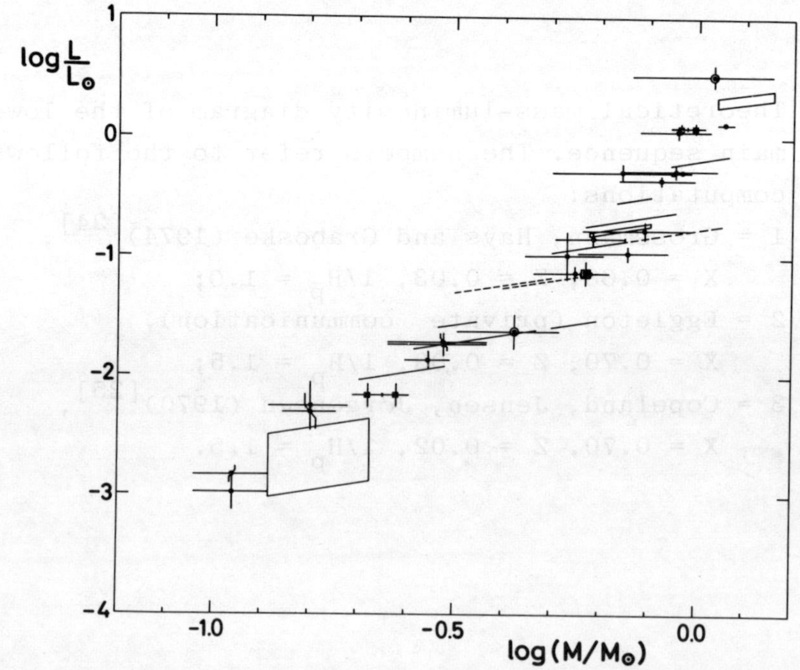

Fig. 5d Mass-luminosity diagram combined from Figs. 5a and 5b.

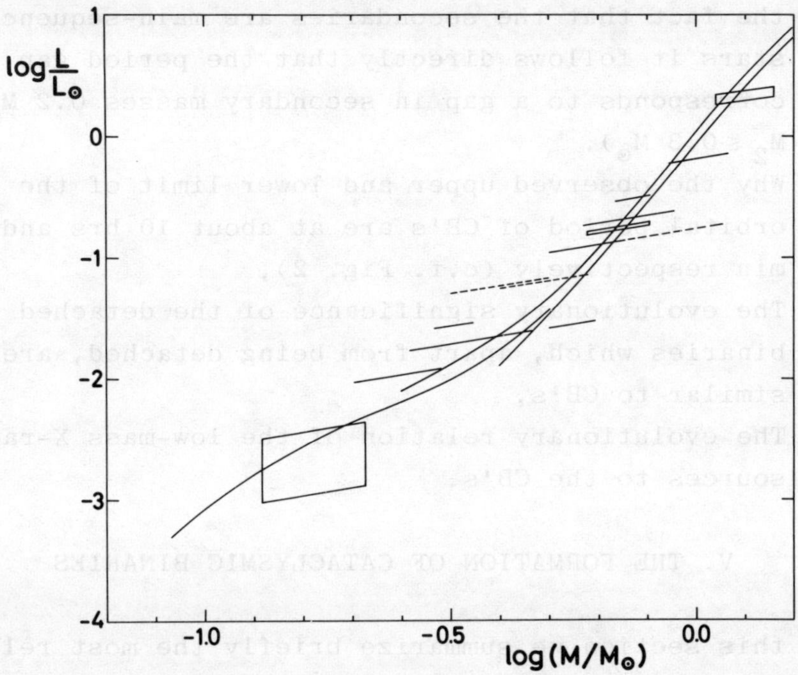

Fig. 5e Mass-luminosity diagram combined from Figs. 5a and 5c.

IV. WHAT HAS TO BE EXPLAINED

Here we give a brief summary of questions which any theory explaining the formation of CB's and of related objects in the context of stellar evolution should be able to answer. In particular we would like to know:

a) How a semi-detached binary consisting of a white dwarf and a Roche-lobe filling low-mass secondary can be formed,

b) Why the secondary components are all unevolved,

c) How the "observed" mass spectrum of the white dwarfs and the "overabundance" of massive white dwarfs can be understood,

d) The significance of the period gap. (N.B. from the fact that the secondaries are main-sequence stars it follows directly that the period gap corresponds to a gap in secondary masses 0.2 $M_\odot \lesssim M_2 \lesssim 0.3\ M_\odot$).

e) Why the observed upper and lower limit of the orbital period of CB's are at about 10 hrs and 80 min respectively (c.f. Fig. 2),

f) The evolutionary significance of the detached binaries which, apart from being detached, are so similar to CB's,

g) The evolutionary relation of the low-mass X-ray sources to the CB's.

V. THE FORMATION OF CATACLYSMIC BINARIES

In this section we summarize briefly the most relevant arguments and ideas about the formation of CB's.

(1) Conservative Evolution

It has been well known for a long time from numerical computations that the evolution of certain types of close binaries results in the formation of a white dwarf and a main-sequence companion[28-30]. However, as a consequence of the assumption that the total mass and the angular momentum are conserved during the evolution, the final systems resulting from these computations always showed two important properties which are not shared by CB's: the mass of the secondary is too large, in most cases exceeding that of the white dwarf, and the orbital separation is too large, so that these systems are far from being semi-detached. Fig. 6 illustrates a typical case: the conserva-

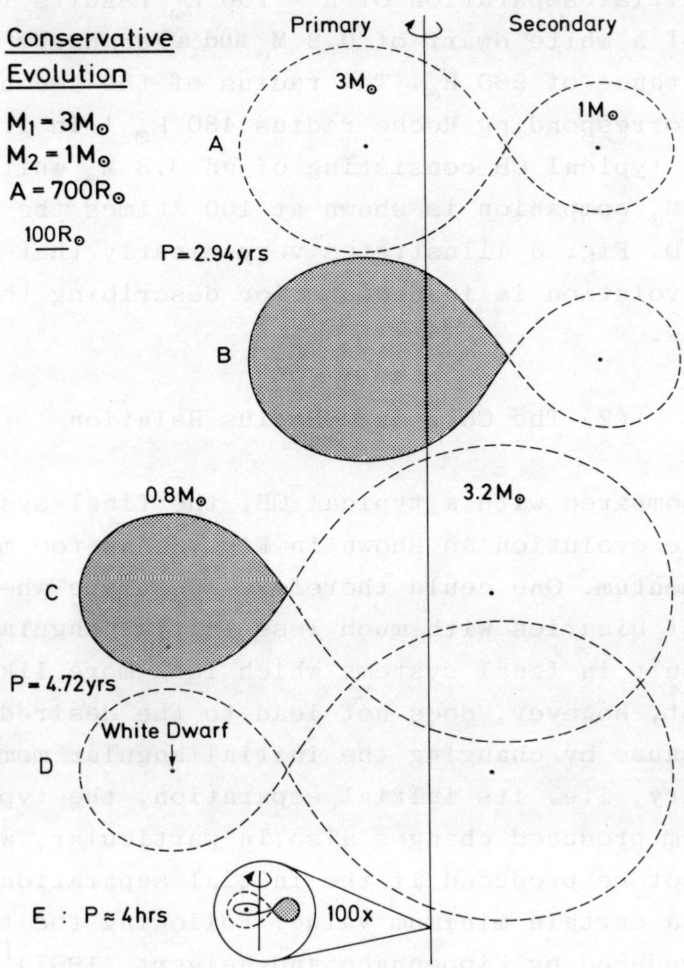

Fig. 6 Conservative evolution of a binary with $M_1 = 3\ M_\odot$, $M_2 = 1\ M_\odot$ and $A = 700\ R_\odot$ drawn to scale.
A) The system on the main sequence, B) at the onset of mass transfer,
C) at the end of mass transfer, D) the final system and E) a typical CB with a white dwarf of the same mass (100 times magnified).

tive evolution of a binary with masses $M_1 = 3\ M_\odot$ and $M_2 = 1\ M_\odot$ at an initial separation of $A = 700\ R_\odot$ results in the formation of a white dwarf of 0.8 $M_\odot$ and a secondary of 3.2 $M_\odot$ at a distance of 960 $R_\odot$. The radius of the secondary is 2 $R_\odot$, the corresponding Roche radius 480 $R_\odot$! In frame E of Fig. 6 a typical CB consisting of an 0.8 $M_\odot$ white dwarf and an 0.4 $M_\odot$ companion is shown at 100 times the scale of frames A - D. Fig. 6 illustrates very clearly that conservative evolution is inadequate for describing the formation of CB's.

(2) The Core Mass-Radius Relation

When compared with a typical CB, the final system of conservative evolution as shown in Fig. 6 has too much angular momentum. One could therefore speculate whether the evolution of binaries with much less initial angular momentum can result in final systems which look more like CB's. This concept, however, does not lead to the desired result. This is because by changing the initial angular momentum of the binary, i.e. its initial separation, the type of final system produced changes also.In particular, white dwarfs cannot be produced if the initial separation does not exceed a certain minimum value. Following the terminology introduced by Kippenhahn and Weigert (1967)[31] and Lauterborn (1970)[29] one can distinguish several types of mass transfer referred to as cases A, B and C. Details are discussed by Thomas (this Proceedings, see also[32]). Helium white dwarfs, with masses $M_{WD} \lesssim 0.45\ M_\odot$, can only be formed by case B mass transfer from primaries with $M \lesssim 2.5\ M_\odot$. Carbon-oxygen white dwarfs, with masses $M_{WD} \gtrsim 0.55\ M_\odot$, can be formed via case C mass transfer from a primary with $M \lesssim 6...8\ M_\odot$. As an example, Fig. 7 shows the evolution of a 5 $M_\odot$ star, Fig. 7a the HR-diagram and Fig. 7b the change of the stellar radius with time. Case C mass transfer is possible only after the radius of the star has increased beyond that at point E, the ignition of central helium burning. The corresponding radius is $\sim 40\ R_\odot$. If the Roche radius of the primary does not exceed this value, no white dwarf will be formed.

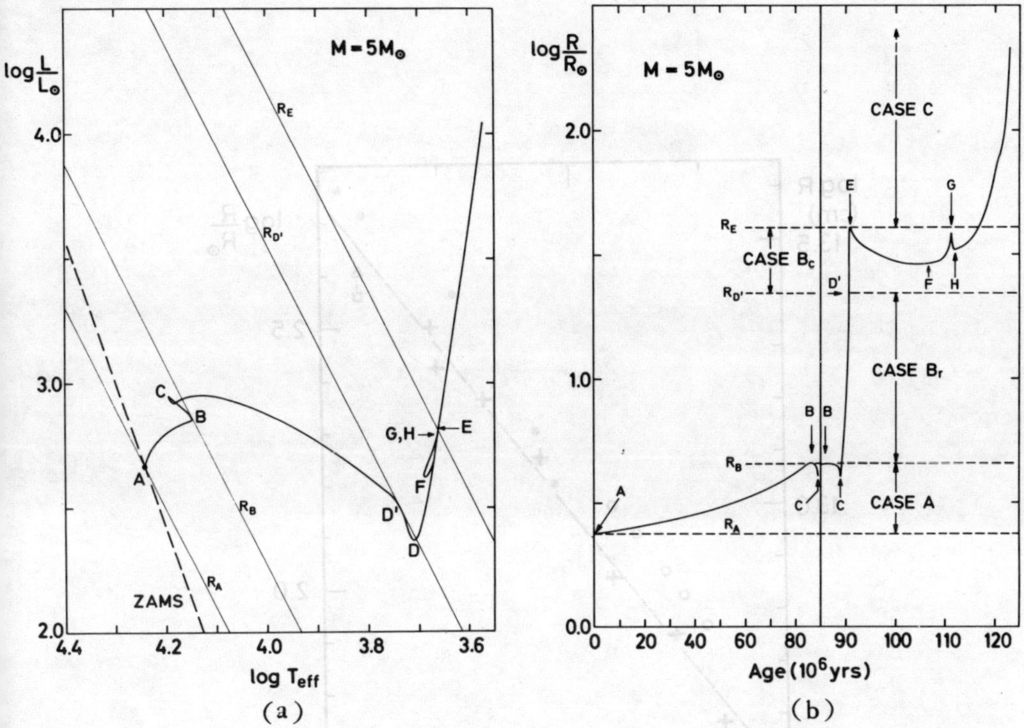

Fig. 7 a) The evolution of a 5 $M_\odot$ star in the HRD

b) The change of the radius of the 5 $M_\odot$ star during its evolution. The minimum and maximum radii corresponding to the cases A,B and C are indicated.

Stars containing degenerate cores, i.e. having white-dwarf precursors, at their centres, are near the Hayashi-line in the HR-diagram. It has long been known from numerical computations that the radius R and the luminosity of these stars are almost independent of the total mass of the star and that they depend almost exclusively on the

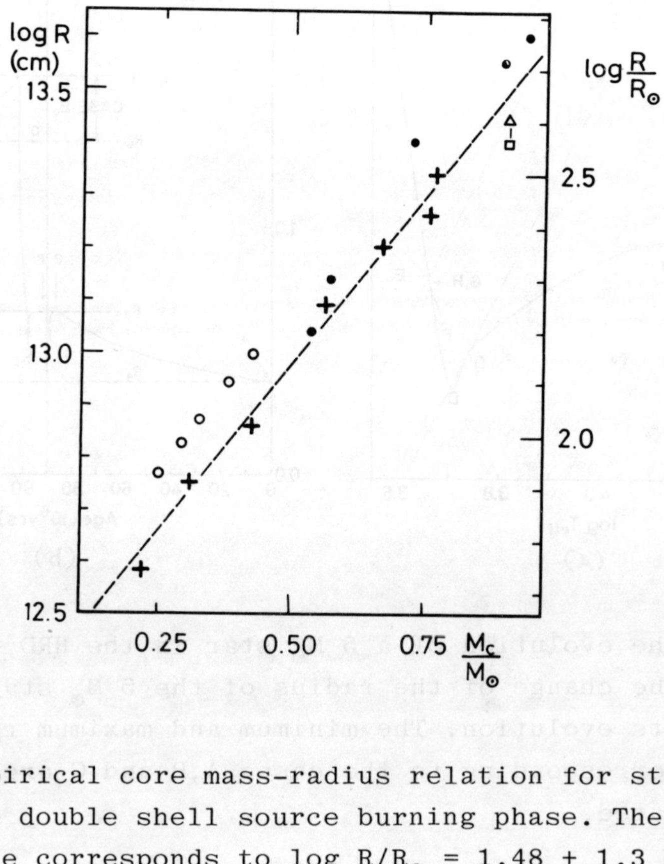

Fig. 8 Empirical core mass-radius relation for stars in the double shell source burning phase. The dashed line corresponds to $\log R/R_\odot = 1.48 + 1.3\ M_c/M_\odot$

mass M_c of the degenerate core [33]. This result is also theoretically understood (see e.g. [34]). Fig. 8 shows the empirical relation between R and M_c derived from numerical computations [21]. It can be approximated by the equation

$$\log \frac{R}{R_\odot} = 1.48 + 1.30 \frac{M_c}{M_\odot} \qquad (2)$$

and is valid for stars containing a degenerate carbon-oxygen core. According to this equation the Roche radius of the primary has to be larger than about $100R_\odot$ in order to form a carbon-oxygen white dwarf ($M_c \gtrsim 0.55\ M_\odot$). Thus under no circumstances can white dwarfs be formed in systems whose initial angular momentum approximates that of typical CB's. The progenitors of CB's have to be very wide binaries with separations of the order $10^2 \ldots 10^3\ R_\odot$ and orbital periods in the range of years [21].

(3) Loss of Mass and Angular Momentum

From the above considerations it appears unavoidable that during its evolution towards the CB stage a binary must lose a considerable fraction of its initial angular momentum and of its mass. It is also clear that the binary cannot lose too much of its initial angular momentum prior to the onset of the mass transfer. By means of the relation between M_c and R it is possible to make a quantitative estimate of the mass and angular momentum excess of possible progenitors over a typical CB. It turns out that a typical progenitor of a CB has to lose 50% - 70% of its mass and up to 99% of its angular momentum in the phases <u>after</u> the onset of Roche-lobe overflow [21]. Thus one of the key problems is to find a mechanism which can account for such enormous losses of mass and angular momentum. The most promising such mechanism is the so-called common-envelope evolution, i.e. the spiraling-in of the secondary in the envelope of the red- giant primary. This mechanism was first proposed by Paczyński (1976)[35]. Before discussing it in more detail we have to explain how the secondary can ever come into contact with the envelope of the primary.

(4) The Evolution into Contact

The crucial phase in the transformation of a wide binary to a CB begins with the onset of Roche-lobe overflow of the primary. At that moment the internal structure of the primary is characterised by its degenerate core and a very deep outer convective envelope. In this stage, the equilibrium radius of the primary depends only on the core mass but not on its total mass. Therefore, if Roche-lobe overflow begins, the equilibrium radius of the primary remains constant. On the other hand, the corresponding Roche radius shrinks since the mass is lost from the more massive component. The fact that the equilibrium radius of the primary cannot adjust to the shrinking Roche radius results in enormous mass-loss rates from the primary, of the order[36]

$$\dot{M}_1 \simeq - \frac{M_1}{P_{orb}} \left(\frac{R_1 - R_{1\,Roche}}{R_{1\,Roche}} \right)^3 . \qquad (3)$$

Accordingly, mass transfer rates as high as 0.1 $M_\odot$ yr^{-1} are possible.

What is the reaction of the secondary which is exposed to such an enormous mass transfer? At the time when the primary starts transferring mass, the secondary is still very close to the main sequence. If the mass of the secondary is $M_2 \gtrsim 1\ M_\odot$, its envelope is mainly radiative and the corresponding thermal time scale is of the order of the Kelvin-Helmholtz time-scale $\tau_{KH} = GM^2/RL$. When the secondary is accreting matter on an accretion time-scale $\tau_{accr} > \tau_{KH}$, the star remains in or near thermal equilibrium and its radius changes according to the main-sequence-mass-radius relation. However, the mass loss rate from the red-giant primary corresponds to an accretion time-scale $\tau_{accr} \ll \tau_{KH}$. In this case the secondary cannot remain in thermal equilibrium. The behaviour of stars with radiative envelopes accreting (spherically symmetrically) on time-scales $\tau_{accr} \ll \tau_{KH}$ has been studied numerically by Kippenhahn and Meyer-Hofmeister (1977)[37]. According to these computations, stars accreting on time scales $\tau_{accr} \ll \tau_{KH}$ expand very rapidly and become red-giant-like objects. As an example, Fig. 9 shows the evolution of a 2 $M_\odot$ star for different

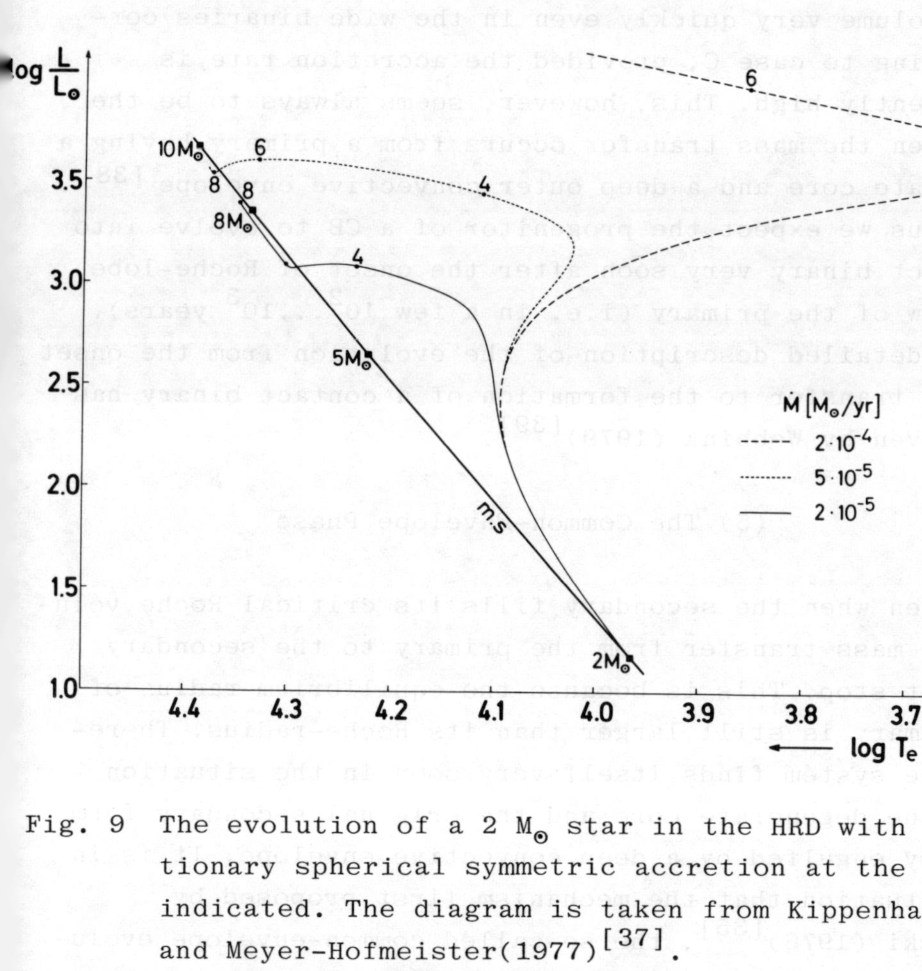

Fig. 9 The evolution of a 2 $M_\odot$ star in the HRD with stationary spherical symmetric accretion at the rates indicated. The diagram is taken from Kippenhahn and Meyer-Hofmeister(1977)[37].

rates of accretion $\dot{M}$. For the highest accretion rate, i.e. $\dot{M} = 2\times10^{-4}$ $M_\odot$ yr^{-1} the star moves very quickly to the Hayashi-line. In reality, the accretion rates are probably even larger. Therefore we can expect a qualitatively similar, but more violent, reaction of the secondary. Fig. 10 shows the evolution of a 2 $M_\odot$ star for different values of $\dot{M}$ in the mass-radius diagram. It shows clearly that the radius of the secondary can increase easily by a factor 100 on time-scales of $\sim 10^3$ years. Also shown in Fig. 10 is the variation of the Roche radius of the secondary (for an assumed primary mass of $M_1 = 5$ $M_\odot$) corresponding to the cases B and C. As can be seen, the secondary will fill its Roche volume very quickly even in the wide binaries corresponding to case C, provided the accretion rate is sufficiently high. This, however, seems always to be the case when the mass transfer occurs from a primary having a degenerate core and a deep outer convective envelope [38].

Thus we expect the progenitor of a CB to evolve into a contact binary very soon after the onset of Roche-lobe overflow of the primary (i.e. in a few $10^2 \ldots 10^3$ years). A more detailed description of the evolution from the onset of mass transfer to the formation of a contact binary has been given by Webbink (1979)[39].

(5) The Common-Envelope Phase

Even when the secondary fills its critical Roche volume the mass transfer from the primary to the secondary does not stop. This is because the equilibrium radius of the primary is still larger than its Roche-radius. Therefore the system finds itself very soon in the situation where the degenerate core and the original secondary form a binary engulfed by a deep convective envelope. It is in this situation that the mechanism first proposed by Paczynski (1976)[35], the so-called common-envelope evolu-

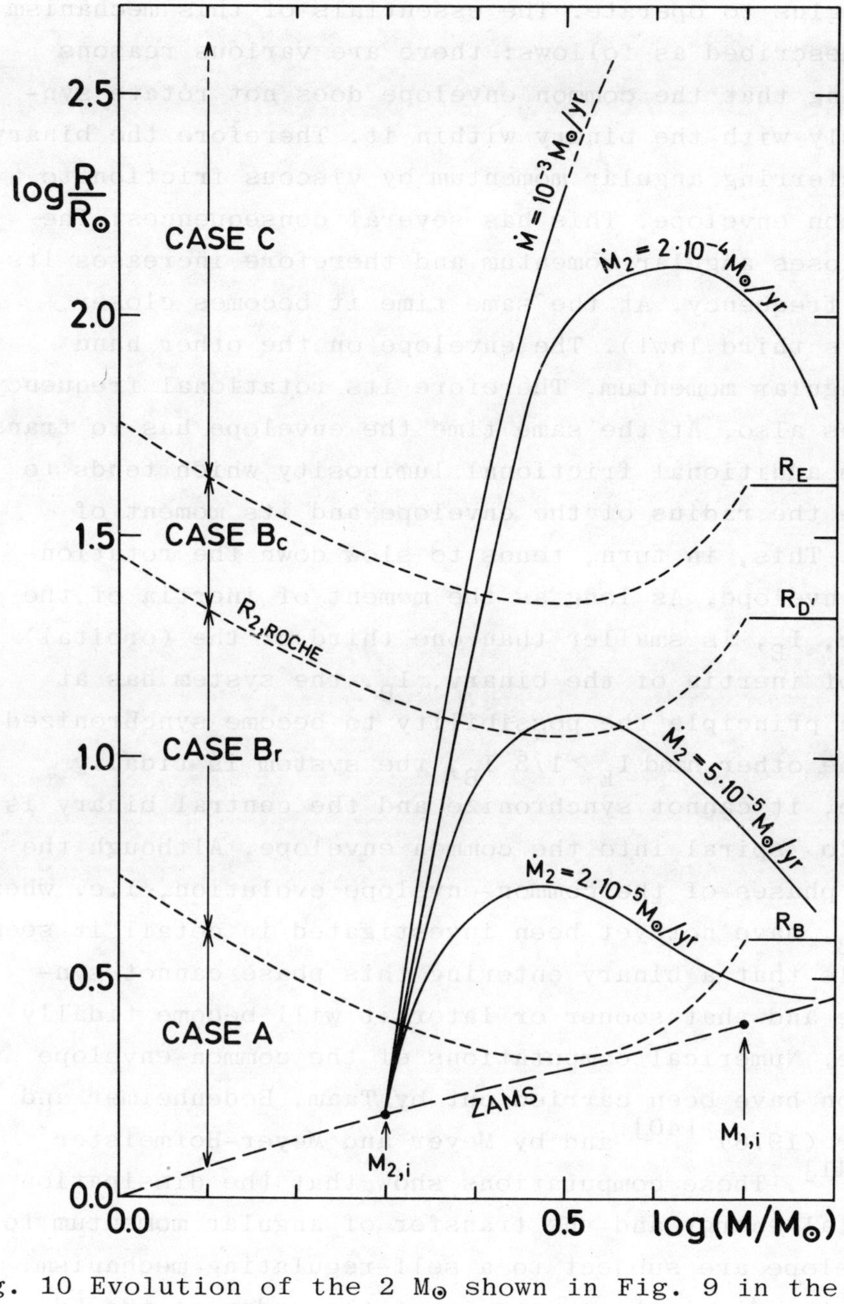

Fig. 10 Evolution of the 2 $M_\odot$ shown in Fig. 9 in the mass-radius diagram. Also indicated is the change of the Roche radius of the secondary in a binary consisting initially of masses $M_1 = 5\ M_\odot$ and $M_2 = 2\ M_\odot$ and which keeps the total mass and angular momentum constant.

tion, begins to operate. The essentials of this mechanism can be described as follows: there are various reasons indicating that the common envelope does not rotate synchronously with the binary within it. Therefore the binary is transferring angular momentum by viscous friction to the common envelope. This has several consequences: The binary loses angular momentum and therefore increases its orbital frequency. At the same time it becomes closer (Kepler's third law!). The envelope on the other hand gains angular momentum. Therefore its rotational frequency increases also. At the same time the envelope has to transport the additional frictional luminosity which tends to increase the radius of the envelope and its moment of inertia. This, in turn, tends to slow down the rotation of the envelope. As long as the moment of inertia of the envelope, I_E, is smaller than one third of the (orbital) moment of inertia of the binary, I_B, the system has at least in principle the possibility to become synchronized. If on the other hand $I_E > 1/3\ I_B$, the system is tidally unstable, it cannot synchronize and the central binary is forced to spiral into the common envelope. Although the initial phases of the common-envelope evolution, i.e. when $I_E < \frac{1}{3} I_B$, have not yet been investigated in detail it seems plausible that a binary entering this phase cannot synchronize and that sooner or later it will become tidally unstable. Numerical computations of the common-envelope evolution have been carried out by Taam, Bodenheimer and Ostriker (1978) [40] and by Meyer and Meyer-Hofmeister (1979)[41]. These computations show that the dissipation of orbital energy and the transfer of angular momentum to the envelope are subject to a self-regulating mechanism. Thereby the dissipation rate is of the order of the Eddington luminosity. This leads to a characteristic time-scale of the spiraling-in of the order of a few thousand years. Because of its comparatively long thermal time-scale, the original secondary is not greatly affected during the few thousand years which the whole evolution, from the onset of mass exchange to the ejection of the common envelope, lasts.

(6) The Ejection of the Common Envelope

From what has been said so far it is now obvious that the formation of CB's proceeds via the ejection of the common envelope at a certain point of the evolution. To the outside observer the ejected envelope would probably appear very similar to an ordinary planetary nebula, while its central star is a close detached or semi-detached binary consisting of a white dwarf and a main-sequence star.

Unfortunately, however, the process which leads to the ejection of a common envelope is not understood. Therefore at present no theoretical predictions about the properties of the emerging binary can be made. Theoretically it is not even clear whether the common envelope can be ejected at all or if the two stars must coalesce. Thus from the purely theoretical point of view the common-envelope evolution as a mechanism for forming CB's is highly speculative.

The reason why the common envelope nevertheless deserves our attention is the observation of a few central stars of planetary nebulae which are very close but detached binaries with orbital periods of $\sim 0.^{d}5$ and which consist of a white dwarf or a pre-white-dwarf primary and of a low-mass secondary. The objects in question are the central stars of Abell 46, Abell 63 and of K 1-2. (also known as V477 Lyr, UU Sge and VW Pyx respectively). It is mainly because these objects correspond so perfectly to the theoretical expectation that the common-envelope concept is now generally accepted. However, from the theoretical side most of the work still remains to be done.

Based on the observation that a planetary nebula is visible for only a rather short time, typically 10^4 years, and that on the other hand the detached binaries are rather long-lived objects we would expect to find many more detached binaries of this type without an associated observable planetary nebula. This is in fact observed. Objects belonging to this class are e.g. Case 1, HZ 9, V471 Tau, GK Vir and maybe AA Dor.

It has also been speculated whether semi-detached binaries can be formed directly in a common envelope. At present we cannot answer the question. Observationally there is no way of distinguishing between very young, newly formed CB's and old ones.

(7) From the Detached to the Semi-Detached State

In order to evolve from the detached to the semi-detached state, i. e. to become a CB, the detached binaries still have to lose some of their angular momentum. Although the secondary could fill its Roche lobe after sufficient nuclear evolution this would not yield a typical CB because of its evolved secondary and its atypical orbital period. In addition the nuclear time-scale of the secondary is uncomfortably long. Angular momentum losses via gravitational waves also yield very long time-scales for the observed binaries, i.e time-scales in excess of 10^{10} years. The most likely process is the loss of angular momentum via a magnetically-coupled stellar wind from the secondary. The secondary of V471 Tau shows star spot activity and the presence of a hot corona. It is therefore very likely that this star has a magnetically-coupled stellar wind. The time-scale for the angular momentum loss would be of the order 10^8 years[42].

VI. THE SECONDARY COMPONENTS

Now that we have sketched the formation of CB's we want to look in some more detail at what can be said about the components. Because there are more observational data for comparison we start with a discussion of the secondaries.

(1) The Mass Loss of the Secondaries

If the evolutionary picture outlined in Sect. V. is correct, then the secondaries have to undergo a substantial mass loss during the formation of a CB. This is easily seen from the following simple argument: let $M_{1,i}$ and $M_{2,i}$ be the initial masses of the CB progenitor on the main sequence. As a result of the CB formation a fraction f of the mass of the primary ends as a white dwarf of mass

$$M_{WD} = M_{1,f} = f \cdot M_{1,i}. \quad (4)$$

Because no secularly stable CB can be formed unless the mass of the secondary $M_{2,f}$ is smaller than the mass of the white dwarf (see Sect. VIII.(2)) the secondary must have lost the mass

$$\Delta M_2 = M_{2,i} - M_{2,f} = M_{2,i} - f \frac{q_i}{q_f} M_{2,i} = M_{2,i}\left(1 - f\frac{q_i}{q_f}\right), \quad (5)$$

where

$$q_i = \frac{M_{1,i}}{M_{2,i}} \quad (6)$$

and

$$q_f = \frac{M_{1,f}}{M_{2,f}}. \quad (7)$$

Since for reasons of secular stability $q_f \gtrsim 1$, the secondary can keep its original mass only if $q_i \gtrsim f^{-1}$. Depending on the initial mass of the primary, f is 0.2...0.5. Thus mass loss from the secondary can only be avoided for rather large initial mass ratios. In most cases, however, the secondary suffers substantial mass loss during CB formation.

When do the secondaries lose their mass? Theory cannot answer this question at present. However, it is very unlikely that it happens before the onset of mass transfer from the primary. From the fact that the observed binary nuclei of planetary nebulae already contain secondaries of very low mass we may then conclude that their mass loss occurs late during the common-envelope evolution. There remains, however, a serious problem: if the mass loss occurred in the late phases of the common-envelope evolution, why are the observed secondaries of the binary nuclei of planetary nebulae apparently main-sequence stars? Since we still see the planetary nebula, the systems could not have emerged from the common envelope more than $\sim 10^4$ years ago. If substantial mass loss occurred such a short time ago, the secondaries cannot yet be in thermal equilibrium again, i.e. they should not look like main-sequence stars. Yet they do.

Incidentally, the requirement of secular stability discussed above and in Sect. VIII. (2), leads to a natural explanation of the observed upper limit of the orbital period of CB's (cf. Fig. 2): According to the period-secondary-mass relation (Eq. (1)) a CB containing a main-sequence secondary of maximum allowable mass $M_2 \simeq M_1 = M_{CHANDRASEKHAR} \simeq 1.4\ M_\odot$ has an orbital period of roughly 10 hours. Only if the secondary is significantly evolved can a CB be secularly stable and have an orbital period in excess of ~ 10 hrs. An example for such a case is GK Per.

(2) The Possibility of Evolved Secondaries

In progenitors of CB's whose initial mass ratio q_i is not too far from unity, the secondary has spent a significant fraction of its main-sequence life-time when the binary enters the common-envelope phase. Making use of the mass-luminosity relation for main-sequence stars, i.e. assuming

$$L = \text{const.}\ M^{3.5}, \tag{8}$$

the fraction of the main-sequence life-time the secondary can spend before its evolution is interrupted by the mass loss during the common-envelope evolution is estimated to be

$$\varepsilon \simeq q_i^{-2.5} \tag{9}$$

Since the same systems which produce the most evolved secondaries are also those where the secondary suffers the largest mass loss there is the possibility that some of the secondaries of CB's now essentially consist of the former hydrogen burning central region and may even expose nuclear processed matter at their surface. Depending on details, the secondary remnants could be chemically homogeneous, having a higher helium content than normal, or they could be inhomogeneous stars of various kinds. In fact these stars are essentially like the remnants of case A mass transfer (see e.g. [43]). Chemically homogeneous stars are expected to lie below the main sequence and inhomogeneous stars above the main sequence in the HRD (see e.g. [44]).

Because of Eq. (9) we do not expect significantly evolved secondaries in systems having $q_i > 2$, even if they lose a large fraction of their initial mass.

Finally we should mention the possibility that the secondaries might be homogenized during the phases of rapid mass accretion or during the final phase of the common-envelope evolution when they probably have to lose significant amounts of mass very rapidly [36]. The strong departure from thermal equilibrium during these phases could result in a significantly larger convective core and thus in the homogenization of those parts of the star which eventually emerge as the secondary. However, it remains to be confirmed whether the essentially adiabatic change of the interior of the secondary really results in a significant change of its chemical stratification.

(3) The Probability of Evolved Secondaries

Since even highly evolved secondaries are possible, the fact that the observed secondaries are all practically unevolved calls for an explanation. It may be found in the distribution of the mass ratios q_i of the progenitors: let $P(q_i)$ be the probability that the progenitor has a mass ratio in the range q_i to $q_i + dq_i$. Accordingly, the probability of finding the relative evolutionary state of the secondary in the interval ε to $\varepsilon + d\varepsilon$ is $P(\varepsilon^{-0.4})$. The probabilities of having $\varepsilon < \varepsilon_0$ or $\varepsilon_1 \leqslant \varepsilon \leqslant \varepsilon_2$ are then given by

$$P(\varepsilon_1 < \varepsilon_0) = \int_1^{\varepsilon_0^{-0.4}} P(q)\, dq \tag{10}$$

and

$$P(\varepsilon_1 \leqslant \varepsilon \leqslant \varepsilon_2) = \int_{\varepsilon_1^{-0.4}}^{\varepsilon_2^{-0.4}} P(q) dq \tag{11}$$

Examples of $P(\varepsilon < \varepsilon_0)$ and $P(\varepsilon_1 \leqslant \varepsilon \leqslant \varepsilon_2)$ for three different functions P(q) are shown in Fig. 11. From Fig. 11 it is very clearly seen that the fraction of significantly evolved secondaries, i.e. $\varepsilon \gtrsim 0.5$, is very small ($\lesssim 10\%$) unless

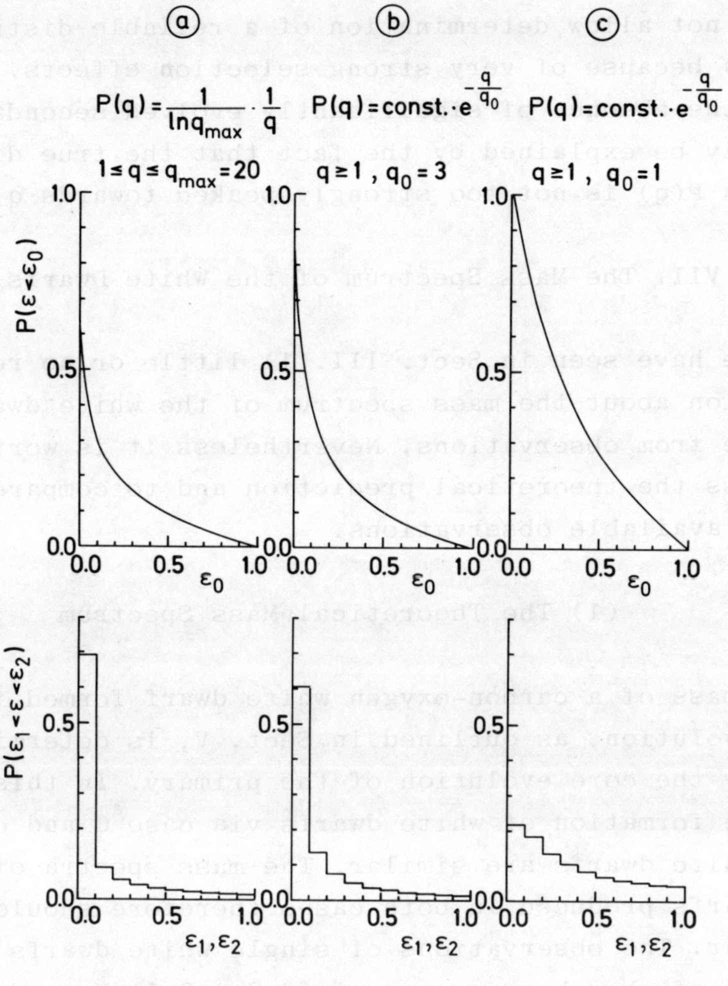

Fig. 11 Probability distributions $P(\varepsilon<\varepsilon_0)$ and $P(\varepsilon_1\leq\varepsilon\leq\varepsilon_2)$ of the relative evolutionary status ε of the secondary for different assumed distributions P(q) of the initial mass ratio.

the distribution of the initial mass ratios is very strongly peaked towards $q_i = 1$ (Fig. 11c). Unfortunately observations do not allow determination of a reliable distribution P(q) because of very strong selection effects. Nevertheless the absence of significantly evolved secondaries can easily be explained by the fact that the true distribution P(q) is not too strongly peaked towards $q_i = 1$.

VII. The Mass Spectrum of the White Dwarfs

As we have seen in Sect. III.(1) little or no reliable information about the mass spectrum of the white dwarfs is available from observations. Nevertheless it is worthwhile to discuss the theoretical prediction and to compare it with the available observations.

(1) The Theoretical Mass Spectrum

The mass of a carbon-oxygen white dwarf formed in a case C evolution, as outlined in Sect. V, is determined mainly by the core evolution of the primary. In this respect the formation of white dwarfs via case C and of single white dwarfs are similar. The mass spectra of the white dwarfs produced in both cases therefore should also be similar. The observations of single white dwarfs show that most of them have a mass of $(0.6 \pm 0.1)\ M_{\odot}$ and that massive white dwarfs with masses above $\sim 1\ M_{\odot}$ are extremely rare[18], see also Fig. 3. The helium white dwarfs formed via a low-mass case B mass transfer have masses in the range $0.2\ M_{\odot} \lessapprox M \lesssim 0.45\ M_{\odot}$. Therefore the theoretically expected mass spectrum should show two pronounced peaks in the mass ranges $0.55\ M_{\odot} \lesssim M \lesssim 0.70\ M_{\odot}$ and $0.20\ M_{\odot} \lesssim M \lesssim 0.45\ M_{\odot}$. Webbink(1979, 1981) [38,45] has investigated the problem in more detail and as a result has obtained the theoretical mass spectrum shown in Fig. 12. Apart from confirming

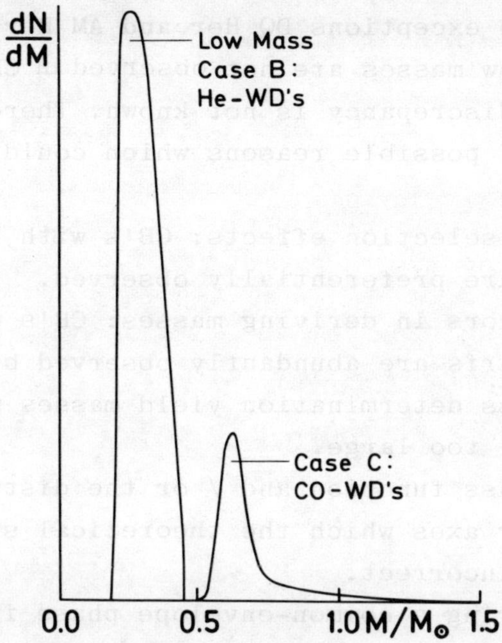

Fig. 12 Sketch of the theoretical mass spectrum of the white dwarfs in CB's as obtained by Webbink(1981)[36]

the expected two peaks this spectrum has a number of other remarkable properties:

a) The number of low mass helium white dwarfs expected is far greater than the number of carbon-oxygen white dwarfs.
b) No white dwarfs in the mass range $0.45\ M_\odot \lesssim M \lesssim 0.5\ M_\odot$ are formed.
c) Only very few white dwarfs with masses $M \gtrsim 0.7\ M_\odot$ are produced.

(2) Comparison with Observations

The most outstanding feature of the mass spectrum

shown in Fig. 12 is the very high number of helium white dwarfs predicted. This is in contrast to what is "observed". With the possible exceptions DQ Her and AM Her, white dwarfs of such low masses are not observed in CB's. The reason for this discrepancy is not known. There are, however, a number of possible reasons which could account for this discrepancy:

a) Observational selection effects: CB's with high-mass white dwarfs are preferentially observed.
b) Systematic errors in deriving masses: CB's with low mass white dwarfs are abundantly observed but the methods of mass determination yield masses which are systematically too large.
c) The initial mass function and / or the distribution of the semi-major axes which the theoretical spectrum is based on are incorrect.
d) Binaries entering a common-envelope phase initiated by a low-mass case B mass transfer cannot eject the common envelope and coalesce.
e) The mass of a white dwarf increases secularly as a consequence of the continuous accretion. This will be discussed in Sect. VIII!

With regard to b), Webbink (1981)[36] has pointed out the striking similarity of the mass spectrum of the white dwarfs (Fig. 12) with the observed period distribution of CB's (Fig. 2). According to Webbink, the ultra-short-period CB's (hereafter USPCB's), i.e. those with $P \lesssim 2^h$, are those containing a helium white dwarf and the CB's having longer periods (i.e. $P \gtrsim 3^h$) contain the carbon-oxygen white dwarfs. Accordingly the period gap would reflect the discontinuity in the core masses associated with a low-mass case B and with a case C evolution respectively. We shall come back to "Webbink's hypothesis" in the subsequent Sects. VIII. and IX.

We have already discussed the possible excess of high-mass white dwarfs in CB's in Sect. III.(1). It becomes even more striking when we compare Fig. 3 and 12. However, we should stress once again that the spectrum shown in Fig. 3 is very uncertain and that some of the above mentioned reasons (in particular a,b, and e) could also be responsible for this discrepancy.

(3) The Formation of Massive White Dwarfs in CB's

Although the observational evidence for an excess of massive white dwarfs in CB's is not very convincing, it has motivated a number of suggestions to account for this fact. The most widely used explanation assumes that the white dwarfs increase their mass secularly as a consequence of the accretion. We shall come back to this point in the next section.

Another possibility arises from the fact that in binaries white dwarfs cannot only be formed via case C or low-mass case B mass transfer but also in a third, completely different, way. Webbink (1979)[38] was the first to suggest this possibility. Accordingly massive white dwarfs of $M \gtrsim 1\ M_{\odot}$ can be explained as the result of a massive case B mass transfer which yields a helium star as an intermediate product. The helium star later undergoes a second case B mass transfer, hereafter called case BB. The ultimate product of such an evolution is either a carbon-oxygen white dwarf with a mass of up to the Chandrasekhar mass or a neutron star. Delgado and Thomas (1981)[46] have investigated numerically the properties of conservative case BB evolution. Law and Ritter (1981)[47] have discussed the same type of evolutión in the context of CB's. As a result they find that the formation of CB's containing massive white dwarfs via case BB evolution requires the binary to undergo at least one, most probably two, separate phases of a spiraling-in in a common envelope. The formation of low-mass X-ray binaries is explained in a similar way, the only difference being that a neutron star results when the mass of the case BB remnant exceeds $\sim 1.4\ M_\odot$. Since the formation of the necessary intermediate helium stars of 2...4 $M_\odot$ requires rather massive primaries on the main sequence, i.e. $M_{1,i} \gtrsim 10...15\ M_\odot$, the relative number of systems of both types, CB's and low-mass X-ray binaries, produced in this way must be small. With regard to the CB's, case BB evolution therefore probably cannot contribute significantly. With regard to the low-mass X-ray sources, case BB also creates a problem: If all the low-mass X-ray binaries, except those observed in globular clusters, are to be formed via case BB, their statistical properties should be those of the extreme population I. This, however, seems not to be the case. From their statistical properties, i.e. belonging to the galactic bulge, one would rather conclude that the progenitors of these objects are binaries of rather low mass. How a neutron star can be formed as a result of the evolution of a low-mass binary is not understood in the context of stellar evolution. Another possible solution of the problem of the low-mass X-ray binaries will be discussed in the next section.

VIII. ACCRETING WHITE DWARFS, NOVAE AND LOW-MASS X-RAY BINARIES

At various points of the foregoing discussion we mentioned the possibility that some of the discrepancies between observations and theory might be explained by the fact that the white dwarfs in CB's are continously accreting. In this section we summarize briefly a number of arguments relevant to this hypothesis.

(1) Accreting White Dwarfs

As a consequence of the continous mass loss of the secondaries, the white dwarfs in CB's are subject to a corresponding accretion. With the exception of the systems AM CVn and GP Com the accreted matter contains hydrogen. From the fact that the secondaries are indistinguishable from main-sequence stars (see Sect. III.(2)) we may even conclude that the transferred matter has approximately solar composition. The properties of white dwarfs accreting hydrogen-rich material have been investigated by many authors (see e.g. [48-52]). The most important results can be summarized as follows:

The behaviour of a white dwarf which accretes hydrogen-rich material depends critically on the mass of the white dwarf M_{WD} and on the accretion rate $\dot{M}_{WD}$. In particular two critical accretion rates $\dot{M}_{crit,1}$ and $\dot{M}_{crit,2}$ can be defined which have the following properties[50]: The higher of the tow, $\dot{M}_{crit,1}$, corresponds to the maximum rate at which hydrogen burns in the shell source of a red giant with a degenerate core. From the core mass-luminosity relation

$$L/L_{\odot} = 2.832 \times 10^{4} + 5.925\left(\frac{M_{WD}}{M_{\odot}} - 1.0\right)\times 10^{4} \quad (12)$$

([33]; see also [34]) it follows that [50]

$$\frac{\dot{M}_{crit,1}}{M_{\odot}/yr} = 2.8\times10^{-7} + 5.9\times10^{-7}\left(\frac{M_{WD}}{M_{\odot}} - 1.0\right). \quad (13)$$

White dwarfs accreting at rates $\dot{M}_{WD} > \dot{M}_{crit,1}$ are very extended objects and if the mass of the hydrogen-rich envelope exceeds a few $10^{-2} M_{\odot}$, the radius is determined by the relation between core mass and total radius (Eq.(2)). Furthermore, if $\dot{M}_{WD} > \dot{M}_{crit,1}$, the mass of the degenerate core M_c and of the hydrogen-rich envelope M_E change according to $\dot{M}_c = \dot{M}_{crit,1}$, $\dot{M}_E = \dot{M}_{WD} - \dot{M}_{crit,1}$. The lower of the two critical accretion rates, $\dot{M}_{crit,2}$, corresponds to the lowest burning rate for which stable hydrogen shell burning is possible. According to Paczyński and Rudak (1980)[50]

$$\dot{M}_{crit,2} \simeq 0.4\,\dot{M}_{crit,1}\,, \tag{14}$$

If a white dwarf accretes at rates $\dot{M}_{crit,2} < \dot{M}_{WD} < \dot{M}_{crit,1}$ its envelope mass, luminosity, effective temperature and radius depend on $\dot{M}_{WD}$. However, for this particular range of $\dot{M}_{WD}$, the radius of the accreting white dwarf is rather small, i.e. $R_{WD} \lesssim 1\,R_\odot$, and correspondingly its effective temperature very high, i.e. $\log T_{eff}(K) \gtrsim 5$. (see e.g.[49]).

If $\dot{M}_{WD} < \dot{M}_{crit,2}$ no stable hydrogen shell-burning is possible. Depending on the thermal history of the hydrogen envelope, i.e. on $\dot{M}_{WD}$, hydrogen burning is ignited in a non-degenerate environment (for high $\dot{M}_{WD}$) which gives rise to ordinary shell flashes[49,52], or in a degenerate state (for low $\dot{M}_{WD}$) which gives rise to a thermo-nuclear runaway (see e.g.[3]). As a consequence of a shell flash the radius of the white dwarf increases very strongly and can reach red-giant size (see e.g.[53,49,52]).

(2) Accreting White Dwarfs in CB's

From the properties of accreting white dwarfs described in Sect. VIII.(1) it seems rather difficult for a white dwarf in a CB to increase its mass via hydrogen burning of the accreted matter. This is due to the particular properties of CB's:

Accretion at rates $\dot{M}_{WD} > \dot{M}_{crit,1}$ is not compatible with the structure of CB's. If $\dot{M}_{WD} > \dot{M}_{crit,1}$, the radius of the white dwarf is so large that it cannot fit within its Roche lobe. An accretion rate $\dot{M}_{WD} > \dot{M}_{crit,1}$ would probably bring the binary back into a common-envelope-like situation. Thereby most of the previously accreted matter is probably lost from the system. It is because of this property that CB's having $M_2 \gtrsim M_1$ cannot exist. If the system is secularly unstable against mass transfer from the secondary i.e. if $M_2 \gtrsim M_1$, the mass-loss rate of the secondary is determined by the thermal time-scale of its envelope. The corresponding mass loss rate is of the order of 10^{-6} $M_\odot$ yr^{-1} and thus larger than $\dot{M}_{crit,1}$.

Stable hydrogen shell-burning at a white dwarf radius smaller than its Roche radius is possible if $\dot{M}_{crit,2} < \dot{M}_{WD} < \dot{M}_{crit,1}$. Mass-loss rates from the secondary in this range are, however, not expected. If the system is secularly unstable, $\dot{M}_{WD} > \dot{M}_{crit,1}$. If, on the other hand, it is secularly stable the mass-loss rates are much smaller than $\dot{M}_{crit,2}$ (see Sect. IX.(1)).

If the accreted hydrogen is burnt via shell flashes, the radius of the white dwarf becomes much larger than its Roche radius during the flash[49,52]. However, it is not clear whether in this way the system will enter a common-envelope phase or only lose a large fraction of the accreted envelope. In either case only a small fraction of the accreted matter could be transmuted into helium and thus remain on the white dwarf.

For accretion rates $\dot{M}_{WD} \lesssim 10^{-9} \ldots 10^{-11}$ $M_\odot$ yr^{-1}, such as are expected to occur in CB's, the hydrogen which accumulates at the surface of the white dwarf becomes degenerate and is burnt in a thermo-nuclear runaway after sufficient hydrogen ($10^{-4} \ldots 10^{-5}$ $M_\odot$) has been accreted.

Depending on the violence of the runaway (which in turn depends on $\dot{M}_{WD}$), more or less of the accreted hydrogen is expelled from the white dwarf. It could even be that the white dwarf itself loses some of its mass as a result of such a runaway (see below). It is now generally accepted that the classical nova outbursts are such thermo-nuclear runaways. For a detailed description of the nova mechanisms, see e.g. Gallagher and Starrfield (1978)[3]. The problem we are interested in is, whether the hydrogen accretion and the subsequent nova outbursts result in a secular change of the mass of the white dwarf and if so, whether it increases or decreases in time.

(3) Arguments Favouring a Significant Secular Increase of M_{WD}

Despite the above-mentioned difficulties the hypothesis that the white dwarfs increase their mass secularly as a consequence of the accretion is currently very popular. The reason is that it allows an explanation for a number of observational facts which otherwise are difficult to understand theoretically.

The most important argument in favour of $\langle \dot{M}_{WD} \rangle > 0$ is based on observational properties of the low-mass X-ray sources. Ignoring the globular cluster sources which probably have a different genesis[5], the low-mass X-ray sources have a spatial distribution corresponding to the old Population I. Therefore they are sometimes referred to as galactic bulge sources. The progenitors of these objects are presumably binaries of rather low total mass. Nevertheless they contain a neutron star. Since in the framework of classical stellar evolution there is no way of forming a neutron star in a binary of low mass, it has been suggested that low-mass X-ray binaries originate from CB's in which the secular increase of the mass of the white dwarf finally resulted in a Type I supernova and in the formation of a neutron star (Woosley, Weaver and Taam, 1980[54]; Fujimoto, 1980[55] and references therein).

At the same time the hypothesis $\langle \dot{M}_{WD} \rangle > 0$ could also explain why fewer low-mass white dwarfs and more high-mass white dwarfs are "observed" than are theoretically predicted (cf. Figs. 3 and 12).

(4) Arguments Against a Significant Secular Increase of M_{WD}

Let us assume for the moment that the mass spectrum of the newly born white dwarfs of CB's is roughly like that shown in Fig. 12. The mean mass of the helium white dwarfs and of the carbon-oxygen white dwarfs is then $\langle M_{He-WD} \rangle =$ $(0.35 \pm 0.10)\ M_\odot$ and $\langle M_{CO-WD} \rangle = (0.60 \pm 0.10)\ M_\odot$ respectively. As outlined in Sect.VIII the mass ratio of a CB has to be compatible with secular stability of the secondary against Roche-lobe overflow, i.e. $M_2 \lesssim M_{WD}$. Thus the total mass of a CB is $= M_{WD} + M_2 \lesssim 2M_{WD}$, i.e. the mean total mass is $\langle \mathfrak{M} \rangle = (0.7 \pm 0.2)\ M_\odot$ and $\langle \mathfrak{M} \rangle = (1.2 \pm 0.2)\ M_\odot$ if the primary is a helium white dwarf or a carbon-oxygen white dwarf respectively. Thus massive white dwarfs cannot result from systems containing a helium white dwarf because there is not enough mass available. The situation is different in the case of CB's containing a carbon-oxygen white dwarf: Provided that all of the transferred matter can be burnt into helium and thus be accepted by the white dwarf, the total mass of a typical CB is sufficient for forming a massive white dwarf and even for inducing a Type I supernova. The question therefore is whether in the real world a sufficiently high fraction of the transferred hydrogen-rich matter can be accepted by the white dwarf. Thus in order to answer the question we must know the efficiency

$$f = -\frac{\int \dot{M}_{WD}\,dt}{\int \dot{M}_2\,dt}\,, \qquad (15)$$

with which matter coming from the secondary is transformed into matter which is accepted by the white dwarf. Clearly, $f < 1$, but by how much?

The value of f depends on M_{WD}, $\dot{M}_{WD}$, the chemical composition of the accreted matter and on the maximum allowable radius of the white dwarf (see e.g. [52]). In CB's the accreted matter has roughly solar composition (except for AM CVn and GP Com which do not contain hydrogen at all) and – because of the secular stability – the accretion rate is low, i.e. $\dot{M}_{WD} \lesssim 10^{-8} M_\odot \, yr^{-1}$. If $\dot{M}_{WD} < 10^{-9} M_\odot \, yr^{-1}$ the accreted hydrogen is burnt via a thermo-nuclear runaway, i.e. in a nova explosion[3]. In this case f is likely to be very small since most of the accreted matter is expelled from the white dwarf and eventually forms a nova shell. In fact from the observations of some nova ejecta one can even conclude that $f < 0$, i.e. that the mass of the white dwarf decreases secularly. This is because abundance determinations of these ejecta show that the CNO elements are over-abundant by a factor ~ 100 as compared to solar composition, as can be seen from the examples given in Table 2. Where do these CNO elements come from? It is

Table 2. Abundance by mass of CNO-elements in the ejecta of different novae

Object	Z_{CNO}	References
DQ Her	0.56	Williams et al. (1978)[56]; Ferland and Truran (1980)[57]
V1500 Cyg	0.30	Ferland and Shields (1978)[58]
T Aur	0.13	Gallagher et al. (1980)[59]
HR Del	0.07	Tylenda (1978)[60]
RR Pic	0.04	Williams and Gallagher (1979)[61]
Sun (for comp.)	0.015	

very unlikely that the matter transferred from the secondary already contains them. Since, on the other hand, the total abundance of the CNO elements is not changed by the CNO burning during the nuclear runaway, it is more likely that the CNO elements come from the interior of the white dwarf. However, it is by no means clear how the carbon and oxygen which are relatively deep in the interior of the white dwarf find their way to the surface in order to be mixed with the accreted matter. A possible mechanism has been suggested by Kippenhahn and Thomas (1978)[62]. Thus, if the white dwarf loses some of its mass with every nova outburst its mass decreases secularly, i.e. $f < 0$! While this might not be true for all novae, f is certainly far from unity. In this context it is interesting to note that according to Young and Schneider (1980)[16] the mass of the white dwarf of the old nova DQ Her is $(0.45 \pm 0.05) M_\odot$. If this value should turn out to be correct it would provide direct evidence for a secular mass loss since carbon-oxygen white dwarfs of such a low mass cannot be formed by normal stellar evolution.

The situation with regard to the value of f is not much better if $\dot{M}_{WD}$ is larger, i.e. $10^{-9} M_\odot\, yr^{-1} \lesssim \dot{M}_{WD} \lesssim 10^{-8} M_\odot\, yr^{-1}$, and the accreted matter is burnt in a hydrogen shell flash rather than in a thermo-nuclear runaway. As outlined in Sect. VIII. (2) most of the accreted matter is probably lost from the system when the white dwarf has attained the size of a red supergiant i.e. $R \gtrsim 10^2 R_\odot$. Although in this case f might still be larger than for novae it is probably not large enough to give rise to a significant increase of the mass of the white dwarf. With the prospect that f is unlikely to exceed 0.1...0.2, less than $(0.1...0.2) M_{WD}$ can be added to the white dwarf.

While adding $\sim 0.1 M_\odot$ of helium to a white dwarf of $M_{WD} \lesssim 0.9 M_\odot$ is of little consequence, it might be different when the white dwarf is already rather massive, i.e. $M_{WD} \gtrsim$

1 $M_{\odot}$. In this case the accumulation of sufficient helium can induce an off-centre helium detonation which results in a Type I supernova even for masses well below the Chandrasekhar mass (see e.g. [63,64,54,55]).

It is therefore possible that the low-mass X-ray binaries are the descendants of CB's which contained an already massive white dwarf. Thus only a small fraction of the CB's would be the progenitors of the low-mass X-ray binaries. This would also be consistent with the relatively small space density of the low-mass X-ray binaries. In fact, if a typical CB eventually ends up as a low-mass X-ray binary, many more of these objects should exist, i.e. the space densities of both types of objects should be comparable. Thus the small number of low-mass X-ray binaries would reflect the inefficiency with which white dwarfs increase their mass via accretion.

IX. THE PERIOD GAP AND THE SECULAR EVOLUTION OF CB's

As the foregoing discussion has shown it is by no means clear whether the mass of the white dwarf changes secularly and whether it is increasing or decreasing. In contrast to the case of the white dwarfs it is clear that the mass of the secondary decreases secularly. The reason for this mass loss and its implications on the evolution of CB's are the topic of this section. Various aspects of the secular evolution of CB's have also been discussed by Eggleton and Whyte (1980)[65].

(1) Mechanisms for Driving the Mass Transfer

We have already pointed out that CB's can only be formed with a mass ratio which is consistent with the secular stability of the secondary against mass loss (see Sect. VIII. (2)). In fact, in all CB's known so far the mass ratio is such that the secondary is secularly stable. Therefore the fact that the secondaries are nevertheless losing mass calls for a mechanism which drives the mass transfer. Possible mechanisms are:

1. Nuclear evolution of the secondary. The increase of the radius of the secondary as a consequence of its nuclear evolution gives rise to a mass transfer on the nuclear time-scale. For secondaries with masses $M_2 \lesssim 1\ M_\odot$, the corresponding mass-transfer rates cannot account for the observations. Since practically all CB's have secondaries with $M_2 \lesssim 1\ M_\odot$, nuclear evolution is relatively unimportant.

2. Loss of angular momentum via gravitational waves. Paczyński (1967) [66] and Faulkner (1971)[67] have suggested that the decay of the orbit via gravitational radiation is driving the mass transfer in CB's and perhaps also in the low-mass X-ray binaries. The corresponding mass transfer rate is of the order $(2\times10^{-11} ... 2\times10^{-10})\ M_\odot\ yr^{-1}$ and is of the right order of magnitude to account for at least some of the observations (see e.g.[68,69]). This mechanism yields larger mass transfer rates than nuclear evolution if $P_{orb} \leqslant 10^h$.

3. Loss of angular momentum via magnetic braking. There are observational indications that gravitational radiation alone cannot account for the mass transfer in all observed cases (see e.g.[70]). Another sink of angular momentum is a magnetically coupled stellar wind emanating from the secondary (see also Sect. V (7)). Although this mechanism is probably very important and effective no reliable quantitative estimates are available. However, see Verbunt and Zwaan (1981)[42].

(2) The Period Gap

The period gap (cf. Fig. 2) ranks among the most significant observational properties of CB's Many attempts have been made to explain the gap (see e.g.[65]). However, no satisfactory explanation has been found so far. Therefore we give here only a brief outline of some of the current ideas on the subject.

1. Webbink's hypothesis. As we have seen in Sect. III.(2) (Figs. 4 and 5), the secondaries of CB's are observationally indistinguishable from main-sequence stars. Let us therefore assume that they obey a unique mass-radius relation. Then from the corresponding $P_{orb} - M_2$ relation (Eq. (1)) it follows immediately that the period histogram (Fig. 2) can be transformed into a histogram of M_2 with a gap in M_2 corresponding to the period gap. The gap in M_2 would be in the mass range $0.2\ M_\odot \lesssim M_2 \lesssim 0.3\ M_\odot$. Thus if the secondaries are mainsequence stars, the period gap implies that no CB's exist with secondaries in the corresponding mass range.

In contrast, Webbink's hypothesis (cf. Sect. VII.(2)) postulates that the period gap separates CB's containing a helium white dwarf, i.e. the USPCB's from those containing a carbon-oxygen white dwarf, i.e. the systems having $P_{orb} \gtrsim$ 3 hrs. Therefore, Webbink's hypothesis is compatible with the mainsequence property of the secondaries only if the CB's are all formed with essentially the same mass ratio $M_1/M_2 \simeq 2$ and, in order to preserve the gap. if the masses of both components remain constant. In the light of the fact that the secondary of a CB loses mass and that therefore the orbital period of the system changes accordingly, it is not clear how the gap. once formed, could be preserved from being destroyed by the secular evolution.

Webbink's hypothesis could be tested observationally by determining the masses of the white dwarfs of the USPCB's. This has in fact already been done for a few systems: although the white dwarfs of at least some of the USPCB's, in particular of Z Cha, OY Car and EX Hya, apparently have masses $M_{WD} \gtrsim 0.6\ M_\odot$ and thus are in contradiction to Webbink's hypothesis, this result – regarding the uncertainties (cf. Sect. III.(1)) – is far from being conclusive.

2. Fast mass transfer - or no mass transfer? If the secondaries of CB's are main-sequence stars, the period gap implies that no CB's exist with secondaries in the mass range $0.2\ M_\odot \lesssim M_2 \lesssim 0.3\ M_\odot$. However, since the secondaries lose mass we expect that sooner or later an originally more massive secondary will have to cross the mass (period) gap, unless the mass loss stops when the upper gap boundary is reached. Let us assume for the moment that the mass transfer does not stop. The observed absence of CB's in the gap could then be explained by assuming that the gap is crossed very quickly, i.e. that the mass-transfer rate becomes much higher when the mass of the secondary has dropped to about $0.3\ M_\odot$. As a consequence of an enhanced mass transfer, i. e. of an increased accretion rate onto the white dwarf, such a binary would not only cross the gap much faster but it would also not look like a CB (cf. Sect. VIII.(2)). Therefore these objects would disappear completely from the period histogram of CB's and eventually "reappear" as USPCB's when the phase of enhanced mass transfer has ended. The obvious difficulty with this hypothesis is that it postulates the existence of a mass-transfer instability which a secondary undergoes as soon as its mass is $\sim 0.3\ M_\odot$ and which ends when the mass is below $\sim 0.2\ M_\odot$. For main-sequence secondaries no such instability is known. Whether such an instability exists if the secondaries are

evolved is also not clear; however, see below.

The existence of the gap could also be accounted for by assuming that the mass transfer stops as soon as the mass of the secondary has dropped to $\sim 0.3\ M_\odot$ and that the CB's cross the gap in a detached state (see e.g.[71]). It is immediately clear that the secondaries cannot be main sequence stars if the gap is to be explained in this way: if a CB crosses the gap as a detached system, the mass of the secondary remains constant;,its radius however has to shrink by

$$\Delta \log R_2 = \frac{2}{3}\ \Delta \log p_{orb} \simeq 0.12,$$

i.e. by about 30%. This, however, is incompatible with the main-sequence mass-radius relation. Therefore crossing the gap in a detached state is not possible if the secondary is a main-sequence star.

3. Inhomogeneous Secondaries. Recently, D'Antona and Mazzitelli (1981)[72] have suggested that the period gap is the result of the secular evolution of chemically inhomogeneous secondaries. In their paper, D'Antona and Mazzitelli have investigated the evolution of low-mass stars which lose mass at a constant rate $\dot{M}$. As an initial model they took stars with $0.6\ M_\odot$ and $0.8\ M_\odot$ and with different degrees of central hydrogen depletion, i.e. with different evolutionary status. The evolution of these stars is characterised by the interaction of three main effects which are all a direct consequence of the secular mass loss. Firstly, with decreasing mass the outer convective envelope penetrates more and more into the star until it becomes fully convective below a certain mass, which is of the order $0.3\ M_\odot \ldots 0.4\ M_\odot$. Secondly, below a certain mass, which depends on $\dot{M}$, the time-scale of the mass loss $\tau_M = M/\dot{M}$

becomes shorter than the thermal time-scale $\tau_{th} \simeq \frac{GM^2}{RL}$ of the star. As a consequence, the mass loss brings the star sooner or later out of thermal equilibrium. Thirdly, hydrogen burning via the p-p chain, in particular the abundance of helium 3, (^{3}He), is not in equilibrium. As a consequence of the mass loss the mass of the hydrogen-burning region decreases. In the wake of the shrinking hydrogen-burning region ^{3}He is accumulated. The amount of ^{3}He which is accumulated in layers in which hydrogen burning has previously occurred depends mainly on the time-scale of the mass loss. As a result, even initially homogeneous stars become inhomogeneous with respect to the ^{3}He abundance.

The crucial phase of the secular evolution of these stars comes when the outer convective envelope reaches the centre of the star, i.e. when the star becomes fully convective and thus chemically homogenized. Thereby the accumulated ^{3}He is mixed into the hydrogen burning region. As a consequence the nuclear burning adjusts to the new ^{3}He abundance and the whole star is restructured on a rather short time-scale. Depending on the initial conditions, i.e. on the mass, the mass-loss rate and on the evolutionary status, the ^{3}He abundance after mixing can be higher or lower than before, which gives rise to a corresponding increase or decrease of the stellar radius as a consequence of the mixing. Thereby the radius can change by a few per cent on a time-scale of the order 10^7 years.

Since in a CB the mass-loss rate of the secondary is not constant, the computations of D'Antona and Mazzitelli (1981) [72] are not directly applicable to CB's. However, they demonstrate that low-mass stars which are losing mass can undergo a rather complicated secular evolution. If the reaction of a secondary of a CB to mass loss is qualitatively similar to that found for constant $\dot{M}$, the period gap could be explained as the result of the interaction of the inhomogeneous ^{3}He distribution and of the penetration of the outer convection zone. If the radius increases or decreases by a sufficient amount on a time-scale which is sufficiently short, a phase of dynamical mass loss or a phase where the secondary is detached from its Roche radius could respectively result. Thus either of the above discussed hypotheses, i.e. fast mass transfer or no mass transfer, could find a physical explanation. Moreover the explanation of the gap as suggested by D'Antona and Mazzitelli (1981)[72] is particularly attractive because it is fully compatible with the observation that the secondaries are mainsequence stars. The inhomogeneity in ^{3}He created as a result of the mass loss is not sufficient to make these stars observationally distinguishable from perfectly homogeneous stars. Finally, however, we should stress that, though the computations by D'Antona and Mazzitelli (1981)[72] are promising, it remains to be seen, whether a self-consistent mass loss from the secondary results in sufficiently strong effects in order to account for the period gap.

(3) From the Red Dwarf to the Black Dwarf

With decreasing mass the thermal time-scale of the secondary becomes longer and longer. On the other hand the time-scale of the mass loss induced by gravitational radiation becomes shorter. Therefore below a mass of $\sim$0.2 $M_\odot$ the mass-loss time-scale becomes shorter than the thermal time-scale and the secondary cannot remain in thermal equilibrium. The consequences of this effect have recently been studied by Paczyński and Sienkiewicz (1981)[73], Rappaport, Joss and Webbink (1982)[74] and by D'Antona and Mazzitelli (1982)[75]. According to their results the secondary evolves from the low-mass main sequence towards the degenerate black-dwarf state as shown in Fig. 13. Thereby the system goes through a minimum of the orbital period.

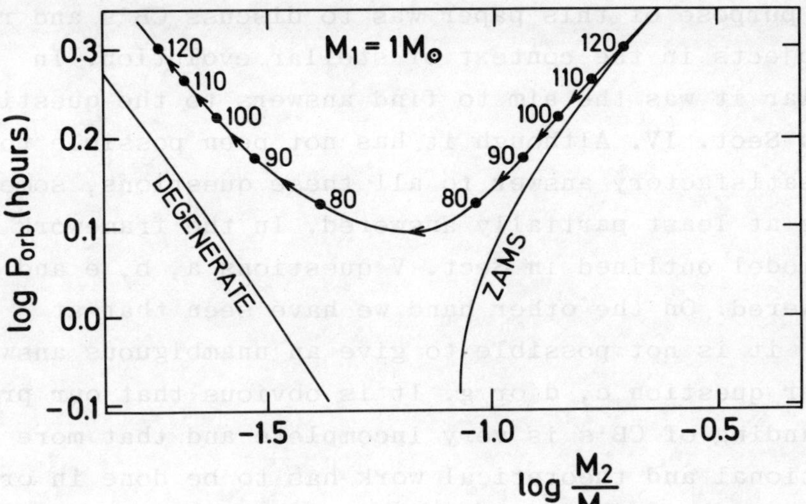

Fig. 13 Secular evolution of the secondaries of CB's under the influence of gravitational radiation in the $P_{orb} - M_2$ diagram. (After [73])

For realistic parameters the observed minimum period of~ 80 min of CB's containing hydrogen-rich secondaries can be reproduced and understood as a direct consequence of the action of gravitational waves. The end of the secular evolution of CB's would accordingly be a binary consisting of a whitedwarf primary and a very low mass degenerate hydrogen secondary (i.e $M_2 \simeq 10^{-2} M_{\odot}$). Systems of this kind could be distinguished from ordinary CB's. However, they are so faint that the chance of observing such a binary is extremely low[76].

X. CONCLUSIONS

The purpose of this paper was to discuss CB's and related objects in the context of stellar evolution. In particular it was the aim to find answers to the questions asked in Sect. IV. Although it has not been possible to give a satisfactory answer to all these questions, some of them are at least partially answered. In the framework of the model outlined in Sect. V questions a, b, e and f are answered. On the other hand we have seen that at present it is not possible to give an unambiguous answer to either question c, d or g. It is obvious that our present understanding of CB's is very incomplete and that more observational and theoretical work has to be done in order to arrive at a better understanding of these objects. At the end of this paper we should also not forget to indicate a number of interesting problems which we have not even mentioned so far:

a) What are the consequences of mass loss from the primary on phases preceding Roche-lobe overflow? How much envelope mass is necessary to allow for the formation of a common envelope?

b) What is the fate of binaries in which the accreting secondary has a low mass ($M_2 < 1\ M_{\odot}$) and therefore an outer convection zone. Can such a system ever evolve into a common-envelope binary or does it remain detached?

c) What is the evolutionary history of CB's containing a helium white dwarf secondary (e.g. of AM CVn and GP Com)?

d) Is a common envelope the only means to explain the formation of CB's? In fact other mechanisms have been suggested [77-79]

e) Why are some binaries such as FF Aqr, Feige 24, BE UMa and the central star of the planetary nebula NGC 2346 relatively wide binaries although they are probably also the result of common-envelope evolution?

The above list is certainly not complete. Nevertheless it indicates very clearly that we are still very far from a complete understanding of the evolution of close binaries and in particular of the formation of CB's and related objects. The obvious incompleteness should however encourage rather than discourage us to invest more work in this exciting field.

ACKNOWLEDGMENTS

It is a pleasure to thank my colleagues Drs. F. D'Antona, Wai-Yuen Law, E. Meyer-Hofmeister, W. Hillebrandt, F. Meyer and H. -C. Thomas for many stimulating discussions on the subject. I am grateful to Dr. J.A. Robertson for correcting the language of the manuscript. I would also like to thank the Chinese Academy of Sciences (Academia Sinica) for the generous invitation which gave the necessary impetus for writing this article.

REFERENCES

[1] Robinson, E. L., 1976, Ann. Rev. Astron. Astrophys., Vol. 14, 119.

[2] Warner, B., 1976, "IAU Symp. No 73", 85.

[3] Gallagher, J.S., Starrfield, S., 1978, Ann. Rev. Astron. Astrophys., Vol. 16, 171.

[4] Cordova, F.A., Mason, K.O., 1981, in "Accretion Driven Stellar X-ray Sources", Cambridge University Press, Cambridge, in press.

[5] Lewin, W.H.G., Joss, P.C., 1981, Space Sci. Rev., Vol. 28, 3.

[6] Van Paradijs, J., 1981, Astron. Astrophys., Vol. 103, 140.

[7] Ritter, H., 1982, Catalogue of Cataclysmic Binaries,

Low-Mass X-Ray Binaries and Related Objects.
[8] Kraft, R.P., 1963, Adv. Astron. Astrophys., Vol. 2,43.
[9] Stover, R.J., Robinson, E.L.,Nather, R.E., 1981, Ap. J., Vol. 248, 696.
[10] Stover, R.J., 1981, Ap. J., Vol. 248, 684.
[11] Patterson, J., 1979, Ap. J., Vol. 234, 978.
[12] Vogt, N., Breysacher, J., 1980, Ap. J., Vol. 235, 945.
[13] Cowley, A.P., Crampton, D., Hutchings, J.B., 1980, Ap. J., Vol. 241, 269.
[14] Walker, M.F., 1981, Ap. J.,Vol. 248, 256.
[15] Stover, R.J., Robinson, E.L., Nather, R.E., Montemayor, T.J., 1980, Ap.J., Vol. 240, 597.
[16] Young, P., Schneider, D.P., 1980, Ap.J., Vol. 238, 955.
[17] Hutchings, J.B., Cowley, A.P., Crampton, D., 1979, Ap.J., Vol. 232, 500.
[18] Koester, D., Schulz, H., Weidemann, V., 1979, Astron. Astrophys., Vol. 76, 262.
[19] Koester, D., Weidemann, V., 1980, Astron. Astrophys., Vol. 81, 145.
[20] Faulkner, J., Flannery, B.P., Warner, B., 1972, Ap. J., Vol. 175, L79.
[21] Ritter, H., 1976, M.N.R.A.S., Vol. 175, 279.
[22] Ritter, H.,1980, The Messenger, Vol. 21, 16.
[23] Ritter, H., 1982 (in preparation).
[24] Grossman, A.S., Hays, D., Graboske, Jr., H.C., 1974, Astron. Astrophys., Vol. 30, 95.
[25] Copeland, H., Jensen, J.O., Jørgensen, H.E., 1970, Astron. Astrophys., Vol. 5, 12.
[26] Lacy, H., 1977, Ap. J., Suppl. Ser., Vol. 34, 479.
[27] Popper, D., 1980, Ann. Rev. Astron. Astrophys., Vol. 18, 115.
[28] Kippenhahn, R., Kohl, K., Weigert, A., 1967, Z. Astrophys., Vol. 66, 58.
[29] Lauterborn, D., 1970, Astron. Astrophys., Vol. 7, 150.
[30] Refsdal, S., Weigert, A., 1971, Astron. Astrophys., Vol. 13, 367.

[31] Kippenhahn, R., Weigert, A., 1967, Z. Astrophys., Vol. 65, 251.
[32] Thomas, H.-C., 1977, Ann. Rev. Astron. Astrophys., Vol. 15, 127.
[33] Paczyński, B., 1970, Acta Astron., Vol. 20, 47.
[34] Kippenhahn, R., 1981, Astron. Astrophys., Vol. 102, 293.
[35] Paczyński, B., 1976, "IAU Symp. No. 73", 75.
[36] Webbink, R.F., 1981, paper presented at the Santa Cruz Summer Workshop on Cataclysmic Variables and Related Objects.
[37] Kippenhahn, R., Meyer-Hofmeister, E., 1977, Astron. Astrophys.,Vol. 54, 539.
[38] Webbink, R.F., 1979, "IAU Coll. No. 53", 426.
[39] Webbink, R.F., 1979, "IAU Coll. No. 46", 102.
[40] Taam, R.E., Bodenheimer, P., Ostriker, J.P., 1978, Ap. J., Vol. 222, 269.
[41] Meyer, F., Meyer-Hofmeister, E., 1979, Astron. Astrophys., Vol. 78, 167.
[42] Verbunt, F., Zwaan, C., 1981, Astron. Astrophys., Vol. 100, L7.
[43] Horn, J., Kříž, S., Plavec, M., 1970 Bull. Astron. Inst. Czl., Vol. 21, 45.
[44] Giannone, P., Kohl, K., Weigert, A., 1968, Z. Astrophys., Vol. 68, 107.
[45] Webbink, R.F., 1981, Bull. American. Astron. Soc., Vol. 12, 848.
[46] Delgado, A., Thomas, H.-C., 1981, Astron. Astrophys., Vol. 96, 142.
[47] Law, W. -Y., Ritter, H., 1981,"IAU Coll. No. 69"(in press).
[48] Sienkiewicz, R., 1975, Astron. Astrophys., Vol. 45, 411.
[49] Paczyński, B., Zytkow, A.N., 1978, Ap.J., Vol. 222, 604.
[50] Paczyński, B., Rudak, B., 1980, Astron. Astrophys., Vol. 82, 349.

[51] Sienkiewicz, R., 1980, Astron. Astrophys., Vol. 85, 295.
[52] Iben Jr., I., 1982, preprint.
[53] Kippenhahn, R., Thomas, H.-C., Weigert, A., 1968, Z. Astrophys., Vol. 69, 265.
[54] Woosley, S.E., Weaver, T.A., Taam, R.E., 1980, in "Type I Supernovae", Wheeler, J.C. (ed.),University of Texas at Austin and McDonald Observatory Austin, Texas, 96.
[55] Fujimoto, M. Y.,1980, in "Type I Supernovae", Wheeler, J.C. (ed.), University of Texas at Austin and McDonald Observatory, Austin, Texas, 155.
[56] Williams, R.E., Woolf, N.J., Hege, E.K., Moore, R.L., Kopriva, D.A., 1978, Ap.J., Vol. 224, 171.
[57] Ferland, G.J., Truran, J.W., 1980, Ap.J., Vol. 240, 608.
[58] Ferland, G.J., Shields, G.A., 1978, Ap.J., Vol. 226, 172.
[59] Gallagher, J.S., Hege, E.K., Kopriva, D.A., Williams, R.E., Butcher, H.R., 1980, Ap.J., Vol. 237,55.
[60] Tylenda, R., 1978, Acta Astron., Vol. 28, 333.
[61] Williams, R.E., Gallagher, J.S., 1979, Ap.J., Vol. 228, 482.
[62] Kippenhahn, R., Thomas, H.-C., 1978, Astron. Astrophys., Vol. 63, 265.
[63] Sugimoto, D., Nomoto, K., 1980, Space Sci. Rev., Vol. 25, 155.
[64] Nomoto, K., 1980, Space Sci. Rev., Vol. 27, 563.
[65] Eggleton, P., Whyte, C.A., 1980, M.N.R.A.S., Vol. 190, 801.
[66] Paczyński, B., 1967, Acta Astron., Vol. 17, 287.
[67] Faulkner, J., 1971, Ap.J., Vol. 170, L99.
[68] Ritter, H., 1980, Astron. Astrophys., Vol. 85, 362.
[69] Ritter, H., 1980, Astron. Astrophys., Vol. 86, 204.
[70] Patterson, J., Nather, R.E., Robinson,E.L., Handler,

F., 1979, Ap.J., Vol. 232, 819.
[71] Robinson, E.L., Barker, E.S., Cochran, A.L., Cochran, W.D., Nather, R.E., 1981, Ap.J., Vol, 251, 611.
[72] D'Antona, F., Mazzitelli, I., 1981, preprint.
[73] Paczyński, B., Sienkiewicz, R., 1981, Ap.J., Vol. 248, L27.
[74] Rappaport, S., Joss, P.C., Webbink, R.F., 1982, Ap.J., Vol. 254, 616.
[75] D'Antona, F., Mazzitelli, I., 1982, preprint.
[76] Ritter, H., 1982 (in preparation).
[77] Livio, M., Salzman, J., Shaviv, G., 1979, M.N.R.A.S., Vol. 188, 1.
[78] Eggleton, P., 1981, paper presented at the Santa Cruz Summer Workshop on Catacalysmic Variables and Related Objects.
[79] Livio, M., 1981, preprint.

R-S MODEL AND RADIO BINARY PULSARS

WANG Zhenru(汪珍如)

(Department of Astronomy, Nanjing University, Nanjing)

It has been 14 years since the first radio pulsar was discovered in 1967. During these years rapid developments have been made in the research of pulsars both in observation and theory.[1,2] There are still many problems to be solved in the field of pulsars. One of the interesting problems is why among more than three hundred radio pulsars all are single pulsars except three—PSR 1913 + 16, PSR 0820 + 02 and PSR 0655 + 64 that are radio binary pulsars.[2] It is a glaring problem that radio binary pulsars are so rare, and it will be probed by use of R-S Model in this paper. Our idea is different from that of A.F. Illarionov et al.[3]

The basic idea of R—S Model is as follows:[4] A polar magnetospheric gap is formed that spans the open field lines from the stellar surface up to an altitude of about 10^4 cm. The electric potential difference between the bottom and top of the gap is about 10^{12} volts. For a neutron star with magnetic moment antiparallel to its spin angular momentum, the direction of electric field is outward from its surface, and positive currents will flow out along the open lines from the star. Owing to the potential difference, the gap continually sparks by forming electron-positron pairs. The gap positrons move out and the electrons flow onto the stellar surface. They all move along the open field lines.

The positrons flowing out emit curvature radiation and the curvature radiation produces electron-positron pairs. The pairs again produce curvature radiation. Such a process repeats over and over again, and a secondary electron-positron plasma, which is nearly neutral, will be formed over the gap. The interaction between primary positrons (or ions) and the secondary $e^{\pm}$ plasma triggers a two-stream instability and results in the coherent curvature radiation. This is the way by which the radio radiation of pulsars is produced. It can be seen from the R—S Model that the existence of the electric field and the potential difference in the gap is the fundamental condition for the radio radiation to be produced. Once this condition is broken up, the radio radiation will disappear.

Let us consider the matter accreted onto the polar region of a neutron star. Electrons flowing into the gap will be accelerated towards the surface of the neutron star because of the electric field of the gap, whereas ions will be decelerated to zero velocity, and then turn to move outward. Hence the separation of charges will be brought about and the electric field or the potential difference opposite to that of the gap will be produced. As a result, if the rate of accretion is greater than a definite value which is necessary for producing an electric field with magnitude equal to the original gap field, the total electric field will disappear and the radio radiation will be extinguished.

Suppose that the accreted ions are protons and they arrive at the top of the gap with freefall velocity.* Let M and R be the mass and radius of a neutron star respectively. Taking $M \simeq 1M_{\odot}$, $R \simeq 10^{6}$ cm, we obtain $v_0 \simeq 1.6 \times 10^{10}$ cm sec^{-1}.

---

* Generally speaking, we have $GM/RC^2 \simeq 0.15 << 1$ for neutron stars and the relativistic effect is thus unimportant. So it is good enough to use Newton mechanics for making magnitude estimation.

Owing to the gap electric field, protons are decelerated to zero velocity. The deceleration time interval $\Delta t \simeq 1.63\times10^{-10}$ sec and the displacement $\Delta s \simeq 1.3$ cm are small enough to be neglected. It means that ions would accumulate on the top of the gap. In order to set up an opposite electric field which can just cancel the primary electric field of the gap, the surface charge density produced by the accumulation of accreted protons must be $\sigma = E/4\pi = \frac{\Delta v}{4\pi h}$, where Δv is the potential difference of the gap and h is the altitude of the gap. Thus we have $\sigma \simeq 2.64 \times 10^4$ esu. and the corresponding number density of protons at the top is $n=5.5\times10^{13}$ cm^{-2}. Suppose that it is provided by the freefall protons in the time interval Δt. Let $n = n_{0c}\, v_0\, \Delta t$, where v_0 is the velocity of protons arriving at the top and n_{0c} is the number density of accreted protons at the top, then

$$n_{0c} = \frac{n}{v_0 \Delta t} \simeq 2 \times 10^{13}\ cm^{-3}. \qquad (1)$$

This is the critical (minimum) value of the number density of accreted protons for the "extinguishment" of the gap electric field.

Suppose that matter is accreted onto a neutron star radially, the equation of continuity can be written as

$$n_r = n_0 (\frac{R}{r})^{3/2}. \qquad (2)$$

If the magnetic field of the neutron star is dipolar we have

$$B = B_0 (\frac{R}{r})^3. \qquad (3)$$

Taking $B_0 \simeq 10^{12}$G, the distance of Alfven surface will be given as

$$r^5 = \frac{R^6 B_0^2}{16\pi G M n_r m_p} \simeq \frac{10^{56}}{n_r}. \qquad (4)$$

Suppose that the density of matter at Alfven surface equals the density of interstellar gas, i.e. $n_r \simeq 0.1\ cm^{-3}$, the distance of such a Alfven surface will be $r \simeq 2.5 \times 10^{11}$ cm. It is easy to obtain the number density of the accreted matter at the top

$$n_0 = n_r \left(\frac{r}{R}\right)^{3/2} \simeq 1.25 \times 10^7\ cm^{-3}. \qquad (5)$$

Obviously, $n_0 << n_{0c}$. This means that the electric field of the gap cannot be cancelled by the accretion of interstellar gas. Hence, single pulsars can maintain their radio radiation.

Generally, in order to cancel out the electric field of the gap, the number density of accreted matter at Alfven surface must satisfy the following condition

$$n_r^{0.7} > 1.6 \times 10^{-8} n_{0c}, \qquad (6)$$

i.e.

$$n_r > 7.2 \times 10^7\ cm^{-3}. \qquad (7)$$

For a binary system containing a neutron star, if there is a lot of matter flowing out from its companion star (in the form of stellar wind or Roche-lobe overflow), Eq. (7) will be satisfied. In such a binary system, the neutron star cannot emit radio radiation and thus cannot be a pulsar. Otherwise, if the companion star is a compact star, there is hardly stellar wind or matter flowing out from it. As a result, the electric field of the gap can maintain and the system appears as a binary pulsar. This conclusion is in agreement with the existing observations made so far. The first observed radio binary pulsar is PSR 1913 + 16, and its companion star is recognized as a compact star, especially a neutron star[5]. Another binary pulsar is PSR

0655 + 64 which was discovered in 1980, and its companion is likely a compact star too[2]. These two binary pulsars support the above argument. As to the third radio binary pulsar PSR 0820 + 02, the observed data are too rare to determine the category of its companion star. Following our discussion here we may expect its companion star to be a compact star or a star nearly without matter flowing out. Whether the above expectation coincides with further observations will be a test of this paper.

REFERENCES

[1] Manchester, R.N., Taylor, J.H., 1981, Preprint No. RPP 2530.

[2] Sieber, W., Wielebinski, R., 1981, "I.A.U. Symposium No. 95", Reidel, Dordrecht.

[3] Illarionov, A.F., Sunyaev, R.A., 1975, Astron. Astrophys. Vol. 39, 185.

[4] Ruderman, M.A., Sutherland, P.G., 1975, Ap. J., Vol. 196, 51.

[5] Taylor, J.H., Fowler, L.A., McCulloch, P.M., 1979, Nature, Vol. 277, 437.

THE FORMATION OF MASSIVE WHITE DWARFS IN CATACLYSMIC BINARIES*

Wai-Yuen LAW, Hans Ritter
(Max-Planck-Institut für Physik und Astrophysik, Institut für Astrophysik, München, Federal Republic of Germany)

ABSTRACT

White dwarfs (WD's) in cataclysmic binaries (CB's) contrast with single white dwarfs in that, while the masses of most single WD's fall into a narrow range of $\sim 0.12\ M_\odot$ around the mean mass of $\sim 0.58\ M_\odot$, the masses of WD's in CB's are distributed quite uniformly over almost the entire permitted mass range. This makes a discussion of the formation of higher mass ($\gtrsim 1\ M_\odot$) WD's of particular interest, and may shed some light on the evolution of CB's in general.

In the calculations of Delgado and Thomas (DT)[1] for conservative Case BB evolution in binary systems, it was

---

*This communication is based on a paper in preparation (Law and Ritter, 1982); an earlier version will appear in: Proceedings of the 5th Bamberg Int. Coll. No. 69 on "Binary and Multiple Stars as Tracers of Stellar Evolution", Bamberg (1982).

found that the resultant system is wide and may consist of a high mass WD and a very high mass main sequence (MS) secondary. An important result of their calculations is that the second Case BB mass transfer (MT) begins with a fast phase and then proceeds to a slow phase, which stops only when almost the entire helium envelope has been lost. We have made use of the results of DT, and also of the computations of the evolution of single helium stars, to investigate the Case BB evolution over the entire probable mass range. We found that a rather high mass helium star of $\sim 2\ M_\odot$ to $\sim 3\ M_\odot$ is needed to form either a CO WD $\gtrsim 1\ M_\odot$ or a neutron star. In order for the helium star to evolve into a WD, it is also necessary for the first Case B MT to be nonconservative in order to produce a short period system, enabling the second Case BB MT to occur before the onset of carbon-burning. This is in accord with the conventional theory which asserts that CB's are descendants of initially very wide systems.

Such an evolution may indeed be nonconservative, since, during its course, favourable circumstances arise for the development of the common envelope (CE), which is an effective mechanism for carrying away mass and angular momentum from the system. A necessary condition for the formation of the common envelope during mass transfer is that the secondary, in response to mass gain, must increase its radius so much that the system comes into contact. This situation can be realized only if the accretion time-scale of the secondary is much less than its thermal time-scale.

If the primary star starts its first Case B MT when it is on the red giant branch, its deep convective envelope induces mass loss on a dynamical time scale, which can easily be overcome by any realistic thermal adjustment time of the secondary. A common envelope almost certainly occurs in this case. Observations of post CE objects then help to provide information on the properties of the systems having experienced CE evolutions. Such a system should be of short period and consists of a low mass MS secondary and a helium star of roughly the same mass as predicted by a conservative evolution.

Due to its very short evolutionary time, the helium star soon starts its second Case BB MT and loses its surface material onto the MS secondary. Extrapolation of the results of Kippenhahn et al[2]. for the mixing of the accreted helium with a hydrogen star gives a mixing time of the order of 10^4 to 10^5 years for the mass range of 1 $M_\odot$ to 4 $M_\odot$. This time scale is somewhat too long compared with the time scale of MT to suggest any significant anomaly of the reaction of the secondary to mass accretion. Therefore, the presence of the fast phase in the Case BB MT and the secondary being of low mass both imply that the occurrence of a second CE phase is very likely.

The nonconservative Case BB evolution (cf. the Sect. VII.(3). of Ritter's paper in this Proceedings) discussed in our work represents a simple method of forming CB's containing high mass WD's, and low mass X-ray binaries. However, the effectiveness of this evolutionary path remains to be tested by the relevant observations. First, the paucity of the ZAMS stars, namely those in the mass range of $\sim$10 $M_\odot$ to $\sim$15 $M_\odot$, which may experience Case BB evolutions seems to question whether the mass distribution of the WD's in CB's is consistent with the mass spectrum of their progenitors on the MS. Second, the concentration of the low mass X-ray binaries at the galactic bulge seems to indicate that their precursors are of low mass; however, the rarity of these X-ray objects, if indeed true, may be an indication that their progenitors also belong to a small population. Finally, we should always keep in mind that the observational data of CB's still do not allow a statistical determination of the mass distribution of their

WD's.

REFERENCES

[1] Delgado, A.J., Thomas, H. -C., 1981, Astron. Astrophys., Vol. 96, 142.

[2] Kippenhahn, R., Meyer-Hofmeister, E., 1977, Astron. Astrophys., Vol. 54, 539.

CASE B
CASE C
Ritter

V. ACCRETION DISKS

ACCRETION DISKS

Friedrich Meyer

(Max-Planck-Institut für Physik und Astrophysik,
München, Federal Republic of Germany)

In High Energy Astrophysics accretion of matter on compact bodies provides powerful sources of radiation by the release of gravitational energy. Systems in which this process plays or may play a role range from cataclysmic variables and X-ray binaries to SS 433, γ-ray burst sources, active galactic nuclei and quasars.

Usually the relative distance over which the accreting matter has to fall covers several powers of ten before it arrives at the gravitational center. Even small amounts of angular momentum, when present, finally prevent the free fall. Instead an innermost radial distance r_i is reached where centrifugal acceleration balances gravity. Only when the material is able to lose its angular momentum will it be able to spiral further in and approach the central body.

The direction of the mean angular momentum of the accreting matter defines a plane through the central body. Motions perpendicular to this plane may all be dissipated by collision and friction between the interpenetrating orbits, the energy of this relative motion being radiated away. The same will hold for any non-circular part of the orbits within this plane. The mean angular momentum, however, cannot thus be removed. Material therefore settles toward this equatorial plan, forming an "accretion disk" in which it performs Kepler orbits around the gravitational center. In the following we limit ourselves to "non-selfgravitating" disks in which the gravity of the central body dominates over the gravity which the disk material exerts on itself. If this latter action is no longer negligible the disks become liable to the Jeans instability and have to be stabilized by some other component in the system (like for example a halo of stars for the case of disks in spiral galaxies).

The Kepler angular frequency

$$\Omega = (GM/r^3)^{1/2} \tag{1}$$

(G gravitational constant, M mass of central body, r distance from center) shows a non-vanishing radial shear, $d\Omega/dr \neq 0$. In the presence of a viscosity μ this shear leads to a radial stress

$$\tau = \mu r \frac{d\Omega}{dr}, \tag{2}$$

which tends to brake the faster interior revolution and transmits angular momentum toward the outer orbits. Thus the material gradually losing its angular momentum will slowly move inward, always orbiting closely to the Kepler law (Eq. (1)). A steady state will be reached when the viscous angular momentum flux outward is exactly compensated by the inward advective flux of angular momentum carried by the matter moving inward. (If the central body is a source or sink of angular momentum for the disk around it, this term has to be added to the balance.)

The viscous heat that is generated in this process has to be brought to the surface of the accretion disk from where it is radiated away into space. In this way the gas of the accretion disk acquires a temperature and pressure which support it against the component of the central gravity perpendicular to the disk which cannot be balanced by the Kepler rotation. If the thermal velocity (sound velocity) of the disk material remains small compared to the Kepler orbital velocity the disk is thin compared to the radial distance r, and the equations for the radial and the vertical structure of the disk may be separated.

Assuming rotational symmetry and calling z the distance from the plane of the disk and ϕ the azimuthal angle of the cylindrical coordinate system r, ϕ, z, one obtains by integrating in z-direction

$$\frac{\partial}{\partial t}(2\pi r\Sigma) - \frac{\partial \dot{M}}{\partial r} = 0 \qquad (3)$$

$$\frac{\partial}{\partial t}(2\pi r^3 \Omega\Sigma) - \frac{\partial}{\partial r}\left[r^2 \Omega \dot{M} + 2\pi r^3 \frac{d\Omega}{dr}\int_{-\infty}^{+\infty} \mu dz\right] = 0 \qquad (4)$$

as the equations for conservation of mass and angular momentum. Here

$$\Sigma = \int_{-\infty}^{+\infty} \rho dz \qquad (5)$$

is the "surface density", $\dot{M}$ is the <u>inward</u> radial mass flow and the friction integral

$$f = \int_{-\infty}^{+\infty} \mu dz \qquad (6)$$

has to be determined from the equations giving the vertical disk structure. Elimination of $\dot{M}$ results in the equation

$$\frac{\partial \Sigma}{\partial t} = \frac{3}{r}\frac{\partial}{\partial r}\left\{r^{1/2}\frac{\partial}{\partial r}[r^{1/2} f]\right\} . \qquad (7)$$

If the time scale of thermal adjustment in the vertical disk structure is short compared to the time scale for the temporal evolution in Eq. (7), f becomes a function of Σ and r alone,

$$f = f(\Sigma, r). \tag{8}$$

Eq. (7) then represents a non-linear diffusion equation for the surface density distribution in the disk, a result of the frictional interaction between neighboring "rings" of matter in the disk.

In the vertical direction the equation of hydrostatic equilibrium is

$$\frac{dP}{dz} = - g_z \rho , \tag{9}$$

(P pressure, ρ density) where for a thin disk

$$g_z = \frac{GM}{r^3} \dot{z} \tag{10}$$

is the "effective" gravity, neglecting terms of order $(z/r)^2$ against 1. For an isothermal ideal gas

$$P = R_g T \rho , \tag{11}$$

(R_g = gas constant, T temperature) the vertical structure becomes

$$P = P_0 \exp(- \frac{z^2}{H^2}) \tag{12}$$

with the scale height H

$$H = (\frac{2R_g Tr^3}{GM})^{1/2} \tag{13}$$

being a measure of the thickness of the disk. In realistic cases the gas will be optically thick and a temperature difference between the interior and the surface has to appear so that the internally produced heat can be conducted to the surface. Also, in hot disks, the radiation pressure may become important. The equation for the vertical heat gradient can be written as

$$\frac{dT}{dz} = -\frac{3\kappa\rho F}{4acT^3}, \tag{14}$$

if the vertical energy flow density F is transported by radiative conduction (κ opacity (in cm^2/g), a radiation constant, c velocity of light). As in stellar structure the law of radiative energy transport is valid if the temperature gradient is <u>subadiabatic</u>:

$$\nabla \leq \nabla_{ad}, \tag{15}$$

i.e. ∇=(d ln T/d ln P) smaller than it's value for adiabatic exchange. In the other case Schwarzschild instability leads to convection and Eq. (14) has to be replaced by

$$\frac{dT}{dz} = \frac{T\rho g_z}{P} \nabla_{conv} \qquad \nabla \geq \nabla_{ad}, \tag{16}$$

where in stellar structure a mixing length theory is used to determine the convective temperature gradient ∇_{con} (see e.g. [1]). We leave aside here the discussion of how the rapid disk rotation interferes with the convection and simply assert that the disk structure will not be much affected by this, in principle important effect. Eqs. (14, 15) or (16) are complemented by the equation of heat generation

$$\frac{dF}{dz} = \frac{9}{4}\frac{GM}{r^3}\mu, \tag{17}$$

where it is assumed that the friction is the only source of heat in the disk.

As in the corresponding stellar case molecular viscosity in accretion disks is so small that it can be neglected. (One may show, that with such small viscosity the disk for the usual mass accretion rates in binaries, becomes so dense and extremely opaque that it will be inflated to a more spherical shape, contrary to the observations, where the usual approximations break down.) Thus there obviously is another sort of effective viscosity, which has been sought in turbulent radial exchange driven by the differential rotation (Eq. (1)) or the effect of small scale magnetic fields, present and wound up in the disk. Both effects have by order of magnitude an upper limit of the velocity of sound for the exchange velocity, which led Shakura and Sunyaev (1973)[2] to parametrize the unknown viscous stress as

$$\tau = \alpha P, \tag{18}$$

which is roughly equivalent to an assumed viscosity

$$\mu = \frac{\sqrt{2}}{3} \alpha \rho H V_s, \tag{19}$$

where $V_s = (P/\rho)^{1/2}$ is the isothermal sound velocity. The non-dimensional parameter α is equivalent to the Mach number of "turbulent exchange" and should remain $\lesssim O(1)$. Comparison of time dependent disk model computations with observations of disks in cataclysmic variables have in general supported Eq. (18) with

$$0.1 \lesssim \alpha \lesssim 1. \tag{20}$$

α has usually been assumed constant in model computations. It may of course depend on other non-dimensional quantities like H/r.

With this law of friction (18) and Eq. (14) the vertical disk structure has been approximately determined by using a "free-free" (Kramers) and electron scattering opacities by Shakura and Sunyaev[2] for stationary accretion disks and by Lightman (1974)[3] and Bath and Pringle (1981)[4] for time dependent disks. For a review and for further literature we refer to Pringle (1981)[5]. Recent numerical computations of the vertical disk structure[6] using stellar structure opacity tables generally confirm the approximation treatments. In particular such analytical or numerical models result in a determination of the function f (Σ, r) in Eq. (7). An example for stationary disk structure solutions is given in Fig. 1, for different accretion rates $\dot{M}$ around a 1 $M_\odot$ central mass. For neutron stars the disks can extend down to inner radii of order 10^6 cm. In the inner regions the slope of the log Σ versus log r - curves changes sign from a negative value at larger radii to a positive value at smaller radii. This is an effect of the importance of the radiation pressure increasing inwards and signals the onset of the Lightman-Eardley(1974)[7] instability. One may see this in the following way. In the stationary disk (with zero net angular momentum flow) the function f must be constant across the disk and then is

$$f = \dot{M}/3\pi. \tag{21}$$

The turning over of the curves in Fig. 1 then signifies that in these interior regions

$$\frac{\partial f}{\partial \Sigma} = \frac{1}{3\pi}\left(\frac{\partial \Sigma}{\partial \dot{M}}\right)^{-1} < 0 \tag{22}$$

for r = const. Multiplying Eq. (7) by $\frac{\partial f}{\partial \Sigma}$ now results in a diffusion equation for f,

$$\frac{\partial f}{\partial t} = \frac{\partial f}{\partial \Sigma} \cdot \frac{\partial \Sigma}{\partial t} = \frac{3}{r}\frac{\partial f}{\partial \Sigma} \cdot \frac{\partial}{\partial r}\left\{ r^{1/2} \frac{\partial}{\partial r}[r^{1/2} f]\right\}, \tag{23}$$

which is <u>unstable</u> whenever the diffusion coefficient, proportional to $\frac{\partial f}{\partial \Sigma}$, becomes < 0. The disk then separates into alternating rings of high and low density. (Shakura and Sunyaev (1976)[8] have shown that at that same time the vertical disk struture becomes thermally unstable, leading to speculations that the disk becomes centrally inflated to

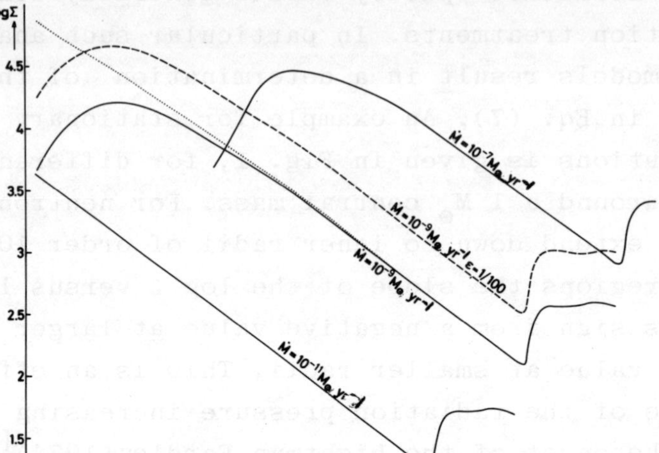

Fig. 1 Surface density Σ versus distance r from 1 $M_\odot$ central object for different accretion rates ($\alpha = 0.071$). Dashed line $\alpha = 0.021$. Dotted line viscosity proportional to gas pressure instead of total pressure. cgs units except M.

a near spherical shape). For models of magnetic friction the total pressure in the law of friction should be replaced by the gas pressure. Then the inner instability disappears, as Lightman and Eardley (1974) have shown[7].

In the case of cataclysmic variables only regions of radii larger than about 10^9 cm, the typical white dwarf radii are important. In these outer regions one sees a typical change of structure in which the surface density sharply increases when convection sets in and keeps the midplane disk matter cooler than a radiative structure would do. Here again, for a limited range in r, Σ-curves for different $\dot{M}$ intersect yielding $\partial f/\partial\Sigma<0$ and instability. This instability is of considerable importance for the cataclysmic variables and appears to explain the dwarf novae outbursts of such systems[9]. Fig. 2 shows the $f(\Sigma)$ curve at a particular r in this region and one sees that at the same surface density there are two stable branches, one of high f, corresponding to a radiative structure and one of low f, corresponding to a convective structure. In between lies an unstable branch, $\frac{\partial f}{\partial \Sigma} < 0$ that corresponds to a transition structure. Imagine now that the stationary accretion rate is such that at this radius its value comes to lie between points B and C. The unstable branch, of course, cannot be realized. Instead a cyclic behaviour must result. If initially the system were in a convective state on the branch between A and B, then the difference between the outside steady accretion rate M and $3\pi f$ (which roughly is the mass flow rate locally passing through) leads to mass accumulation and an increase in Σ until point B is reached. Any further increase in Σ then enforces a transition to the radiative branch at point D. Now the local mass flow rate $3\pi f$ is larger than the stationary value $\dot{M}$ and the surface density Σ decreases rapidly until point C is reached, from where the solution transits back to the convective state. A more quantitative treatment of this cyclic behaviour provides

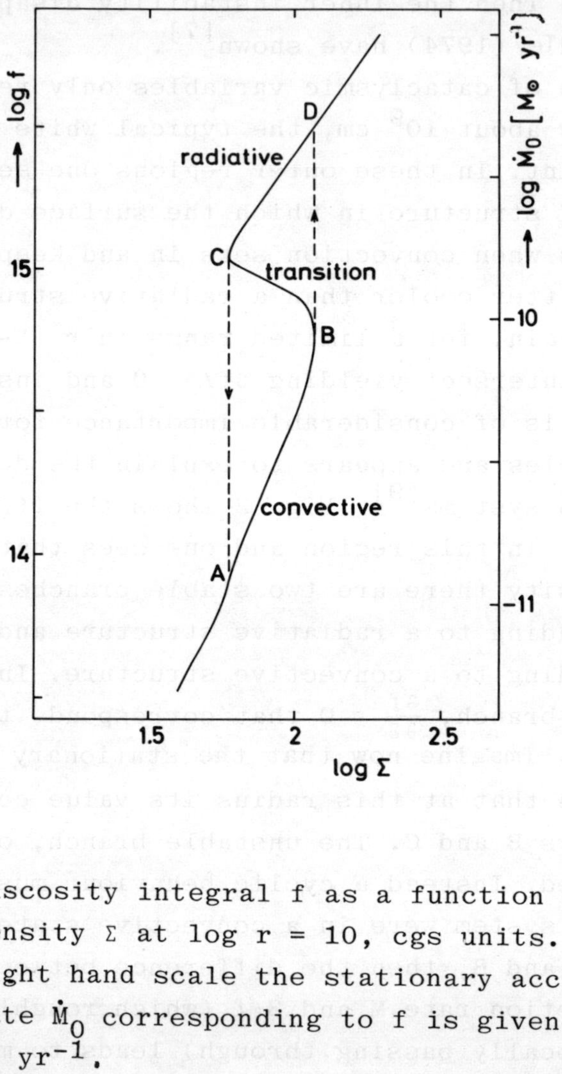

Fig. 2 Viscosity integral f as a function of surface density Σ at log r = 10, cgs units. On the right hand scale the stationary accretion rate $\dot{M}_0$ corresponding to f is given in units $M_\odot$ yr^{-1}.

good agreement with the dwarf nova observations, where the quiescent stage corresponds to the long time of "convective" build-up of the surface density and the outburst corresponds to the rapid release of this "stored" matter while this disk area is in its radiative state. Further interesting results from this model can be expected.

REFERENCES

[1] Kippenhahn, R., Weigert, A., Hofmeister, E., 1967, "Methods in Computational Physics", Vol. 7, Academic Press Inc., New York, 129.

[2] Shakura, N.I., Sunyaev, R.A., 1973, Astron. Astrophys., Vol. 24, 337.

[3] Lightman, A.P., 1974, Ap. J., Vol. 194, 429.

[4] Bath, G. T., Pringle, J.E., 1981, M.N.R.A.S., Vol. 194, 967.

[5] Pringle, J.E., 1981, Ann. Rev. Astron. Astrophys., Vol. 19, 137.

[6] Meyer, F., Meyer-Hofmeister, E., 1982, Astron. Astrophys., Vol. 106, 34.

[7] Lightman, A.P., Eardley, D.N., 1974, Ap. J., Vol. 187, L1.

[8] Shakura, N.I., Sunyaev, R.A., 1976, M.N.R.A.S., Vol. 175, 613.

[9] Meyer, F., Meyer-Hofmeister, E., 1981, Astron. Astrophys., Vol. 104, L10.

INTERACTION OF ACCRETION DISKS AND ROTATING MAGNETIC NEUTRON STARS

U. Anzer, G. Börner

(Max-Planck-Institut für Physik und Astrophysik, Institut für Astrophysik, München, Federal Republic of Germany)

It is now generally believed that pulsating X-ray sources are powered by the accretion of gas onto neutron stars. Since these neutron stars are very compact ($r_* \simeq 10$ km for $M \simeq 1M_\odot$) a sizeable fraction of the rest energy of the infalling gas can be released when this material hits the surface of the neutron star. Typical values for the efficiency are $\eta \simeq 0.1$. Far away from the neutron star two different flow patterns can be envisaged: spherical flow and disk flow. The first will be encountered when a single neutron star accretes from a surrounding gas cloud. For binary systems the assumption of spherical flow is obviously a poor one. Only in systems where the primary emits a strong wind one may use spherical accretion as a first approximation. But if the primary star looses its mass through the inner Lagrangian point ("Roche overflow"), see Fig. 1, then the material carries so much specific angular momentum that a disk will be formed. The flow in the disk will be approximately along circular Kepler orbits ($\vec{v} = \vec{v}_\phi \simeq r^{-\frac{1}{2}}$). In this case two types of disk geometry may occur: thin disks and thick disks, i.e. either $d \ll r$ or $d \lesssim r$. Since d is usually of the order of the scale height H, and $H/r \simeq v_{th}/v_k$, the distinction between the different geometries is given by the thermal velocity in the disk (relative to the Kepler velocity), and therefore by the energy balance inside the disk. The assumption of a rotationally symmetric disk will break down for radii comparable to the Roche radius. Most models which have been considered so far also assume that the disk normal and the spin axis of the neutron star are parallel to each other.

Next we want to discuss the magnetospheres of neutron stars. The "pulsations" of X-ray sources are interpreted theoretically as being due to magnetic poles which rotate with the neutron star. So far for one object one also has direct observational evidence for the existence of strong surface fields: Trümper et al. (1978)[1] measured the spectrum of Her X-1 and found a line which can be interpreted as a cyclotron line originating in a field of 5×10^{12} gauss. In the following investigation we shall assume that this magnetic field is of the dipole type (higher order multipoles may also play some role at the surface of the neutron star but fall off much more rapidly than the dipole). There will also be some modification of the dipole field because of interaction with the disk. The dipole field is given by

$$\vec{B} = \frac{\vec{\mu}}{r^3}(2\cos\theta, \sin\theta, 0), \tag{1}$$

where spherical coordinates around the dipole axis are used. We shall consider two particularly simple cases:

a) $\vec{\mu}$ parallel to the plane of the accretion disk and b) $\vec{\mu}$ perpendicular to it.

The first configuration was studied by Anzer and Börner (1980)[2], see Fig. 2. They consider a geometrically thin disk with very large electrical conductivity. In that case the diamagnetic disk plasma will be shielded from the magnetosphere without distorting the dipole field. Inside the disk we take hydrostatic equilibrium with

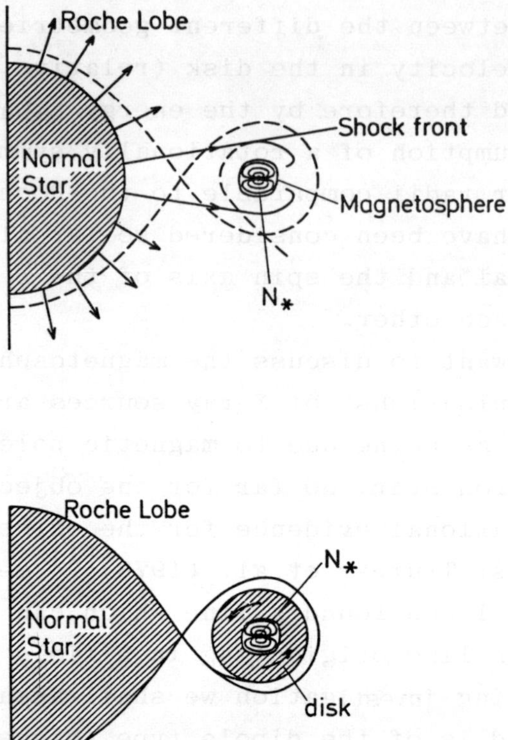

Fig. 1 Mass flow in binary systems with a neutron star.

$$p = p_0 \exp(-z^2/2H^2), \tag{2}$$

$$H/r = \sqrt{RTr/GM} = v_{th}/v_k \tag{3}$$

and the surface of the disk is defined by the balance between the gas pressure inside it and the magnetic pressure of the dipole outside.

The perpendicular dipole has been proposed by Ghosh and Lamb (1978)[3]. They also assume a geometrically thin disk extending radially to an inner edge $r_1 >> r_*$. If the conductivity in the disk is infinite then all the flux of the dipole has to go through the hole in the disk ($r<r_1$). If the conductivity is finite then some fraction of the magnetic flux can also penetrate the disk itself. Such a configuration is shown in Fig. 3. Ghosh and Lamb (1979)[4] argue that rapid reconnection of antiparallel fields (Petschek mechanism) and turbulence inside the disk will reduce the electrical conductivity sufficiently.

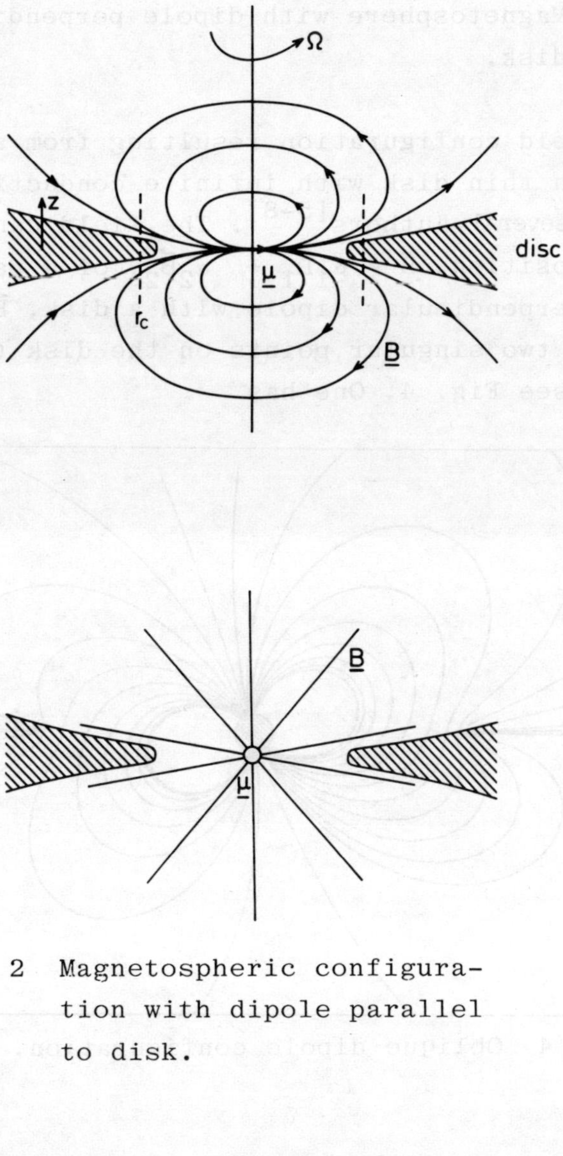

Fig. 2 Magnetospheric configuration with dipole parallel to disk.

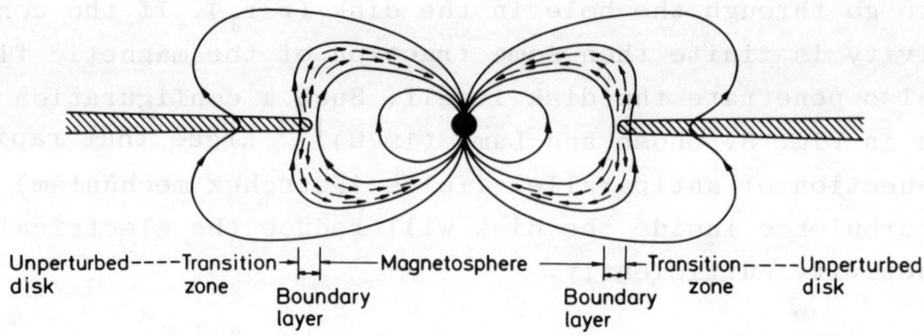

Fig. 3 Magnetosphere with dipole perpendicular to disk.

The field configuration resulting from an oblique dipole and a thin disk with infinite conductivity has been derived by several authors[5-8]. The field can be obtained as a superposition, $\vec{B} = \alpha_1\vec{B}_1 + \alpha_2\vec{B}_2$, of a parallel dipole, $\vec{B}_1$, and a perpendicular dipole with a disk, $\vec{B}_2$. These fields have two singular points on the disk (at $r = r_s$, $\phi = 0, \pi$), see Fig. 4. One has

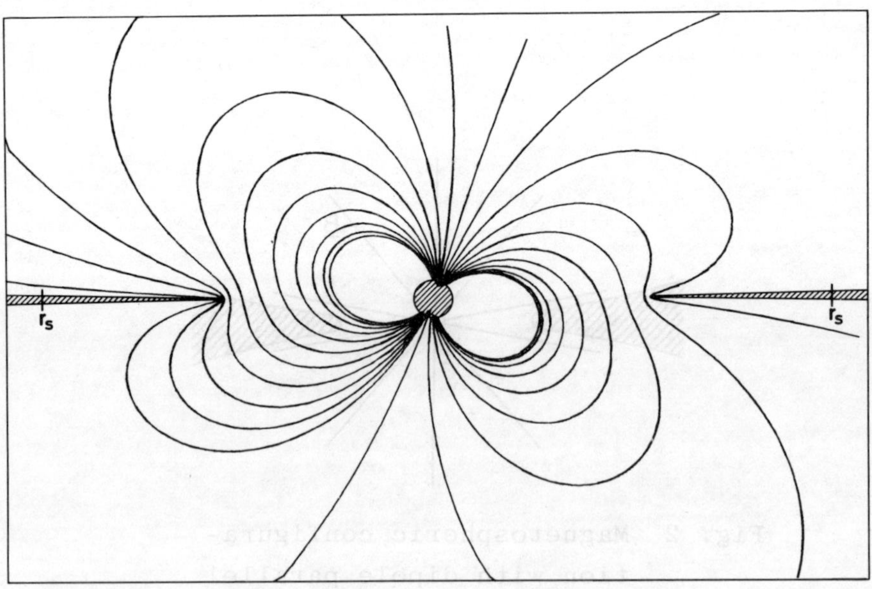

Fig. 4 Oblique dipole configuration.

$$r_s/r_1 = (1 + \frac{4}{\pi^2} \mathrm{tg}^2\theta)^{\frac{1}{2}}. \tag{4}$$

One can also see that for $r > r_s$ the field topology near the disk resembles that of the parallel dipole, whereas for $r < r_s$ it is closer to that of the perpendicular dipole. From Eq. (4) one deduces $r_s/r_1 < 1.03$ for $\theta < 20°$; therefore even a sizeable deviation from the parallel alignment will lead only to a negligible modification of the special field structure which was considered by Anzer and Börner.

Next we want to discuss a possible mechanism for mixing disk material into the magnetosphere. At a first glance it might seem that in the case of parallel alignment and high conductivity no mixing would occur, disk and magnetosphere appear to be completely separated. But since the rotation laws in both regimes are different, there will be a velocity difference at the interface. One has

$$v = \sqrt{\frac{GM}{r}} - \Omega r. \tag{5}$$

There is a certain radius, r_c, where $\Delta v = 0 (r_c = (GM/\Omega^2)^{\frac{1}{3}})$, everywhere else v differs from zero. Such a situation will lead to a Kelvin-Helmholtz instability.

Here we shall discuss this instability for an especially simple geometry. We take two homogenuous media, separated by a discontinuity at a planar interface and moving against each other with a constant velocity Δv. If small perturbations of the form

$$\vec{\xi} = \vec{\xi}_0 \, e^{i(\vec{k}_\mu \vec{x} - \omega t)} \tag{6}$$

are studied, one obtains the following dispersion relation[9]:

$$\frac{\dfrac{\rho_1^2[v_1^2 \cos^2(\alpha - \beta_1) - w^2]^2}{w^4}}{w^2(u_1^2 + v_1^2) - u_1^2 v_1^2 \cos^2(\alpha - \beta_1)} =$$

$$\frac{\rho_2^2[v_2^2 \cos^2(\alpha - \beta_2) - (w - \Delta v \cos\alpha)^2]^2}{1 - \frac{(w - \Delta v \cos\alpha)^4}{(w-\Delta v \cos\alpha)^2(u_2^2+v_2^2)-u_2^2 v_2^2 \cos^2(\alpha-\beta_2)},} \tag{7}$$

where the symbols have the meaning: $v_i = B_i / \sqrt{4\pi\rho_i}$ Alfven velocities, u_i sound velocities, β_i angles between $\vec{B}_i$ and $\Delta\vec{v}$, angle between $\Delta\vec{v}$ and wave vector $\vec{k}_t$ and $w = \omega/k_t$, with $i = 1,2$. We shall identify medium 1 with the disk where we neglect the magnetic field ($v_1 = 0$), and medium 2 with the magnetosphere where we assume $u_2 \simeq 0$. With the following notation $v_A = v_2$, $c_s = u_1$, $\beta = \beta_2$, $x = \Delta v \cos\alpha / v_A$ and $m = w/c_s$ one then obtains from Eq. (7).

$$\rho_1^2 \, w^4[1 - (m \frac{c_s}{v_A} - x)^2] = \rho_2^2 \, v_A^4[\cos^2(\alpha - \beta)-(m \frac{c_s}{v_A} - x)^2]^2(1-m^2). \tag{8}$$

pressure equilibrium at the interface leads to

$$\rho_1 c_s^2 = \frac{1}{2} \rho_2 v_A^2$$

and hence

$$m^4[1-(m \frac{c_s}{v_A} - x)^2] = 4[\cos^2(\alpha - \beta)-(m \frac{c_s}{v_A} - x)^2]^2(1-m^2). \tag{9}$$

A sufficient condition for stability is that all roots of Eq. (9) are real. In the following discussion of this equation we shall assume that $c_s/v_A \ll 1$ holds. To find the roots of Eq. (9) we now consider various regimes:

a) $m\, c_s/v_A \ll x$ then Eq. (9) simplifies to

$$m^4(1-x^2) = 4[\cos^2(\alpha-\beta)-x^2]^2(1-m^2). \tag{10}$$

The roots of Eq. (10) are

$$m^2 = \frac{2}{x^2-1} \Big\{ [\cos^2(\alpha-\beta)-x^2]^2 \qquad (11)$$

$$\pm | \cos^2(\alpha-\beta) - x^2 | \sqrt{(\cos^2 (\alpha-\beta) - x^2)^2 - (x^2-1)} \Big\} .$$

For stability we must have $m^2 > 0$ which requires $x^2 > 1$ and in addition the expression under the square root of Eq. (11) must be positive. Hence

$$[x^2 - \cos^2(\alpha-\beta)]^2 > x^2-1. \qquad (12)$$

For $\cos^2(\alpha-\beta) < \frac{3}{4}$ this is fulfilled for all values of x and for $\cos^2(\alpha-\beta) > \frac{3}{4}$ it requires

$$\text{either } x^2 > \cos^2(\alpha-\beta) + \frac{1}{2} + \sqrt{\cos^2(\alpha-\beta) - \frac{3}{4}}$$

$$(13)$$

$$\text{or } \quad x^2 < \cos^2(\alpha-\beta) + \frac{1}{2} - \sqrt{\cos^2(\alpha-\beta) - \frac{3}{4}}$$

In Fig. 5 we have shaded the region of stability. To have stability for all possible values of α and β the inequality $x^2 > 2$ must hold.

b) Next consider the regime $m \frac{c_s}{V_A} >> x$ (which implies $m >> x$): since we already know from a) that only $x > \sqrt{2}$ could be stable we consider only $x > \sqrt{2}$ leading to $m >> 1$. Then Eq. (9) can be approximated by

$$m^4 \left(1 - m^2 \frac{c_s^2}{V_A^2} \right) = 4m^4 \frac{c_s^4}{V_A^4} - 4m^6 \frac{c_s^4}{V_A^4} , \qquad (14)$$

dividing by m^4, we find

$$m^2 \frac{c_s^2}{V_A^2} \left(1 - 4 \frac{c_s^2}{V_A^2} \right) = \left(1 - 4 \frac{c_s^4}{V_A^4} \right) .$$

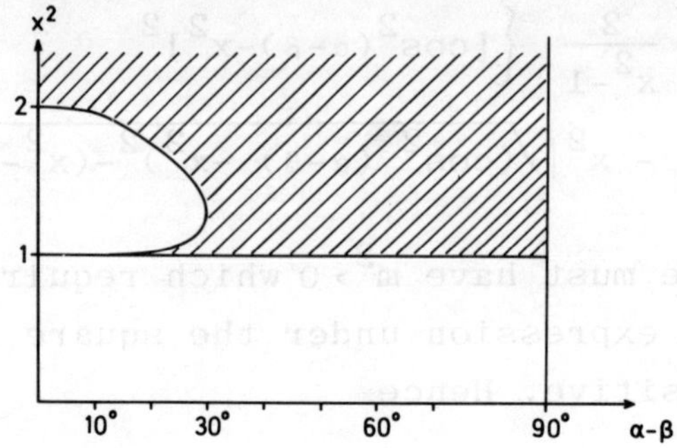

Fig. 5 Region of stable modes (where ω is real).

This would require $m^2 \frac{c_s^2}{V_A^2} \simeq 1+4\frac{c_s^2}{V_A^2} - 4\frac{c_s^4}{V_A^4} \lesssim 2$, and therefore in regime b) there are no solutions.

c) Finally let $m \frac{c_s}{V_A} \simeq x$ (leading also to $m \gg x$) and $x > \sqrt{2}$. Set $y \equiv \frac{mc_s}{V_A} - x$, this leads to $\frac{V_A^2}{c_s^2} (y + x)^2(y^2 - 1) = 4[\cos^2(\alpha-\beta)-y^2]^2$. (15)

Since $\frac{V_A}{c_s} \gg 1$, (15) can be approximated by

$$(y + x)^2(y^2 - 1) = 0 .$$

$y = -x$ gives $m = 0$, which is unacceptable, but $y = \pm 1$ are 2 possible solutions, corresponding to

$$m \frac{c_s}{V_A} = x \pm 1. \tag{16}$$

Thus (11) and (16) give 6 real roots of Eq. (9) as long as $x > \sqrt{2}$ holds, because then (13) is also satisfied for all values of α and β.

The condition for stability $x^2 > 2$ requires

$$\Delta v \cos\alpha > \sqrt{2}\, V_A .$$

We can rewrite this by using the condition of pressure equilibrium $\rho_1 c_s^2 = \frac{1}{2} \rho_2 V_A^2$, as

$$\Delta v \cos\alpha > 2 \sqrt{\frac{\rho_1}{\rho_2}}\, c_s . \tag{17}$$

In general $\rho_2 << \rho_1$. If we take for example $\rho_1 = 100\rho_2$, $\Delta v \cos\alpha > 20\, c_s$ must hold for stability.

The inequality (17) will be violated if $\Delta v < 2\sqrt{\rho_1/\rho_2}\, c_s$ for all values of α, i.e. for all orientations of the wave vector with respect to $\Delta\vec{v}$. For $\Delta v > 2\sqrt{\rho_1/\rho_2}\, c_s$ only modes which fulfil $\cos\alpha < 2\sqrt{\rho_1/\rho_2}\, c_s/\Delta v$ may become unstable.

We have seen from Eq. (11) that for $x^2 < 1$ exponentially growing modes will occur. In addition to the exponential growth of the modes one also has to request that the perturbations are localized around $z = 0$, i.e. $\xi \to 0$ for $z \to -\infty$ (region 1) and $z \to +\infty$ (region 2). This is equivalent to

$$\mathrm{Im}\,(k_{z1}) < 0 \text{ and } \mathrm{Im}\,(k_{z2}) > 0. \tag{18}$$

Following Fejer (1964)[9] one has

$$\left(\frac{k_{z1}}{k_t}\right)^2 = m^2 - 1 \tag{19}$$

and

$$\left(\frac{k_{z2}}{k_t}\right)^2 = -1 + \left(m \frac{c_s}{v_A} - x\right)^2 \simeq x^2 - 1. \tag{20}$$

From Eq. (11) one obtains in the case $x^2 < 1$

$$m = -i\gamma \tag{21}$$

for the growing mode. Conditions (18) together with Eqs. (19) and (20) lead then to

$$k_{z1} = -i\sqrt{1+\gamma^2}\; k_t \quad \text{and} \quad k_{z2} = i\sqrt{1-x^2}\; k_t . \tag{22}$$

The condition for continuity of the perturbations at the interface z = 0 reads

$$-\frac{m^2}{k_{z1}} \simeq 2\,\frac{\cos^2(\alpha-\beta)-x^2}{k_{z2}}, \tag{23}$$

and therefore

$$\frac{\gamma^2}{-i\sqrt{1+\gamma^2}} = 2\,\frac{\cos^2(\alpha-\beta)-x^2}{i\sqrt{1-x^2}} . \tag{24}$$

This then implies that a necessary and sufficient condition for instability in the regime $x^2 < 1$ is given by

$$x^2 > \cos^2(\alpha-\beta). \tag{25}$$

Instability therefore occurs for

$$\left|\frac{\cos(\alpha-\beta)}{\cos\alpha}\right| < \frac{\Delta v}{v_A} < \left|\frac{1}{\cos\alpha}\right| . \tag{26}$$

The lower limit is given by 1 for $\beta = 0$ and $|\mathrm{tg}\,\alpha|$ for $\beta = \frac{\pi}{2}$. We show the region of instability in Fig. 6 for the cases $\beta = 0$ and $\beta = \pi/2$. One sees that as Δv increases a smaller and smaller interval of angles becomes unstable.

Now in most parts of the unstable region the kinetic energy density of the matter $\frac{1}{2}\rho_1(\Delta v)^2$ exceeds the energy density in the magnetic field:

$$\frac{1}{2}\rho_1(\Delta v)^2 \gg \rho_1 c_s^2 = \frac{B^2}{8\pi} . \tag{27}$$

Therefore the disk material which diffuses into the magnetosphere because of the instability will lead to a strong winding-up of the field, a large azimuthal component B_ϕ will then develop. Since pressure equilibrium with the dipole field outside the interaction region has to be maintained, B_ϕ will be limited by the dipole value of B. If the field is wound up to greater strength, it will just expand in the vertical direction. One can imagine that the final result will be the formation of a layer above the unperturbed disk surface.

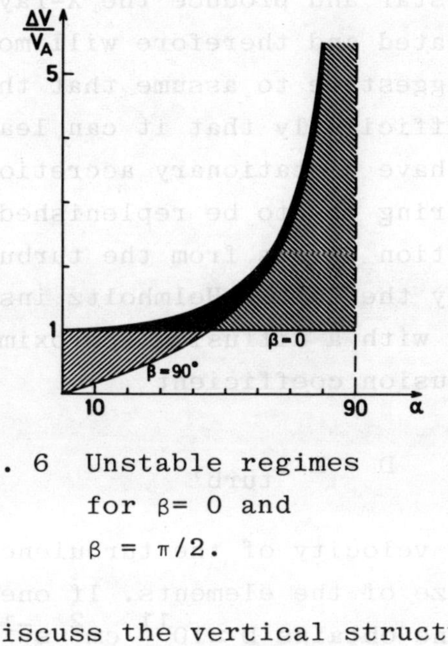

Fig. 6 Unstable regimes for β= 0 and β = π/2.

Let us now discuss the vertical structure of this layer. Underneath of it we have the unperturbed disk region, $z<z_0$; here we took $\rho=\rho_0 \exp(-z^2/2H^2)$ and $\vec{B} = 0$. At larger values of z the Kelvin-Helmholtz instability leads to an efficient mixing between disk material and the dipole field. In the innermost part of this region the kinetic energy dominates and one finds $\vec{B} \simeq \vec{B}_\phi$. Since here the magnetic field will increase with z one has an inward directed magnetic pressure gradient which gives a more rapid pressure fall-off (compared to $\exp(-z^2/2H^2)$). Here one still has $\vec{v} \simeq \vec{v}_k$. When the gas density has become small enough, at $z = z_1$, then the

magnetic field can be sufficiently strong to take up angular momentum from the gas leading to a gradual transition from the Keplerian motion to solid body rotation. Here also the field will change from an azimuthal field to the dipole configuration. The region around r_c where the instability occurs can be divided into two parts. For $r<r_c$ one has $v<v_k$, therefore a net inward force (in radial direction) will act on the gas and leads to an inward flow of the material. The gas originating from this region will fall onto the neutron star and produce the X-rays. For $r>r_c$ the plasma is accelerated and therefore will move radially outward. It is suggestive to assume that this material will be accelerated sufficiently that it can leave the system.

In order to have a stationary accretion flow the material in this ring has to be replenished by vertical motion. Such a motion arises from the turbulent mixing which can be produced by the Kelvin-Helmholtz instability. We model this mixing with a diffusion approximation by taking an effective diffusion coefficient

$$D = L\, v_{turb},$$

here v_{turb} is the velocity of the turbulence and L a characteristic size of the elements. If one takes $v_{turb} \simeq 0.1\, c_s$ and $L = 0.1H$ one obtains $D \simeq 10^{11}\ cm^2\ s^{-1}$. The diffusion velocity u is given by $u = D/\Delta z \simeq 0.1 c_s\, L/\Delta z$ where $L/\Delta z<1$ must hold.

The theoretical considerations presented above of the interaction between disk matter and the neutron star's magnetic field have some immediate consequences: Matter which gets onto the magnetic field lines exchanges angular momentum with the neutron star. A detailed quantitative model—which does not exist yet—would determine the amount δJ of angular momentum transferred to the star. Now, if the neutron star reacted as a rigid body, its own angular momentum would change corresponding to

$$I(\frac{2\pi}{P})\,(\dot{P}/P) = \delta J, \tag{28}$$

where I is the moment of inertia of the neutron star, P its period, $\dot{P}$ the time derivative of the period.

In our model the Kelvin-Helmholtz instability leads to a turbulent diffusion of disk matter onto the magnetic field lines in a region around the radius of corotation r_c. Inside r_c this matter, $\dot{M}_{in}$, accelerates the star, while the mass flow $\dot{M}_{out}$ into the magnetosphere for $r>r_c$ is responsible for the braking. For $r>r_c$ corotating matter experiences an effective outward acceleration $g_{eff} = \frac{3GM}{r_c^2}(r-r_c)$, and at a certain radius $r_{eq} = r_c\, x_{eq}$ it leaves the magnetic field lines. The braking torque depends on this radius r_{eq}, and the angular momentum exchange is

$$\delta J_b = \dot{M}_{out}\ \Omega r_c^2\ (x_{eq}^2 - 1). \tag{29}$$

The acceleration of the star is due to

$$\delta J_{acc} = \dot{M}_{in}\ \Omega r_c^2. \tag{30}$$

Here we have assumed that the matter enters the magnetosphere close to r_c. If we take $\dot{M}_{in} = \dot{M}_{out} = \dot{M}$, then the rate of angular momentum change can be written as

$$\delta J = \dot{M}\Omega r_c^2\ (2 - x_{eq}^2). \tag{31}$$

A complete balance ($\delta J = 0$) can be achieved by setting $x_{eq} = 1.4$.

If we set $\alpha = 2 - x_{eq}^2$, then the observations give spin-up time scales which are consistent with values of α around 0.05, if we take typical values for the neutron star of

$M = 1M_\odot$, $I = 10^{45}$ g cm^2 assuming $\dot{M} = 10L_x/c^2$. Such a low value of α indicates that the loss and the gain of angular momentum are well balanced in most sources.

A good fit to the data can be obtained by setting $\delta J = \alpha\Omega r_c^2 \dot{M}$ with $\alpha = 0.05$. According to Eq. (28) this gives a relation

$$\dot{P} = \alpha P^{7/3}\dot{M}\frac{(GM)^{\frac{2}{3}}}{(2\pi)^{\frac{4}{3}}\, I} . \tag{32}$$

We have plotted the observational data points, and the theoretical relation of $\dot{P}$ vs. $(P^{\frac{7}{3}}\dot{M})$ in Fig. 7.

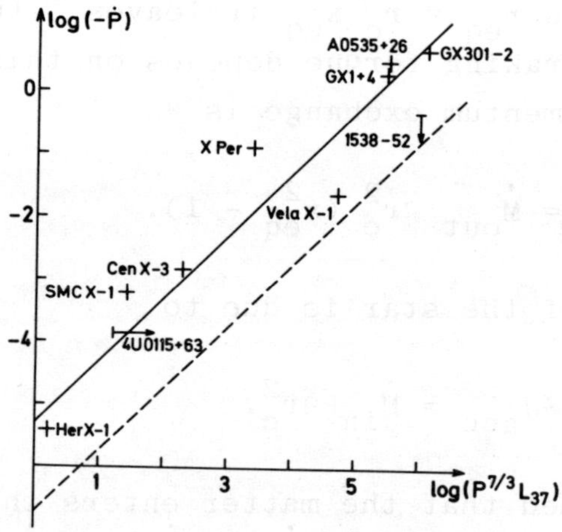

Fig. 7 Plot of $\dot{P}$ vs $(P^{7/3}L)$ for the model of Anzer and Börner, dashed line shows acceleration of a white dwarf with $r = 5 \times 10^8$ cm and $M = 1\ M_\odot$.

The data points representing the observations show a large scatter which is only partially due to the uncertainties of the observations. There is certainly also an intrinsic scatter in the parameters of the X-ray sources: The neutron stars may well have different values of M, I, $\vec{B}$, etc. These obvious uncertainties prohibit definite conclusions from the observations. To demonstrate how unconvincing these fitting procedures are in distinguishing between different models, we have considered the following very unrealistic situation: Assume that the accretion disk reaches all the way to the neutron star, and transfers only its Keplerian angular momentum at the stellar radius:

$$\delta J = \dot{M} \sqrt{GMr_*} \tag{33}$$

$$\text{i.e. } \dot{P} = \frac{\dot{M}}{2\pi I} \sqrt{GMr_*}\, p^2 . \tag{34}$$

Then a curve is obtained which does fit the data not any worse than the other models (see Fig. 8).

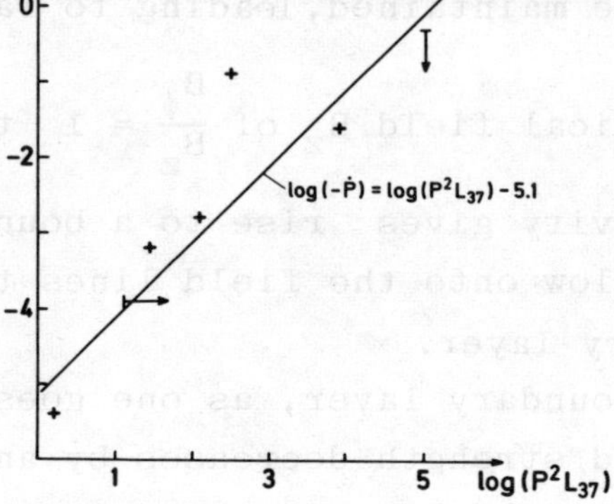

Fig. 8 Plot of $\dot{P}$ vs (P^2L) for simplified model.

The alignment of rotation and dipole axis of the neutron star leads to an axisymmetric configuration. This looks simpler at a first glance, and has therefore received considerable attention[10,11]. The details of the interaction between disk matter and stellar fields are, however, as intricate problems as in the less symmetric perpendicular case. One basic difficulty is, of course, that accretion onto the magnetic poles of the neutron star does not lead to pulsed emission in the axisymmetric case.

In this case the diamagnetic properties of the plasma would prohibit the penetration of field lines through the disk, and the field lines near the magnetic equator would bend in and follow the disk surface tangentially. It has been argued[10] however, that such a magnetosphere with strong, oppositely directed magnetic fields above and below the disk could not be maintained. Instead, rapid reconnection of field lines through the disk is advocated, introduced in the model by an effective electrical resistivity. The ratio of the reconnection speed to the Alfven velocity is adopted as 0.1 to 1. It is then assumed that a balance between shear amplification and reconnection of magnetic field lines can be maintained, leading to ratios of azimuthal field B_ϕ and vertical field B_z of $\frac{B_\phi}{B_z} = 1$ to 10. The assumed effective resistivity gives rise to a boundary layer structure. Mass flow onto the field lines takes place in this thin boundary layer.

Across the boundary layer, as one goes into the disk, the magnetic field strength decreases by an order of magnitude, the azimuthal motion changes from nearly corotational to nearly Keplerian. The accreting matter, and the magnetic torque inside the corotation radius $r_c \left(r_c = \left(\frac{GM}{\Omega^2}\right)^{\frac{1}{3}} \right)$ accelerate the stellar rotation; while the torque from the disk beyond r_c leads to braking.

Various aspects of the model have remained open to doubt and criticism:

a) The aligned rotator model cannot explain the X-ray pulsations. A regular time variation in the radiation pattern must be due to an inclination of the dipole axis to the rotation axis.

b) The reconnection speed is given a rather high value. Ghosh and Lamb[10] use an estimate based on X-type magnetic field configurations. The physical situation in the disk geometry may, however, reduce this mechanism. In X-type reconnection processes the field is dissipated rapidly when the magnetic field energy can be taken up as kinetic energy of plasma motion. In a Keplerian disk, however, the motion of the material is restrained.

c) In their paper Ghosh and Lamb[10] want to make use of the concept of small scale magnetic fields in the disk which connect to the large scale stellar field. Then convection driven by a temperature gradient should transport the magnetic fields through the disk. It is by no means clear, however, that such a convective pattern will produce the desired rapid field connection across the disk.

d) Finally, their estimates give a magnetic field in the disk which is considerably weaker (by a factor $\frac{1}{5}$) than the undisturbed dipole field. Therefore about $\frac{4}{5}$ of the magnetic flux which would pass through the disk in the unperturbed dipole case, now has to be squeezed through the hole in the disk ($r \leq r_M$). In Fig. 9 we have tried to illustrate this situation by superimposing $\frac{4}{5}$ of the field of a dipole plus disk solution[5,6,8], where the disk is taken to be ideally conducting, and $\frac{1}{5}$ of an unperturbed dipole solution. The resulting field deviates significantly from a dipole, and it generates a severe problem for the accretion flow: the field lines threading the boundary layer all extend to large radial distances before they approach the magnetic poles. Therefore a mass flow along $\vec{B}$ would at first have to flow outwards in the gravitational potential, which seems impossible.

Thus, even if we accept the general features of the construction by Ghosh and Lamb[10], we should remark that an essential problem of the accretion flow has not been solved in their model. Probably additional Kelvin-Helmholtz or Rayleigh-Taylor instabilities of the matter in the magnetic field have to be considered in the aligned rotator model. It seems to us that in this case even the question whether accretion occurs at the magnetic poles or at the equator cannot be easily decided.

In the model of Ghosh and Lamb[10] the acceleration is due to the torque of the accreting matter, and the braking occurs through a magnetic torque acting in an extended transition zone outside of r_c. This transition zone, where field lines thread the disk, but do not control the motion, must reach very far out to achieve a close balance between acceleration and braking. Ghosh and Lamb take α as a function of the period P and the magnetic field B, and they obtain a good fit to the observations with $B \simeq 5 \times 10^{11}$ gauss ($M=1.3\ M_\odot$) for all sources (Fig.10). There is a problem here, because Her X-1 has a measured value[1] of $B = 5 \times 10^{12}$ gauss. Ghosh and Lamb[10] try to reconcile their model with this measurement by invoking a strong quadrupole component near the stellar surface — which represents the measured field — and a much weaker dipole component corresponding to their theoretical prediction.

In our opinion the fits to the data are at present only consistency checks on the models. No strong conclusions about the nature of the physical processes involved are possible.

To summarize let us say that our understanding of the interaction between disks and magnetospheres is still very fragmentary. Even the two simplest configurations of the aligned and the perpendicular magnetic dipole still pose many unsolved problems. In our opinion one should try to extend the investigation of MHD instabilities (both Kelvin-Holmholtz and Rayleigh-Taylor instabilities), one also needs an understanding of the finite amplitude perturbations and their influence on enhanced diffusion rates, furthermore a model for a self-consistent global magnetic field structure is in need. A further step then would have to be to attack the problem of the oblique rotator to make the model as general as possible.

Fig.9 Magnetosphere with part of the flux penetrating the disk.

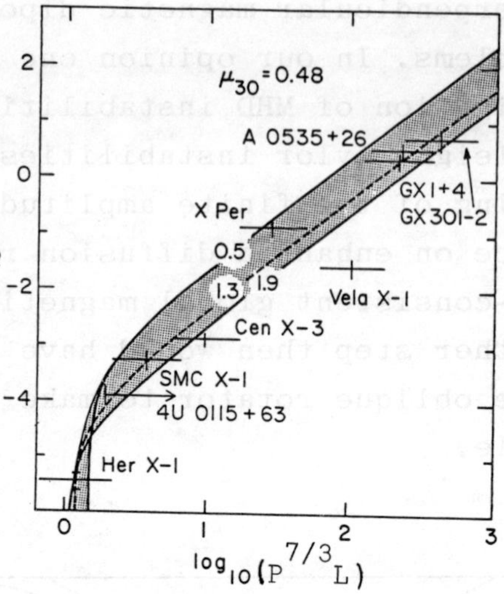

Fig.10 Plot of $\dot{P}$ vs $(P^{7/3}L)$ for the model of Ghosh and Lamb[10].

REFERENCES

[1] Trümper, J., Pietsch, W., Reppin, C., Voges, W., Staubert, R., Kendziorra, E., 1978, Ap. J., Vol. 219, L105.

[2] Anzer, U., Börner, G., 1980, Astron. Astrophys., Vol. 83, 133.

[3] Ghosh, P., Lamb, F.K., 1978, Ap. J., Vol. 223, L83.

[4] Ghosh, P., Lamb, F.K., 1979, Ap. J., Vol. 232, 259.

[5] Aly, J. J., 1980, Astron. Astrophys., Vol. 86, 192.

[6] Kundt, W., Robnik, M., 1980, Astron. Astrophys., Vol. 91, 305.

[7] Lipunov, V. M., 1978, Sov. Astron., Vol. 22, 702.

[8] Riffert, H., 1980, Astrophys. Space Sci., Vol. 71, 195.

[9] Fejer, J. A., 1964, Phys. Fluids, Vol. 7, 499.

[10] Ghosh, P., Lamb, F. K., 1979, Ap. J., Vol. 234, 296.

[11] Ghosh, P., Lamb, F. K., Pethick, C. J., 1977, Ap. J., Vol. 217, 578.

FURTHER INVESTIGATION OF RAPID X-RAY BURSTS

WANG Deyu (王德焴)
(Purple Mountain Observatory, Academia Sinica, Nanjing)

ABSTRACT

A model has been suggested to explain the observation of the rapid X-ray burster MXB1730-335. The period and the energy of the bursts are distinctly different between the initial phase and active phase. Calculation shows that the bursts with energy less than 5×10^{38} erg are caused only by the tearing mode instability; but bursts with energy more than 8×10^{38} erg are due to the tearing mode instability which in turn brings about Kruskal-Schwarzschild instability at the inner boundary of the accretion disc.

I. INTRODUCTION

The rapid X-ray burster MXB1730-335 can be divided into three sub-periods in the observational period: initial phase, active phase, and decay phase. The period and the energy of the burst are different between the initial phase and the active phase. These phenomena are divided into mode I and mode II by Marshall et al.[1](as shown in Fig. 1) The same phenomena were also observed by the X-ray astronomy satellite "Hakucho" of Japan[2]. However, there is not yet any theory to explain it.

II. THE EQUILIBRIUM AND THE INSTABILITY OF THE NEUTRON STAR'S ACCRETION DISC

When the rotation axis and the axis of dipolar magnetic field of neutron star are aligned, there appears a neutral sheet configuration in the magnetic field of the accretion disc of the neutron star. (as shown in Fig.2) The inner boundary of the accretion disc, namely the boundary between the accretion disc and the magnetosphere of neutron star, would satisfy the pressure equilibrium relation

$$[\vec{B}_n(R_m)+\vec{B}_i(R_m)]^2=8\pi(P_m+\frac{1}{2}\rho_m v_r^2), \quad (1)$$

where $\vec{B}_n(R_m)$ and $\vec{B}_i(R_m)$ are the natural magnetic field of the neutron star and induction magnetic field at the boundary equilibrium radius R_m respectively, P_m and ρ_m are the pressure and the density at R_m respectively, and v_r is the flow velocity in the radial direction. Moreover, it would satisfy the radial force equilibrium relation

$$\frac{GM_x M_m}{R_m^2}=M_m R_m \Omega_s^2+\frac{2\pi R_m}{c} I_i B_n+\frac{2\pi^2}{c^2} I_i^2 L+\frac{2\pi I_i}{c^2} I_j L_{ij}+2\pi a^2 P_m, \quad (2)$$

where a is the thickness of the accretion disc, I_i is the induced current of the accretion ring with cross section a^2 at R_m, L is the self-induction coefficient of the accretion ring, L_{ij} is the mutual-induction coefficient between the ring and other part of the accretion disc and Ω_s is the rotation frequency of the accretion disc.

The third and fourth term in Eq. (2) will be negligible when $R_m >> a$ and $B_n(R_m) > B_i(R_m)$. Therefore, the equilibrium radius R_m and induced-current I_i can be calculated from Eqs. (1) and (2), and $v_\phi = R_m \Omega_s$, while the M_x, μ, v_ϕ, P_m and ρ_m are known.

The tearing mode instability can take place in the neutral sheet of the accretion disc. The growth rates of the tearing mode instability can be expressed as

$$p_a \simeq \left(\frac{2}{3}\right)^{4/5} \tilde{\eta}^{3/5} \tilde{\rho}^{-1/5} (2)^{2/5} \alpha^{-2/5} S^{2/5} (F')^{2/5} \quad (3)$$

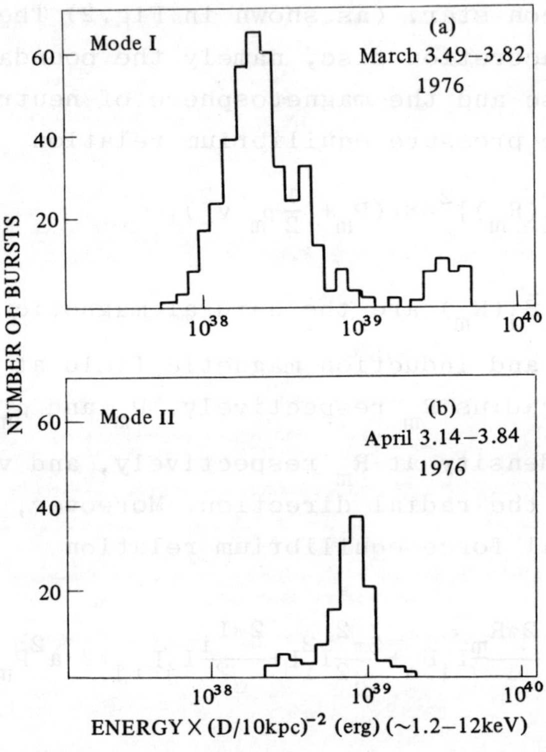

Fig. 1 Distribution of Burst energies from MXB 1730-335

and the thickness of the resistant layer in the accretion disc as

$$\varepsilon_a \simeq \frac{1}{3}\,\frac{a\tilde{n}}{\alpha p_a}\,. \quad (4)$$

The meaning of the symbols is as shown in [3]. When the tearing mode instability takes place in the accretion disc,

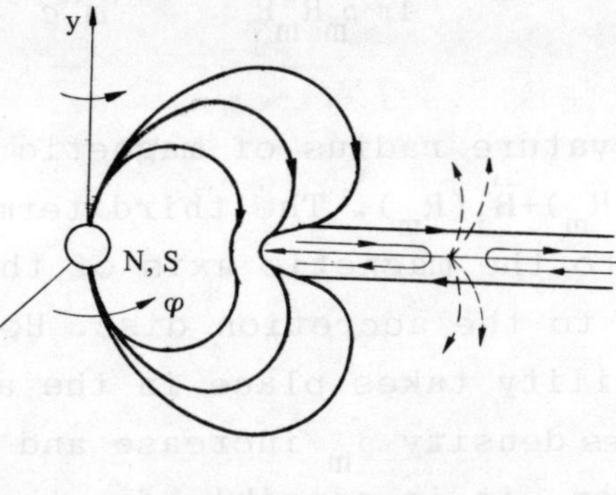

Fig. 2 Magnetic field profile in the accretion disc of neutron star.

and the magnetic field line reconnection appears in the tearing place, it not only causes a part of plasmas to tear out the neutral sheet, but will also make the resistant layer in the neutral sheet accelerate and heat. (as shown in Fig. 2). The numerical calculation was completed by Schnack and Killeen [4]. It is shown that the magnetic energy which is released by tearing mode instability transforms about 10% into the kinetic energy of plasma and the other into the thermal energy of plasma. It was shown by Yang and Sonnerup [5], that an additional radial velocity v_r in the resistant layer can be given while magnetic field line reconnection occurs. The value of v_r is about $\rho_m v_r^2 = \rho_t (2.3)^2 v_A^2$, where ρ_t and v_A are the density and Alfven velocity at the tearing place, respectively.

The inner boundary of accretion disc would be stable for Kruskal-Schwarzschild instability, while the tearing mode instability does not take place and $v_r = 0$. The criterion of K-S instability is

$$g_{eff} = \frac{GM_x}{K_m^2} - \Omega_s^2 R_m - \frac{(\vec{R}\cdot\vec{\mu})^2}{4\pi \rho_m R_m^6 R} - \frac{B^2(R_m)}{4\pi \rho_m R_c} > 0, \qquad (5)$$

where R_c is the curvature radius of magnetic field line at R_m, and $\vec{B}(R_m) = \vec{B}_n(R_m)+\vec{B}_i(R_m)$. The third term of Eq.(5) is equal to zero, due to the magnetic axis of the neutron star being perpendicular to the accretion disc. However, when the tearing mode instability takes place in the accretion disc, then $v_r \neq 0$. It makes density ρ_m increase and equilibrium radius decrease. Then it is possible for the K-S instability to occur throwing out some of the particles of the inner boundary to the magnetic pole of the neutron star, and thus a strong X-ray burst occurs.

Another important factor is the pressure P_m, which is related to the temperature of accretion disc. In the case of the initial phase of burst, the accretion disc is lower in temperature ($\sim 3\times10^8$ K). The equilibrium radius R_m is larger. It is not assured to activate the K-S instability after every tearing mode instability. Perhaps, the K-S instability may take place only when more bursts accumulate. Therefore, in the initial phase, bursts may be divided into two types. One type is caused only by the tearing mode instability, and in this case fewer particles inject and smaller burst energy is produced. Another type is causred by the K-S instability which is activated by the tearing mode instability, and in this case more particles inject and larger burst energies are produced. However, the accretion disc has been heated to high temperature state ($\sim 5\times10^8$ K) due both to the reconnection of magnetic field lines and to the X-ray emitted from the neutron star, while the rapid burst source is in the stage of active phase. It makes the inner boundary equilibrium radius of the accretion disc less than that in the initial phase. Therefore, it is easy to activate the K-S instability when the tearing mode instability takes place.

III CALCULATION AND CONCLUSION

According to Eqs. (1)-(5), the equilibrium radius of accretion disc, the growth times of tearing mode instability, the particle number N_t flowing into the surface of neutron star, and the X-ray energy E_{xt} of the burst can be calculated. It is also possible to calculate whether or K-S instability takes place not, and the particle number N_R and the burst energy E_{xK} after K-S instability occurs. The principal differences between initial phase and active phase of bursts are the temperature of the accretion disc and the equilibrium radius. The calculated results are given in the following table:

	Mx	μ	v_ϕ (cm s^{-1})	ρ_m (g cm^{-3})	T(K)	R_m(cm)
Initial Phase	$1M_\odot$	3×10^{30}	1.95×10^9	$(5-8)\times10^{-3}$	$\sim3\times10^8$	3.5×10^7
Active phase	$1M_\odot$	3×10^{30}	2.04×10^9	$(5-8)\times10^{-3}$	$\sim5\times10^8$	3.2×10^7

a (cm)	$<\eta>/c^2$(s)	T_t(s)	N_t(g)	E_{xt}(erg)	R_c(cm)
$(3-5)\times10^6$	$10^{-9}-10^{-8}$	<1	$(1-5)\times10^{18}$	$(1-5)\times10^{38}$	$\sim1.5\times10^7$
$(3-5)\times10^6$	$10^{-9}-10^{-8}$	<1	$(1-5)\times10^{18}$	$(1-5)\times10^{38}$	$\simeq1.8\times10^7$

V_r (cm s^{-1})	R'_m (cm)	R'_c (cm)	g_{eff}	N_K(g)	E_{XK}(erg)
2.5×10^8	$\sim3.4\times10^7$	$\sim1.6\times10^7$	<0		
2.7×10^8	$\sim3.1\times10^7$	$\sim1.9\times10^7$	>0	$(0.8-4)\times10^{19}$	$(0.8-4)\times10^{39}$

where R'_m and R'_c are the equilibrium radius and curvature radius while the first tearing mode instability has been taken place, respectively. The calculation shows that the bursts with energy less than $\sim 5 \times 10^{38}$ erg are caused only by the tearing mode instability. But the bursts with energy more than $\sim 8 \times 10^{38}$ erg are due to the tearing mode instability which, in turn, brings about the K-S instability at the inner boundary of the accretion disc.

REFERENCES

[1] Marshall, H.L. et al., 1979, Ap. J., Vol. 227, 555.

[2] Inoue, H. et al., 1980, Nature, Vol. 283, 358.

[3] Wang Deyu, Qu Qinyue, 1981, Scientia Sinica, Vol. 24, 1400.

[4] Schnack, D., Killeen, J., 1977, UCRL-79261.

[5] Yang, C.K., Sonnerup, B.U.ö., Ap. J., Vol. 206, 570.

[6] Wang Deyu et al., 1981, Acta Astronomica Sinica, Vol. 22, 364.

VI. BLACK HOLES

Freihoffer
The topic, I am going to talk about today, is MODERN PINHEADLOGY

BLACK HOLES AND SPHERICALLY SYMMETRIC ACCRETION

D. Freihoffer
(Max-Planck-Institut für Physik und Astrophysik,
Institut für Astrophysik, München, Federal
Republic of Germany)

Black holes are one of the most obscure and horrible things in astrophysics. They are not even things in an ordinary sense: Being merely centers of gravitational attraction, they possess no material surface; they consist only of empty space and time twisted in a strange manner, with a singularity in the middle. This space-time singularity is so singular that even physicists have been till now completely unable to describe it. But as this central mystery is hidden behind a so-called 'horizon', physicists are usually not too disturbed about it. The horizon has only a geometrical meaning: Everything, light included, that crosses it, is condemned to fall into the singularity. Nothing that happens beyond the horizon can influence the outer world; nobody can see the interior of a black hole. Black holes are thus like incorporeal monsters, insatiable and invisible, swallowing everything that has the misfortune to approach too near to them. Perhaps still more frightening is the fact that every star which a cruel fate has equipped with more than 3 solar masses may become a black hole at the end of its evolution. Thus the universe may be crowded with black holes, and perhaps everything that exists is condemned to find its ultimate grave in one of these monsters[1].

Possibly you are inclined to think now that black holes only originated in the dark fantasy of certain astronomers, who, instead of doing serious astrophysics, wanted to create a symbol for their pessimistic world view. But unfortunately there are very profound reasons why black holes exist - in theory. Actually one can show under rather general assumptions that space-time singularities are inevitable. Many calculations have been performed — what happens if a black hole is member of a binary system[2]? — what happens if two black holes collide? what about black holes and gravitational waves, charged and rotating black holes? — and so on [3,4]. People have also studied thermodynamics and quantum mechanics of black holes[5-9] and find that black holes with a mass of 2×10^{32} g should evaporate within 8×10^{70} years. Thus, theoretically, things have become rather clear. But astronomers may be reminded to a problem which was intensively discussed in Europe during the end of the Middle Ages: How many angels can sit on a pinhead? The medieval philosophers found the number 17, but just as is the case with black holes, they were unable to perform any experiments. So it is natural that we ask the question:

Are there any reasons to believe that black holes really also exist in reality, not only in theory?

It is a great pity for theoretical physicists that nobody is quite sure about the answer. On the other hand, even experimental astronomers often have a remarkable interest in black holes: Black holes play an important part in astrophysics despite the fact that there is no absolutely compelling proof for their existence. But if they exist they may be the clue to the explanation of some of the most spectacular phenomena which have been detected in recent decades[10,11].

Thus, in part 1, let us have a very short and by no means complete look at how black holes can manifest themselves, if they are not to be only ghosts wandering in astronomy books.

In the second part I will then discuss a special problem concerning the interaction of black holes with their environment: I will speak about matter falling spherically symmetrically into a black hole, i.e. about spherically symmetric accretion.

PART 1

The man who in 1795 considered for the first time the possibility of black holes — though he didn't call them by that name —, was the French astronomer Laplace. He argued that if a star with the same density as the earth were to become larger and larger, it "would not, in consequence of its attraction, allow any of its rays to arrive at us; it is, therefore, possible that the largest luminous bodies in the Universe may, through this cause, be invisible"[12].

General relativity shows that such heavy stars could not be stable. If their masses are higher than some critical mass, the value of which is about 3 solar masses, they cannot find a stable configuration at the end of their evolution and have to collapse[13]. There exists no pressure which could prevent this catastrophe, because any pressure is connected with energy and energy itself has gravity. The result is the formation of a black hole. So if one wants to find black holes it is natural to look for (spectroscopic) binary systems, which have optically invisible components with masses larger than the critical value. In this way Cyg X-1 came under suspicion of being a black hole[14]. Cyg X-1 is a binary system. The X-ray source, which was detected in 1966, shows no regular time variability. The optical counterpart was detected five years later. The hypothesis that Cyg X-1 is a black hole is based on the analysis of the binary elements, which gives a lower mass of $M = 5\ M_\odot$ for the compact object. An ordinary star with this mass would be quite visible in the optical and quite invisible in the X-ray spectrum. You may wonder why the black hole now suddenly should be visible, and even in the X-ray spectrum. I will explain this in a few moments.

There are also some hints that supermassive black holes, i.e. black holes with 10^6 solar masses or more, exist[15,16]. Oort[17] showed that there is some evidence to think there may be such an object with about $4\times10^6\ M_\odot$ in the center of our own galaxy. Of course this center is optically quite invisible. But one can analyse the velocities of those stars which are nearest to it and are still observable. Because of their high velocities, as measured by their Doppler-shifts, one concludes that the mass in the center of the galaxy is higher than would be inferred by measuring the luminosity of this region. The difference could be due to a black hole.

There is a large number of other situations in which black holes could be engaged: Active galaxies, quasars, and radio galaxies for example[18,19]. Black holes could also have played (and still be playing) a major part in cosmology, in the formation and evolution of galaxies[20,21]. This is because there could be primeval black holes which were already formed at the beginning of the universe by perturbations of the initial density distribution[22].

It is quite obvious that, if black holes really do exist, they only can be observed very indirectly. (We disregard here the possibility of detecting them by gravitational waves, which should be emitted by a collapsing star.)

The only way black holes can interact with their environment is by gravitational attraction. If the black hole is a member of a binary system, matter can flow from the normal star into the black hole, either because the normal star fills its Roche lobe or because it loses matter through a stellar wind. If a black hole is embedded in a gaseous cloud, it will tend to absorb the gas gradually. Supermassive black holes in the center of galaxies can even eat entire stars without chewing them before.

Now if matter falls into black holes, gravitational energy will be transformed into internal energy of the infalling matter, either by compressional work or by turbulence and other dissipative processes. Part of this energy will be radiated away before the matter crosses the horizon and the black hole will become 'visible'. What we really see is of course the infalling matter.

Let us see how this could work for a quasar. If quasars really have cosmological distances, as is inferred from their high redshifts, one runs into the problem of looking for a mechanism which can deliver up to 10^{47} erg s^{-1} and which is not larger than a few light-hours or light-days. The source must be so small, since its luminosity can vary within a few hours or days. Of course, a high mass concentration is needed to produce such amounts of energy. The typical length scale of a black hole is the Schwarzschild horizon $r_s = 2GM/c^2 = 3\times10^5\ M/M_\odot$ cm, where M is mass of the black hole. The significance of this horizon was already mentioned in the introduction. Let us assume the region where the infalling matter produces the bulk of radiation lies at $5\ r_s$ [18]. We then have

$$5\ r_s \equiv 5\times3\times10^5\ \frac{M}{M_\odot} = \text{a few light-days} = 10^{16}\ \text{cm}$$

or

$$M = 10^9 - 10^{10}\ M_\odot .$$

On the other hand, the luminosity L produced by accretion is limited (at least in the spherically symmetric case) by the Eddington-luminosity

$$L_E = \frac{4\pi G M m_p c}{\sigma_{Th}} = 1.26\times10^{38} \frac{M}{M_\odot} \text{ erg s}^{-1} \tag{1}$$

(where G is the gravitational constant, M is the mass of the accreting object, m_p is the mass of the proton, $c = 3\times10^{10}$ cm s^{-1} and $\sigma_{Th} = 6.625\times10^{-25}$ cm^2), if Thomson scattering is the dominant opacity.

Inserting 4×10^{47} erg s^{-1} (which is the optical luminosity of 3C 273) we again have

$$M = 10^9 - 10^{10}\ M_\odot .$$

The amount of matter which has to be accreted to produce this luminosity is given by

$$L = \eta \dot{M} c^2, \tag{2}$$

where $\dot{M}$ is the accretion rate, i.e. the amount of matter falling into the black hole per second, and η is the unkown efficiency of conversion of gravitational energy into radiation.

Putting η = 10 per cent, we arrive

$$\dot{M} = 10 - 20\ M_\odot \text{ per year,}$$

a very reasonable value.

This black hole model is of course a very rough picture, but anyway it demonstrates that in principle a black hole with 10^9 solar masses, which eats a few solar masses per year, can both supply the energy required and be consistent with the observed time variability of quasars. Of course one still has to work out the model in detail to explain the spectral features, which probably arise further away from the black hole.

Does matter, which is going to fall into a black hole but still has not crossed the horizon, have any chance of avoiding its fate? Matter always has internal energy, and any form of internal energy corresponds to a pressure which works against gravitation. Also the luminosity coming from the inner parts of the flow provides a radiation pressure which opposes the infalling matter. If, in addition, matter possesses angular momentum, it is forced to spiral around the black hole and must lose its angular momentum, before it can sink into it. In this case a disk develops and the accretion rate depends on the efficiency of angular momentum transport[2]. Within 3 Schwarzschild radii, no stable circular orbits are possible.

If the total angular momentum per gramm of accreted matter is less than $\sqrt{3}\ (\frac{26M}{c^2})\ c$, the accretion can be spherical[23]. Spherical accretion also could take place, if the inner part of a disk is supported by radiation.

As spherically symmetric accretion is the subject of part 2, let me finish here with the overview of part 1 and let us come to

PART 2,

where we will have a closer look at spherical accretion. Consider a black hole embedded in a fully ionized hydrogen plasma. We assume that the black hole does not rotate and has no electric charge. What will happen? Matter will begin to fall into the black hole, certainly. But there are also some processes which oppose the infall. As we deal with spherical symmetric accretion, we can neglect angular momentum. Let us also neglect magnetic fields, which could be frozen into the matter, as well as turbulence[24,25]. Then the only way matter can fight against its downfall is by raising its thermal energy, which works against compression and by exploiting the radiation, which exerts a pressure on the matter.

Before going into further details, I want to make some basic remarks about accretion:

a) The first investigations concerning accretion onto normal stars were motivated by the hope that accretion could augment the luminosity produced inside stars by nuclear fuel[26]. But since the gravitational potential at the surface of normal stars is only of the order $GM/r \simeq 10^{-6} c^2$ the energy due to accretion of matter is usually very small. In contrast to this, the figure is $GM/r \gtrsim 0.1 c^2$ for neutron stars and black holes, so we can expect much higher energies. The maximum luminosity, which can be attained by spherical accretion, is given for Thomson scattering opacity by the Eddington luminosity Eq. (1). When $L = L_E$, gravitation is just counterbalanced by radiation pressure.

b) If all the potential energy of matter falling from infinity into a black hole could be entirely converted into radiation, we would have

$$L = \dot{M} c^2, \tag{3}$$

(where $\dot{M}$ is the accretion rate). But since the efficiency by which gravitational energy is transformed into radiation is less than 1, we arrive at the necessary condition

$$\dot{M} \gg \dot{M}_{CR} \equiv L_E / c^2 \tag{4}$$

for the luminosity to approach the Eddington luminosity L_E.

If this condition is fulfilled, the accretion is called supercritical, otherwise subcritical. Substantial accretion was studied by Shapiro[27,28] and others.

In the following only supercritical accretion will be discussed.

c) The condition $\dot{M}>>\dot{M}_{CR}$ has two consequences. First, high plasma densities, say $10^4 - 10^8$ particules per cm^3, are needed far away from the black hole, in order to ensure a supercritical accretion. Second, if $\dot{M}>>\dot{M}_{CR}$, a radius r_0 exists where matter becomes optically thick for Thomson scattering. This is given approximatley by[29]

$$\left(\frac{r_0}{r_s}\right)= \frac{(\dot{M}/M_{CR})^2}{4(1-L/L_E)} . \tag{5}$$

For $r<r_0$ matter and radiation can come into local thermodynamic equilibrium (LTE) (As LTE cannot be established solely by Thomson scattering, which exchanges no energy between matter and radiation, a better formula for r_0 can be obtained by using an effective optical depth[30][31]).

If supercritical accretion into a black hole could in fact produce luminosities near L_E, the bulk of the radiation would naturally have to emerge from the LTE-region $r<r_0$. To study this region, one has to use a general relativistic theory. The hydrodynamical equations are derived by contracting the stress-energy tensor

$$T^{ab} = (\varepsilon+p)u^a u^b+pg_{ab}+u^a q^b+u^b q^a \tag{6}$$

where ε and p are the total (matter and radiation) energy-density and pressure respectively, (u^a) is the four-velocity of matter, (g_{ab}) the Schwarzschild metric tensor and (q^b) the Poynting four-vector of energy flux[11][32].

By putting $T^{ab}_{\ ;b} = 0$ and projecting the equation

onto the four-velocity of matter and the three-space of matter respectively, one obtains two equations for energy and momentum conservation. Since we assume the matter flows steadily, all partial time derivatives vanish, so all functions simply depend on the radial Schwarzschild co-ordinate r. One thus obtains

$$\frac{\varepsilon+p}{2}\frac{d}{dr}(u^2-\frac{1}{r}) = -\Gamma^2\frac{dp}{dr}+\frac{\Gamma\rho}{\dot{M}}\frac{d}{dr}(u^2L) \qquad (7)$$

(momentum conservation) and

$$u(\frac{d\varepsilon}{dr}-\frac{p+\varepsilon}{\rho}\frac{d\rho}{dr}) = \frac{1}{4\pi r^2\Gamma}\frac{d}{dr}(L\Gamma^2) \qquad (8)$$

(energy balance), where we have used L_E, r_s and c as units. ρ is the density of matter and $\Gamma^2 = 1-\frac{1}{r}+u^2$. u is the radial part of the four-velocity (u^a).

Radiation transport is described by the equation

$$\frac{L}{4\pi r^2}+\frac{1}{2\pi\dot{M}}\frac{d}{dr}(u^2L) = -\frac{4}{6\pi\rho}(\frac{U}{2\Gamma r^2}+\frac{U}{2\Gamma}\frac{d}{dr}(u^2)+\frac{\Gamma}{4}\frac{d}{dr}U) \qquad (9)$$

where U is the energy density of radiation. This can either be derived by using the relativistic Fourier equation[33] or by using a moment method and neglecting the non-LTE terms[34].

The last equation is the law of baryon conservation[11]

$$4\pi\rho r^2 u = \dot{M}. \qquad (10)$$

Now the problem of spherically, symmetric, time independent accretion can be formulated in the following way:

Given the density and temperature of hydrogen plasma at the accretion radius far away from the black hole (i.e. at the radius where the thermal velocity of matter equals

its free-fall velocity), what is
a) the accretion rate $\dot{M}$? and
b) the luminosity L produced in the optically thick part of the flow?

The first question can easily be answered as shown by Bondi[35]. At the accretion radius r_A, where the gas is still non-relativistic, matter crosses its sonic point and becomes supersonic. (We can assume that for $r > r_0$ matter falls isothermally[24], because the energy loss due to bremsstrahlung is compensated by the radiation of nearby hot stars, which also keeps the plasma ionized at $r < r_A$.) Now, as Bondi showed, if we demand matter to flow steadily through its sonic point, we get a condition which fixes the accretion rate $\dot{M}$. Mathematically this condition is due to the fact that the hydrodynamic equations become singular at the sonic point. They can only be solved if an inner boundary condition is fulfilled. The value one obtains for $\dot{M}$ is approximately given by the equation of continuity

$$\dot{M} = 4\pi\rho_A v_A r_A^2, \tag{11}$$

where ρ_A is the density of matter and v_A the isothermal sound speed at r_A.

The second question, concerning the value of the luminosity L, arises only for stationary accretion into a black hole. If matter accretes onto a normal star with a hard surface, one always has a definite relation between the total luminosity and the accretion rate, because one has to require that the surface of the star be not heated by the infalling matter. So the matter has to radiate away all of its potential energy, either during the infall or after the impact on the surface, whereby one gets

$$L = \frac{G\dot{M}M}{r}, \tag{12}$$

where M is the star's mass and r the radius of the star. For black holes it is not obvious whether a corresponding relation also exists, since matter can drag energy beyond the Schwarzschild horizon.

What then will determine the luminosity? We are faced with a second difference between black holes and other compact stars: whereas matter can land with an arbitrary velocity on the surface of a normal star, matter has to cross the Schwarzschild horizon with a speed β which is larger than any propagation velocity of energy within the matter. 'Speed of matter' here means the value, which a stationary observer, i.e. one at rest with respect to the black hole, would assign to the flow.

As regards to our problem, there are two velocities of energy propagation within the matter. The first is that of ordinary sound waves and is connected with the thermal energy of matter. The second is due to the diffusion of photons, and its value is approximately given by $\beta_D = 1/\sqrt{3}$, which is the same as the ultra-relativistic sound speed. (Remark: If we were to use the equation of Eckart[36] as a general relativistic version of the Fourier equation, we would have an infinite value for β_D. This happens because Eckart's equation leads to a parabolic equation of heat conduction, whereas the equation we used leads to hyperbolic equation with $\beta_D \simeq 1/\sqrt{3}$.[33])

Now in the same way that the accretion rate $\dot{M}$ was determined by the requirement that matter flows steadily through the sonic point r_A, where $\beta = v_A$, the luminosity is determined at the radius where β equals β_D.

At both points the system (7,8,9), which can be written in the form

$$A \frac{d}{dr} \begin{pmatrix} u \\ T \\ L \end{pmatrix} = b$$

becomes singular, i.e. the determinant det A of the matrix

A vanishes. To solve the system there, one has to demand det A* = 0, where the matrix A* is obtained from A by replacing one column by the vector b. So we get two additional conditions, one fixing the accretion rate $\dot{M}$, the other the luminosity L.

The rest is only numerics. (For details, see [32].) Fig. 1 shows the results for $1 - 10^3$ solar mass black holes and various accretion rates $\dot{M}$. The temperature at $r = r_A$ is assumed to 10^4 K; the matter densities at r_A can easily be calculated by using Eq. (11). In all models the luminosity L produced in the optical thick part of the flow is much

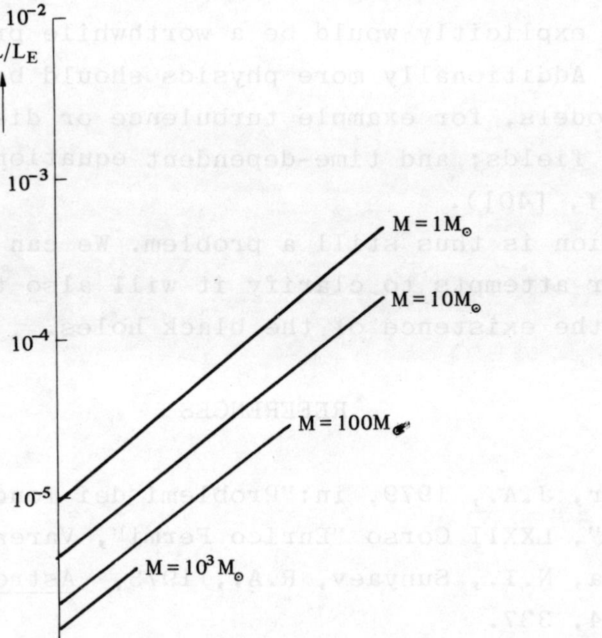

Fig. 1 Luminosity as function of the accretion rate. (M: Mass of the black hole.)

smaller than Eddington luminosity. For M = 1 and $n = 10^8$

particles cm^{-3} at $r = r_A$ one obtains a model which has already been studied in Tamazawa et al.[37]. Although their method of calculation differs in some aspects from that we used, the results coincide. Other calculations of optically thick accretion have been performed by Kafka and Mészáros[38] and also by Begelman[39]. These authors used non-relativistic equations and therefore could not find definite results.

For several reasons the procedure we have used in our calculations cannot be extended to supermassive black holes ($M > 10^6\ M_\odot$). This is mainly because the transition region between the optically thick outer part of the flow and the inner part can no longer be neglected. To study this region more explicitly would be a worthwhile project for the future. Additionally more physics should be incorporated in the models, for example turbulence or dissipation of magnetic fields; and time-dependent equations should be used (c.f. [40]).

Accretion is thus still a problem. We can only hope that further attempts to clarify it will also throw some light onto the existence of the black holes.

REFERENCES

[1] Wheeler, J.A., 1979, in:"Problemi dei fondamenti della fisica", LXXII Corso "Enrico Fermi", Varenna, 396.

[2] Shakura, N.I., Sunyaev, R.A., 1973, Astron. Astrophys., Vol. 24, 337.

[3] Eardley, D.M., Press, W.H., 1975, Ann. Rev. Astron. Astrophys., Vol. 13, 381.

[4] Rees, M., Ruffini, R., Wheeler, J. A., 1974, "Black Holes, Gravitational Waves and Cosmology: An Introduction to Current Research". Gordon and Breach Science Publishers, New York.

[5] Bardeen, J.M., Carter, B., Hawking, S.W., 1973, Commun.

Math. Phys., Vol. 31, 161.
[6] Hawking, S.W., 1977, Sci. Am., 236.
[7] Hawking, S.W., 1975, Commun. Math. Phys., Vol. 43, 199.
[8] Bekenstein, J., 1973, Phys. Rev., D7, 2333.
[9] Davies, P.C.W., 1978, Rep. Prog. Phys., Vol. 41, 1313.
[10] Avrett, E.H. (ed.), 1976, "Frontiers of Astrophysics". Havard Univ. Press, Cambridge, Massachusetts, London.
[11] Misner, C.W., Thorne, K.S., Wheeler, J.A., 1973, "Gravitation". San Francisco, Cal.
[12] Laplace, P.S., 1796, "Expositions des Systéme du Monde", Vol. 2, Paris (the relevant page is reproduced photographically in [11]).
[13] Nauenberg, M., Chaplin Jr., G., 1973, Ap.J., 179, 277.
[14] Giacconi, R., 1973, Phys. Today, Vol. 26, No. 5, 38.
[15] Young, P., Westphal, J. A., Kristian, J., Wilson, C. P., Landauer, F. P., 1978, Ap. J., Vol. 221, 721.
[16] Sargent, W.L.W., Young, P.J., Boksenberg, A., Shortridge, K., Lynds, C.R., Hartwick, F.D.A., 1978, Ap. J., Vol. 221, 731.
[17] Oort, J. H., 1977, Ann. Rev. Astron. Astrophys., Vol. 15, 295.
[18] Strittmatter, P.A., 1979, In: "Extragalactic High Energy Astrophysics". 9th Advanced Course Swiss Society of Astronomy and Astrophysics; Bléchá, A., Maeder, A.(eds.), SAAS-FEE, 153.
[19] Lynden-Bell, D., 1969, Nature, Vol. 223, 690.
[20] Carr., B. J., 1977, M.N.R.A.S., Vol. 181, 293.
[21] Mészáros, P., 1975, Astron. Astrophys., Vol. 38, 5.
[22] Zel'dovich, Ya, B., Novikov, I.D., "Relativistic Astrophysics". Vol. I (1971), Vol. II (1972). University of Chicago Press, Chicago.
[23] Illarionov, A.F., Sunyaev, R.A., 1975, Astron.

Astrophys., Vol. 39, 185.

[24] Mészáros, P., 1975, Astron. Astrophys., Vol.44, 59.

[25] Mészáros, P., 1975, Nature, Vol. 258, 589.

[26] Rees, M.J., 1974, in: "Gravitation and Radiation and Gravitational Collapse", IAU Symp. No. 64.

[27] Shapiro, S.L., 1973, Ap. J., Vol. 180, 531.

[28] Shapiro, S.L., 1973, Ap. J., Vol. 185, 69.

[29] Kafka, P., Meszaros, P., 1976, Gen. Relativ. Gravitation, Vol. 7, No. 11, 841.

[30] Felton, J.E., Rees, M.J., 1972, Astron. Astrophys., Vol. 17, 226.

[31] Vitello, P.A.J., 1978, Ap. J., Vol. 225, 694.

[32] Freihoffer, D., 1981, Astron. Astrophys., Vol. 100, 178.

[33] Freihoffer, D., 1980, Gen. Relativ. Gravitation, Vol. 12, No. 6.

[34] Thorne, K. S., Flamming, R.A., Zytkow, A.N., 1981, M.N.R.A.S., Vol. 194, 475.

[35] Bondi, H., 1952, M.N.R.A.S., Vol. 112, 195.

[36] Eckart, C., 1940, Phys.Rev., Vol. 58, 919.

[37] Tamazawa, S., Kiyotaka, T., Noboru, K., Yôrô,Ô., 1975, Astrophys. Space Sci., Vol. 32, 403.

[38] Kafka, P., Mészáros, P., 1976, Gen. Relativ.Gravitation, Vol. 7, 841.

[39] Begelman, M.C., 1979, M.N.R.A.S., Vol. 187, 237.

[40] Gilden, D.L., Wheeler, J.A., 1980, A.J., Vol. 239, 705.

THE HAWKING EVAPORATION OF DIRAC PARTICLES IN GENERAL KERR-NEWMAN BLACK HOLE BACKGROUND

XU Chongming(须重明), SHEN Yugung(沈有根)
(Department of Physics, Fudan University, Shanghai)

In 1974, Hawking[1] proposed that there might be some quantum processes, in which the irreducible mass of a black hole is radiated away. Such a quantum vacuum polarization process is named Hawking evaporation.

In 1976, Damour and Ruffini[2] discussed the Hawking evaporation of Klein-Gordon particles background Schwarzschild and Kerr metric through Klein-Souter-Heisenberg-Euler formalism. Recently, Liu et al.[3,4] have found the solution of the Dirac equations just outside the horizon for quasi-extreme Kerr and Kerr-Newman black holes by means of the decoupled Dirac equations with non-zero mass derived by Chandrasekhar[5] and Page.[6] They also derived the corresponding Hawking thermal spectrum formula.

However, it is unnecessary to impose quasi-extreme conditions on the black holes for solving the equations. In this paper, we discuss the general Kerr-Newman black hole directly.

The decoupled and separated Dirac equations background Kerr-Newman metric first were derived by Page.[4,6] The radial equation is

$$\sqrt{\Delta}\frac{d}{dr}\sqrt{\Delta}\frac{dR}{dr} - \frac{i\mu\Delta}{\lambda+i\mu r}\frac{dR}{dr} + \left[\frac{K^2+i(r-M)K}{\Delta} - 2i\omega r+ieQ - \frac{\mu K}{\lambda+i\mu r} - \mu^2 r^2 - \lambda^2\right]R = 0, \quad (1)$$

where R is the radial wave function

and

$$\Delta = r^2 + a^2 + Q^2 - 2Mr; \ K = (r^2 + a^2)\ \omega - am - eQr,$$

where M, Q and a are the mass, the electric charge and the angular momentum per unit mass of the black hole (B.H.), respectively; m is the magnetic quantum number; μ, e and ω are the mass, the electric charge and frequency of the Dirac particle, respectively; λ is the eigenvalue of the angular Dirac equation.

Eq. (1) is solved at the outer horizon. Tortoise coordinate is introduced

$$\hat{r} = r + \frac{Mr_+ - Q^2/2}{\sqrt{M^2-a^2-Q^2}} \ln(r-r_+) - \frac{Mr_- - Q^2/2}{\sqrt{M^2-a^2-Q^2}} \ln(r - r_-). \quad (2)$$

At r_+, there are $\hat{r} \sim \frac{Mr_+ - Q^2/2}{\sqrt{M^2-a^2-Q^2}} \ln(r-r_+)$, where r_+ and r_- are the radius of the outer horizon and inner horizon, respectively. Multiplying (1) by $\frac{\Delta}{(r^2+a^2)^2}$, let $r \to r_+$, then Eq. (1) at r_+ becomes

$$\frac{d^2R}{d\hat{r}^2} - \eta \frac{dR}{d\hat{r}} + \left[\frac{K_+^2}{(r_+^2 + a^2)^2} + \frac{i\eta K_+}{(r_+^2+a^2)}\right] R = 0, \quad (3)$$

where K_+ is equal to K, when r_+ is substituted into K. Therefore,

$$K_+ = (r_+^2+a^2)\left(\omega - \frac{am+eQr_+}{r_+^2+a^2}\right) = (r_+^2+a^2)(\omega - \omega_0), \quad (4)$$

where ω_0 is defined as

$$\omega_0 = m\,\Omega + e\,V;$$

$$\Omega = \frac{a}{r_+^2+a^2} = \frac{a}{2M^2+2M\sqrt{M^2-a^2-Q^2}-Q^2} \tag{5}$$

is the angular velocity of the outer horizon;

$$V = \frac{Q\,r_+}{r_+^2+a^2} = \frac{Q(M+\sqrt{M^2-a^2-Q^2})}{2M^2+2M\sqrt{M^2-a^2-Q^2}-Q^2}$$

is the potential of the outer horizon at $\theta = 0$. In Eq. (3) η is the gravitational acceleration at r_+

$$\eta = \frac{1}{2}\left(\frac{r_+-r_-}{r_+^2+a^2}\right) = \frac{\sqrt{M^2-a^2-Q^2}}{2Mr_+ - Q^2}\,. \tag{6}$$

Eq. (3) is a second-order constant coefficient differential equation. The general solution is

$$R = C_1 e^{(\eta-i(\omega-\omega_0))\hat{r}} + C_2\, e^{i(\omega-\omega_0)\hat{r}}. \tag{7}$$

Only the outgoing wave of the situation $\omega>\omega_0$ is considered. (Actually, the situation $\omega<\omega_0$ corresponds to the another kind of black hole radiation, i.e. the so-called Starobinsky-Unruh radiation, which is not considered in this paper.) So $C_1 = 0$, we obtain

$$R \simeq e^{i(\omega-\omega_0)\hat{r}}. \tag{8}$$

Counting the time oscillation factor, the outgoing wave function in region $r \geqslant r_+$ is

$$\phi_\omega^{out} \simeq e^{-i\omega t} e^{i(\omega-\omega_0)\frac{1}{2\eta}\ln(r-r_+)} = e^{-i\omega t}(r-r_+)^{i(\omega-\omega_0)/2\eta} \tag{9}$$

As in [2], we make some analytic continuation from $r>r_+$ to $r<r_+$ about the solution (9). In the complex plane,

a half cycle in center of r_+ is taken with a radius $\varepsilon = |r-r_+|$ which is an arbitrary infinitesimal.

The analytic continuation is along the lower part of circumference. The whole wave function outside and inside r_+ is given as:

$$\phi_\omega = N_\omega[Y(r-r_+)e^{-i\omega t}(r-r_+)^{i\frac{\omega-\omega_0}{2\eta}}$$

$$+Y(r_+-r)e^{\frac{\pi(\omega-\omega_0)}{2\eta}}(r_+-r)^{i\frac{\omega-\omega_0}{2\eta}}e^{-i\omega t}]$$

$$= N_\omega[Y(r-r_+)\phi_\omega^{out}(r-r_+)+Y(r_+-r)\,e^{\frac{\pi}{2\eta}(\omega-\omega_0)}\phi_\omega^{out}(r_+-r)], \quad (10)$$

where $Y(x)$ is the step function; ϕ_ω^{out} has been normalized and N_ω is the normalization constant of ϕ_ω.

Outside the horizon, Eq. (10) represents an outgoing positive-energy flux of particles coming from the horizon. Inside the horizon, space coordinate and time coordinate are reversed; so r represents the time coordinate. Eq.(10) describes a positive-energy flux of particles traveling backward in time, i.e., describes a negative-energy flux of antiparticles traveling forward in time to the singularity. Therefore, the physical meaning of analytic continuation is that particle-antiparticle pairs are created near the horizon, where the antiparticle with negative energy fall in singularity, then the irreducible mass of a B.H. is reduced, and the particles with positive energy cause the radiation of B.H., i.e., the Hawking evaporation.

Noting that $\phi_\omega^{out}(r-r_+)$ and $\phi_\omega^{out}(r_+-r)$ are normalized and orthogonal, the condition of normalization of ϕ_ω yields

$$(\phi,\phi) = N_\omega^2\ [1+\exp(\frac{2\pi(\omega-\omega_0)}{2\eta})] = 1, \quad (11)$$

then

$$N_\omega^2 = \frac{1}{1+e^{(\omega-\omega_0)/kT}} , \tag{12}$$

where

$$T = \eta/\pi k \tag{13}$$

is the temperature of B. H.; k is the Boltzmann constant. Eq. (12) is the Hawking thermal spectrum formula about Dirac particles for the general Kerr-Newman B. H.. The wave function ϕ_ω^{out}, ω_0, η and thermal spectrum formula in[2-4] are the special cases of our results. Now we discuss as follows:

a) For the case of quasi-extreme Kerr-Newman B. H., $M^2 - a^2 - Q^2 = \delta^2$, our formula (3) reduces to formula (14) of [4], and thermal spectrum formula (11) is (21) of [4]. If $r_+ = M + \sqrt{M^2 - a^2 - Q^2} = M + \delta$ is considered, Eqs. (6) (9) and (10) reduce to the case of quasi-Kerr-Newman black hole[4].

b) When Q = 0, these are the formulae in general Kerr B. H.. If the quasi-extreme B. H. is considered ($M^2 - a^2 = \delta^2$, δ^2 is a small quantity), the formulae also reduce to the case of [3].

c) When $Q \neq 0$, a = 0, the case reduces to Hawking radiation background Reissner-Nordstrom B. H.

d) When Q = 0, a = 0, the formulae in this paper reduce to Hawking evaporation of the Dirac particles out of Schwarzschild B. H.[2].

Acknowledgements:

We are much obliged to Prof. Liu Liao for his enlightening discussions. We should also like to thank Prof. Xie Xide very much for her encouragement and for revising the manuscript.

REFERENCES

[1] Hawking, S.W., 1975, Commun. Math. Phys., Vol. 43, 199.

[2] Damour, T., Ruffini, R., 1976, Phys. Rev., Vol. D14, 332.

[3] Liu Liao, Xu Dianyau, 1980, Acta Physica Sinica, Vol. 29, 1617.

[4] Zhao Zheng, Guei Yuanxing, Liu Liao, 1981, Acta Astrophysica Sinica, Vol. 1, No. 2, 141.

[5] Chandrasekhar, S., 1976, Proc. Roy. Soc. London, Vol. A349, 571.

[6] Page, D.N., 1976, Phys. Rev., Vol. D14, 1509.

VII. QUASARS AND GALAXIES

QUASAR AND RELATED PROBLEMS
—SOME RECENT DEVELOPMENTS

Sachiko Tsuruta

(Department of Physics, Montana State University,
Bozeman, Montana, U.S.A.)

I. INTRODUCTION

Almost 20 years since the discovery of quasi-stellar objects (quasars)[1], a consensus is yet to be reached concerning the true nature of these spectacular objects. There is now, however, wider agreement that they are a member of a family of objects known as Active Galactic Nuclei (AGN). Among other members are: Seyferts, BL Lacs, some of strong radio galaxies, N galaxies, and narrow emission line galaxies. They are common in the sense that the presence of an active compact nuclear region in a galaxy is somehow responsible for their peculiar energetic phenomena.

Three representative models have been proposed for active galactic nuclei. These are: a) dense star clusters (including multiple supernovae and pulsars), b) supermassive stars (including "spinars" and "magnetoids"), and c) accretion onto supermassive black holes. See, for instance, reference[2], for general review. It is quite understandable if different phenomena are responsible for different kinds and degrees of various nuclear activities in galaxies, and at the present stage, all of the above models are still possible. However, it appears that an impressive body of observational data which have been accumulated during the last few years have already tilted the balance heavily toward the third possibility, i.e.c) accretion onto supermassive black holes. In the present review, we shall concentrate on these new exciting developments. Here we admit that the selection of topics was somewhat influenced by our own interest, and in this sense this report by no means is meant as a complete review. However, they are interesting especially because investigation along this line may give us some clues to the most fundamental aspect of the quasar phenomenon; that is, what is the central engine powering various kinds of super-energetic phenomena in these AGN's?

Before introducing recent observational developments, we shall briefly describe these AGN's. First, optically a Seyfert is a galaxy (most of them a spiral) with an intense, compact nucleus from which are emitted intense, broad emission lines (both permitted and forbidden lines). A compact non-thermal continuum radiation is superimposed on the stellar contribution. In type I Seyferts the permitted lines show very broad wings, while these broad wings are absent in type II Seyferts. Narrow emission line galaxies have active nuclei with sharp emission lines. A quasar is a bright object stellar in appearance, but with continuum radiation of power-law spectra (the energy spectral index $\alpha = \sim 0 - 2$) and Seyfert-I-like strong emission lines. A major difference between a quasar and a Seyfert I is that the surrounding galaxy is not observed in quasars. However, accepting the cosmological origin of quasar redshifts and noting that the redshift of quasars is generally larger than that of Seyferts, it is quite reasonable to assume that the apparent absence of a parent galaxy around a quasar is due to the prominence of the bright nucleus and greater distance. In this sense, it is generally considered that quasars and type I Seyferts are similar objects. An N galaxy is an object intermediate between a Seyfert and a quasar, in the sense that the nucleus is brighter and more conspicuous than a Seyfert nucleus but the parent galaxy is still detectable. BL Lacs are similar to quasars, except that emission lines are absent or at most very weak, and that the slope of the power-law continuum spectra is steeper ($\alpha > \sim 2\text{–}3$). The red-shifts of BL Lacs (whenever known) are about those of near-by quasars, and very often a surrounding "fuzz" is identified with a parent galaxy (apparently an elliptical). Intrinsically, quasars are the brightest of all, ranging in luminosity from $\sim 10^{43}$ to $\sim 10^{47}$ erg s^{-1} or more (in the infrared-optical band). Seyferts are less luminous by one to two orders of magnitude[1–3].

Strong radio galaxies are either a) powerful very extended (mostly double) radio sources with steep spectra, or b) powerful compact radio sources with flat or inverted spectra, the radio emission originating from a small region in the nucleus. The brightest example of a) is Cygnus A. A most well known example of b) is 3C 273. Very often both extended and compact components are seen in the same object (e.g. Cen A). Most of these galaxies are giant ellipticals, but a very small fraction ($< \sim 10\%$) of quasars (e.g. 3C 273) and Seyferts (e.g. 3C 382 for a) and 3C 120 for b))are strong radio sources. In most of quasars (radio-quiet) and Seyferts ($> \sim 90\%$), optical emission far exceeds radio emission (if present at all), but in strong radio galaxies their radio luminosities can be comparable with or exceed optical emission, with luminosity going up to $\sim 10^{46}$ erg s^{-1} in some cases[1].

In recent years, Seyfert nuclei (especially type I) and BL Lac objects have been found to be strong X-ray emitters. Until recently, most quasars were too distant to be observed by existing X-ray detectors. However,mainly through the greatly improved sensitivities of the Einstein and other recent satellites, it is now known that quasars are also strong X-ray emitters[4]. Their luminosities in the X-ray band are comparable with or exceed those in the lower energy bands. X-rays have been detected from some other AGN's, too, -radio galaxies, N galaxies, etc[5-7]. A typical example is Cen A, a huge double radio source with a giant elliptical as a parent galaxy.

II. SOME RECENT OBSERVATIONAL DEVELOPMENTS

It is generally considered that a best way to learn about the central "power house" of quasar-like objects is through studies of their X-ray properties. Therefore, we shall first look at developments in this area.

(1) Time Variability

Variations in time scales of years to months are common in radio and optical emission from quasars and other AGN's. Sometimes optical variations of weeks (down to days) have been seen. Similar variations of years to days are recently observed in X-ray bands also(Figs. 1a,1b). In addition to these longer term variations, however, recent X-ray observations revealed very short time (minutes to hours) and very large amplitude variations of X-ray intensities in several AGN's[5–7,10]. These are summarized in Table 1. The first column gives the name and kind of AGN's and the next three columns give X-ray luminosity (typically in the range 2-10keV), relative amplitude, and time scale of variations. The last column gives an order of magnitude of the emission region estimated from the "light-travel-time" consideration[2]. We note that this fast variation is independent of the luminosity of the source. For instance, 3C 273 is a strong quasar, while NGC 4151 is a weak Seyfert. Fig. 1c shows the most striking case, NGC 6814 (type I Seyfert)[10]. With a time scale of only ~ 100 sec, it gives a severe constraint on the size of the emitting region, and thus the nature of the central engine (see Sect. III).

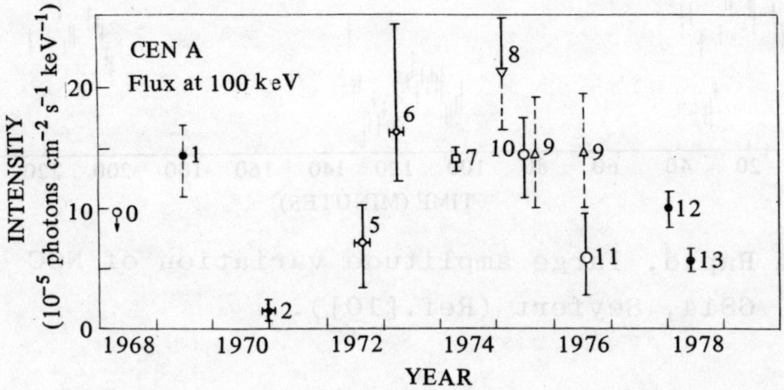

Fig. 1a Time variation of Cen A (Ref.[8]).

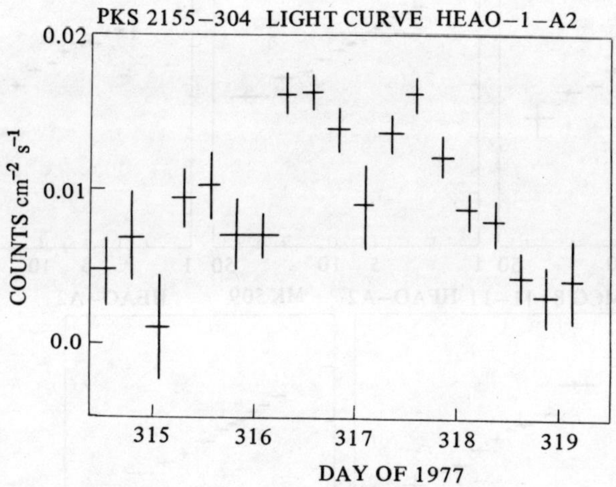

Fig. 1b Time variation of PKS 2155−304, BL Lac (Ref.[9]).

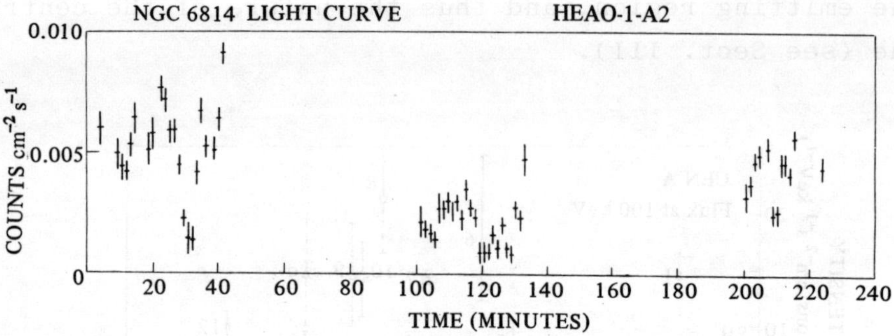

Fig. 1c Rapid, large amplitude variation of NGC 6814, Seyfert (Ref.[10]).

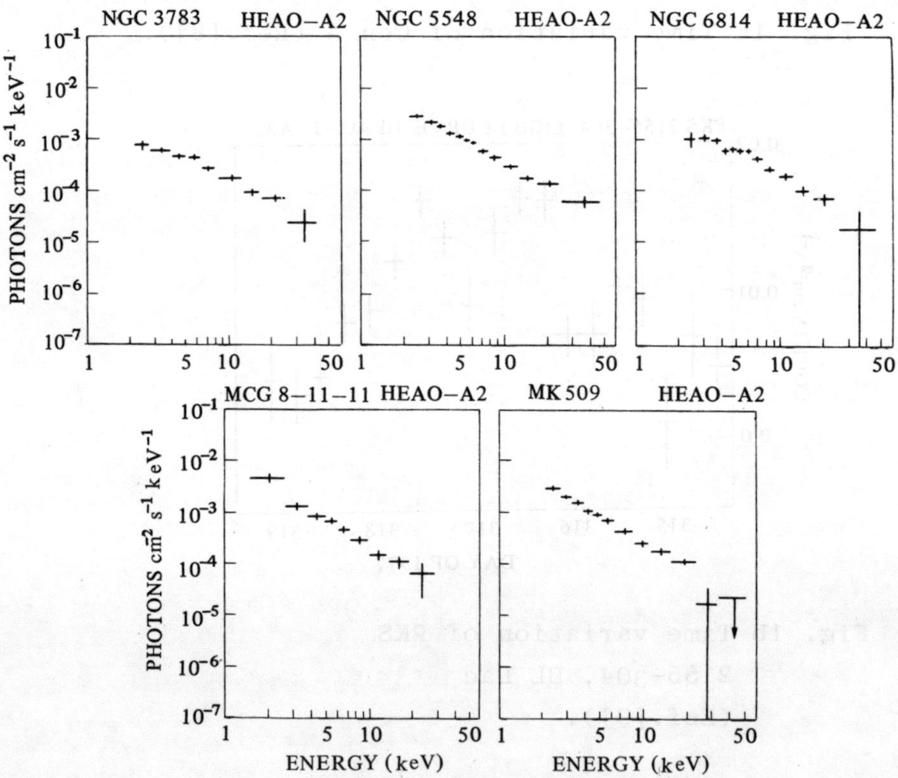

Fig. 2 Typical X-ray spectra of Seyferts (Ref.[14–17]).

(2) Spectra

The first detailed information on the X-ray spectra of AGN's has become available through recent results from Ariel V, HEAO-A, etc. The NASA group reports that the 2–50 keV power law spectra from HEAO-A are relatively flat with energy index $\alpha = \sim 0.2 - 1$, and that α stays relatively constant during intensity variations[14–17]. They note that this appears to be a universal feature common to various kinds of AGN's, including Seyferts, quasars, N galaxies, and radio galaxies. Some examples are shown for Seyferts in Fig. 2. Similar results are reported by some other groups[18,19].

For BL Lac objects, the power-law spectra are steeper ($\alpha > \sim 2$) and softer[14–17]. A typical example is shown in Fig. 3a, for H 2156–304. It is generally highly variable, both in intensity and slope, e.g. MK 501 (Fig. 3b). For some BL Lacs, in addition to the steep soft component, a flatter, high energy tail is sometimes observed, e.g. PKS 0548–322 (Fig. 3c). On the other hand, steep, soft excess was sometimes detected for some Seyferts, e.g. ESO 141–G55 (Fig. 3d)[14–17] and a newly discovered Seyfert E 1616 + 061[20].

For some well-studied sources, e.g. Cen A, NGC 4151, 3C 273, the flat component extends to a few hundred keV[21–23], but there is some indication of breaks beyond a few hundred keV[8,24].

From the Ariel V data, Hayes et al[25] report on observation of a variety of X-ray spectral slopes for various Seyferts and narrow emission line galaxies. However, their error bars are larger than the HEAO-A data. Also, their energy window is 2 - 10 keV, lower than the 2 - 50 keV window of HEAO-A. When we take these into consideration and note the high degree of intensity and spectral variations often observed in soft X-ray regions, it appears that the results from Ariel V and HEAO-A do not contradict with each other. A significance of the Ariel V results[25] is the detection of iron lines and variable soft X-ray absorption in some of these sources (e.g. NGC 4151, NGC 3783), because that may suggest the presence of dense, hot moving clouds near the X-ray source[26].

Table 1. Rapid, large amplitude variability of AGN's. (R), (B), (S), and (Q), are radio galaxy, BL Lacs, Seyferts, and quasars (Ref.[5–7][10–13]).

SOURCE	L_x (erg s^{-1})	$\Delta\bar{L}_x/L_x$	Δt (s)	$c\Delta t$ (cm)
(R) CEN A	1×10^{43}	0.25	7×10^{3}	2×10^{14}
(B) MK 421	1×10^{44}	2.0	1×10^{5}	3×10^{15}
(S)NGC 6814	1×10^{43}	1.5	100	3×10^{12}
(Q) OX 169	1×10^{44}	1.5	6×10^{3}	1.8×10^{14}
(Q) 3C 273	1.7×10^{46}	0.1	6×10^{3}	1.8×10^{14}
(S) NGC 4151	5×10^{42}	3.0	700	2×10^{13}
(Q) 1515+227	3.3×10^{44}	–	200	6×10^{12}

Composite spectra of AGN's are quite complicated[9,18,19,22]. An example is shown in Fig. 4a, for 3C 273, a powerful variable quasar. Another is shown in Fig. 4b, for a BL Lac object PKS 2155–304. Both of these objects are strong radio sources. Qualitatively similar spectra are obtained for other AGN's, except that the radio portion is missing or much weaker for most of Seyfert nuclei, and that gamma rays have been detected from only a few of these objects.

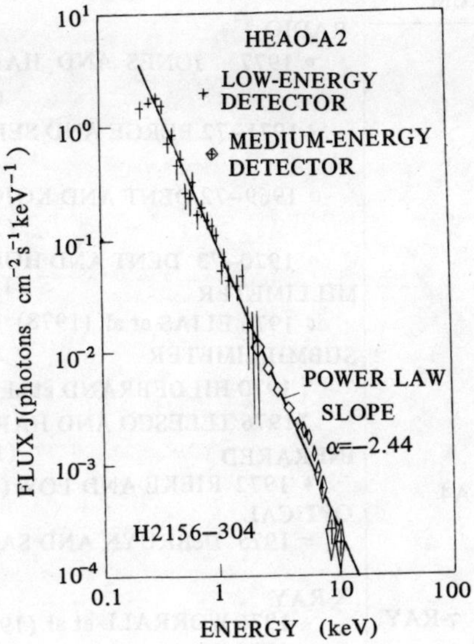

Fig. 3a X-ray spectrum of H 2156−304, BL Lac (Ref. [14−17]).

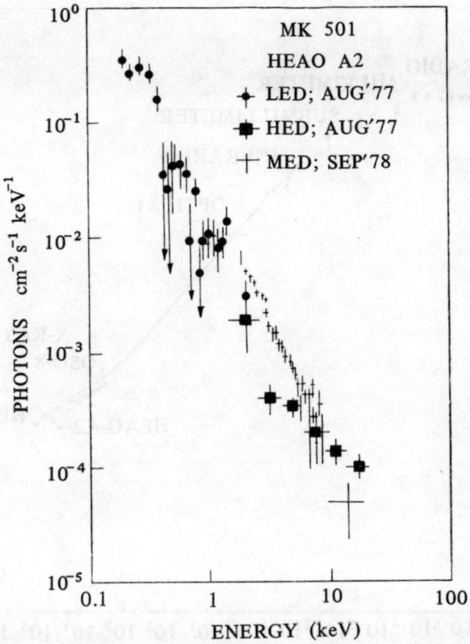

Fig. 3b X-ray spectrum of MK 501, BL Lac (Ref. [14−17]).

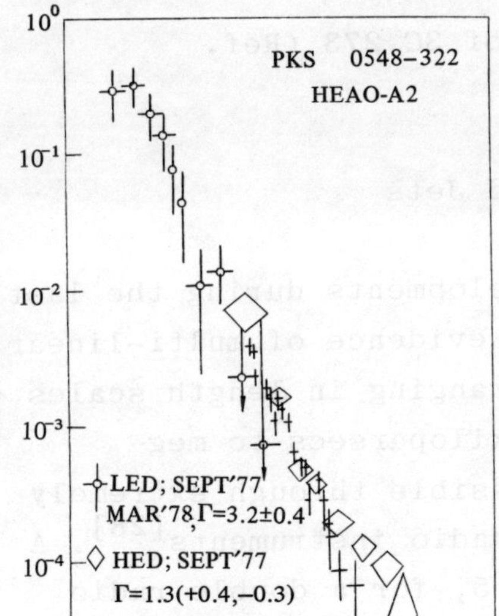

Fig. 3c X-ray spectrum of PKS 0548−322, BL Lac (Ref. [14−17]).

Fig. 3d X-ray spectrum of ESO 141−G55, Seyfert (Ref. [14−17]).

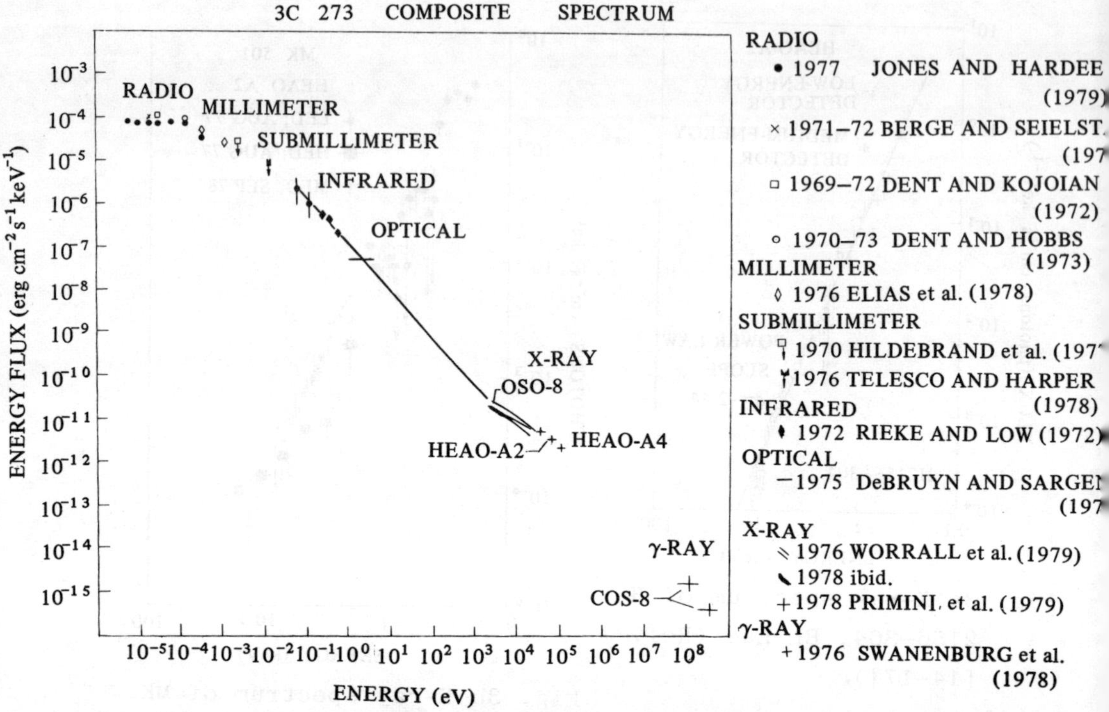

Fig. 4a Composite spectrum of 3C 273 (Ref. [18,19]).

(3) Beams and Jets

Among the most exciting developments during the last few years is the ever increasing evidence of multi-linear structure associated with AGN's ranging in length scales from only a few parsecs through kiloparsecs to megaparsecs[27]. This has become possible through extremely high spatial resolution of VLBI radio instruments[28]. A typical example is shown in Fig. 5, for a double radio source NGC 6251. A striking discovery is the beautiful alignment of various structure of very different scales. Moreover, the smallest scale structure of a few parsecs is located near the nucleus of the parent galaxy.

Through VLBI studies, it is now known that radio jets

are associated with many double extended radio sources (e.g. Cen A, NGC 6251, 3C 66B, 3C 219, NGC 315) and many compact radio sources (e.g. 3C 273, 3C 345, 3C 147, 3C 380). VLBI analysis shows that compact variable radio sources are generally of the "core-jet" type in which a jet emerges from an optically thick nucleus. Apparent "superluminal" expansion (v>c) is often observed between the jet and the core[28]. Optical jets are also observed from many of AGN's (e.g. Cen A, 3C 273, 3C 66, 2C 31, M 87, DA 240). Recent X-ray picture of Cen A from the Einstein Observatory[5–7] shows an X-ray jet located a few parsecs from the nucleus, just inside optical jets (Fig. 6). The X-ray, optical and radio jets of Cen A are roughly aligned in the same general direction of both compact and extended radio sources, which coincides with the direction of the rotation axis of the parent galaxy.

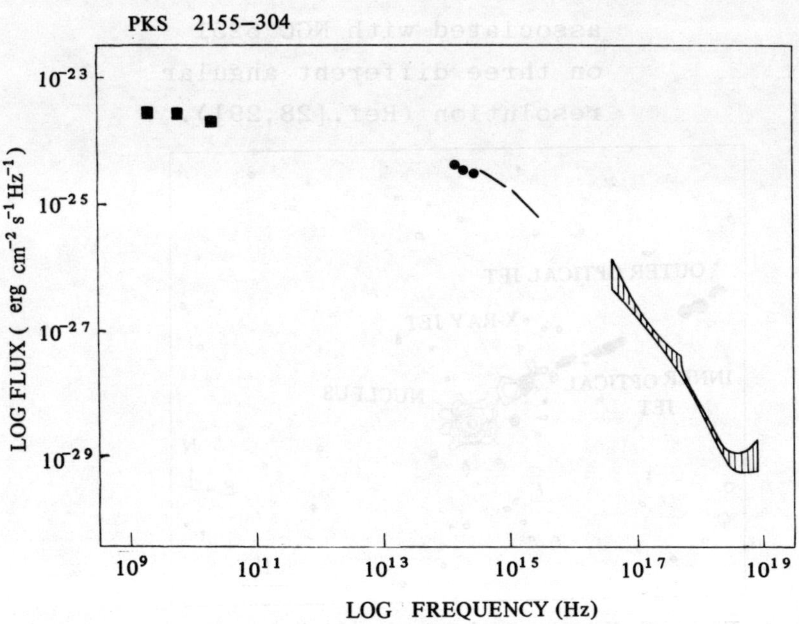

Fig. 4b Composite spectrum of PKS 2155–304 (Ref.[9]).

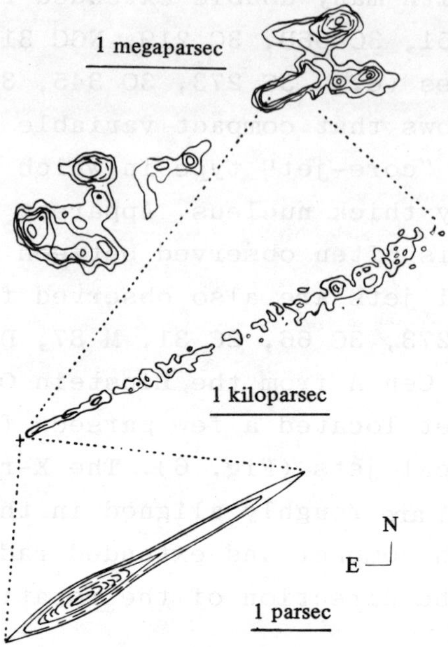

Fig. 5 Radio structure of source associated with NGC 6251 on three different angular resolution (Ref.[28,29]).

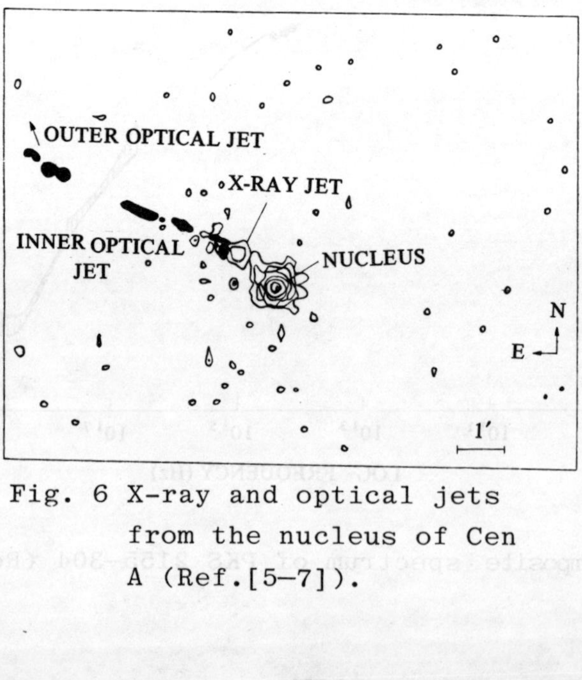

Fig. 6 X-ray and optical jets from the nucleus of Cen A (Ref.[5–7]).

In the next sections we shall consider possible implications of these new observational developments.

III. SUPERMASSIVE BLACK HOLE MODEL

Accepting a cosmological origin of quasar redshifts, quasar-like-objects are losing energy at typical rates of ~ 10^{43} –10^{47} erg s^{-1}. The total mass efficiently converted at these rates into radiation energy during a typical life time of ~ 10^{6} – 10^{8} years amounts to ~ 10^{6} – 10^{9} $M_{\odot}$, which is comparable with or larger than a mass of galactic nuclei. A similar amount of mass is needed if we apply a commonly used constraint that the observed luminosity does not exceed the Eddington limit. On the other hand, the dimension of the emitting region estimated from rapid intensity variations is ~ 10^{12} –10^{15} cm (Table 1), which is less than ~ 0.001 pc. (Note that a typical size of a galactic nucleus is ~ 1 pc.) The use of a deep potential well around a supermassive black hole will be a most efficient and natural way of trapping a huge amount of energy and extracting it from such an extremely compact region. For instance, the gravitational radius of a supermassive black hole of ~ 10^{6} – 10^{9} $M_{\odot}$ is ~ $3(M/M_{\odot})$ km = ~3×10^{11} –3×10^{14} cm. In this sense, we emphasize the significance of the reported ~100–200 sec large amplitude variations of X-rays (Table 1, Fig. 1c).

In a typical black hole model, the energy lost by radiation is supplied by accreting gas, with a rate $\dot{M}$. The total luminosity L_t and $\dot{M}$ are related by $L_t = \dot{M} c^2 \eta$, where η is efficiency. With an efficient conversion with η= ~10%, the accretion rate of $\dot{M}$ = ~0.001 –1 $M_{\odot}$ yr^{-1} is sufficient to explain ordinary AGN phenomena. There are various possibilities of supplying this gas, through natural evolution processes of surrounding ordinary stars, tidal disruptions, stellar collisions, capture of intergalactic clouds, etc.

(see, e.g., references[2,3,29]). For luminous quasars, larger amount of $\dot{M}$ is required. That may be supplied, for instance, through stellar collisions and subsequent disruptions when the surrounding galactic nuclei are extremely dense (denser than observed in Seyfert galaxies). Another possibility is that real luminosities are less than the observed "Apparent" luminosities. That can happen in various ways, for instance, when the radiating source is approaching nearly in the direction of the observer with relativistic speeds[30].

It has been suggested for some time that the energy contained in the extended radio lobes of giant radio galaxies had been supplied by beams of ralativistic particles ejected from the galaxy[27]. As an example, let us consider Cen A. In order to supply the energy contained in the huge extended double radio source, a total mass of at least $\sim 10^7\ M_\odot$ is required. The alignment of the large scale (kiloparsecs to megaparsecs) and small scale (parsecs) multi-linear structure recently discovered in many of radio galaxies (e.g. NGC 6251, etc., see Figs. 5,6) is a strong indication that the preferred direction is well remembered for millions of years. We note that small structures of parsecs are observed right near the galactic nuclei. These considerations strongly suggest the presence of a collapsed center in a form of a supermassive black hole. Then, in this picture, these radio sources are energized by activity involving the supermassive black hole, through beams and jets of relativistic particles ejected from the center. It is interesting to note that in spite of the tremendous amount of power apparently supplied to the extended radio lobes, the nuclei of many of radio galaxies emit little or relatively low degree of detectable radiation. This has led some authors to a scinario in which the energy carried by beams and jets has been extracted from rotational energy of the black hole[31]. They may be regarded as hungry black holes because very little amount of $\dot{M}$ is required. In this picture, matter is ejected along the axis of rotation, and therefore the extended radio lobes and other multi-linear structure lie along the rotation axis of the hole.

Radio, infrared-optical, and X-ray emission from AGN's is generally well fit by power law spectra (Sect. II). A power law spectrum is naturally produced by synchrotron radiation, and it is generally agreed that synchrotron process is responsible for radio emission from these sources (both extended and compact, as well as various jets). Various problems have been posed for synchrotron models of higher frequency radiation (optical through gamma rays)[1–3]. It was noted recently, however, that these problems can be avoided if the source is expanding relativistically or it forms thin layers, e.g. filaments, shock fronts[27]. Successful, though phenomenological, synchro-Compton models have been constructed for violently variable BL Lacs (e.g. PKS 2155–304), where the observed radio through soft X-ray emission is due to synchrotron process involving electrons in a single region, while the high energy tail is the result of inverse Compton scattering of lower energy synchrotron photons by the same electrons[9]. However, that is not the only possibility. According to recent theoretical calculations[32–35], unsaturated inverse Compton scattering of soft photons also produce naturally a power law spectrum. Various possible sources of seed soft photons have been suggested[3,36,37]. Recently, we have carried out detailed calculations to show that cyclotron higher harmonics are a natural source of these seed photons, and that the result is consistent with various observed properties such as rapid, large amplitude variations, relatively flat power law spectra, etc.[38]. The relative constancy of the hard X-ray spectral index is consistent with the model if time-dependent approach is adopted in the treatment of comptonization[39].

IV. DISCUSSION AND CONCLUDING REMARKS

Until recently it seemed hopeless to distinguish between different models of quasar-like objects. The situation has become brighter after the first series of temporal and spectral information has become available for X-rays, -through Ariel V, HEAO-A, Einstein Observatory, etc. A large number of similar or more improved data, however, may be needed, before we can draw statistically significant conclusion. For instance, it may be that the rapid variability (reported in Table 1) is a sporadic event. If so, long term systematic observation of candidate objects with high temporal resolution detectors may lead to discovery of more of NGC 6814's (fast variations of order of minutes). If this is a fairly common phenomenon, it may become hard not to accept black hole models.

It appears that already available spectral data suggest the existence of at least two X-ray components, a soft steeper one and a hard flatter one. It seems that the former is predominant in one type of objects (e.g. BL Lacs), while the latter is a more or less universal character of AGN's. It may be that the flat X-ray tail is not seen in NL Lacs often, because it is relatively weak as compared with the steep component[38]. However, such a hypothesis should be checked carefully through further spectral analysis of high quality. If the existence of these components is confirmed, it is significant because it poses various interesting questions. For instance, if soft excess is occasionally seen in Seyferts, could it correspond to a high flare state as seen in some of galactic X-ray sources, -or is it something else?

There may be several ways of distinguishing between different emission mechanisms. For instance, according to a synchro-Compton model[9], radiation from different wavebands (radio, infrared-optical,UV, X-rays and gamma rays)

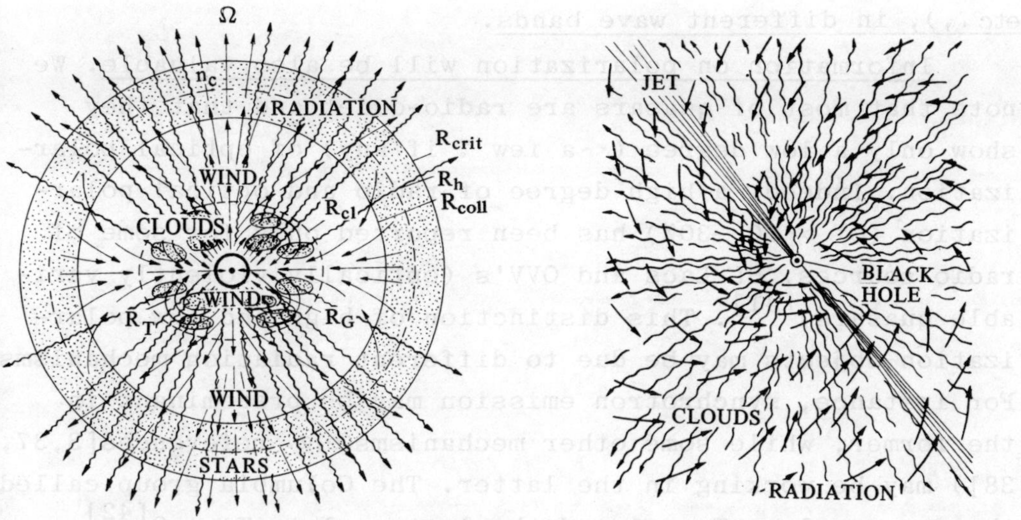

(a) FOUNTAIN MODEL (Ref.[3]) (b)

Fig. 7 Hypothetical supermassive black hole models of quasar-like objects (not drawn to scale).

should all come from the same physical region. On the other hand, in our cyclo-Compton model[38], different mechanisms may be responsible for emissions in different wave bands. They can also come from different physical regions. According to data from Kriss et al.[40], correlation between optical continuum luminosity and X-ray luminosity is poor for Seyferts. Also, it may be noted that the large amplitude rapid variability reported for X-rays (Table 1) is absent in radiation from lower energy bands (radio, optical, etc.). These data may suggest that at least for these objects, the observed lower energy photons come from regions different from (most likely larger than) the X-ray emitting region, and that different mechanisms are responsible for them. In connection with these problems, we strongly emphasize the importance of <u>simultaneous observations of variations of various properties (intensity, hardness ratio, polarization,</u>

etc.,), in different wave bands.

Information on polarization will be also valuable. We note that most of quasars are radio-quiet and that they show only a low degree (<~a few % if any) of optical polarization. Indeed, a high degree of radio and optical polarization (of >~10 –30%) has been reported only for some of radio sources, BL Lacs and OVV's (optically violently variable quasars)[41]. This distinction of high and low polarization objects may be due to different radiation mechanisms. For instance, synchrotron emission may be predominant in the former, while some other mechanisms(e.g. references[3,37, 38]) may be working in the latter. The Columbia group called the former class Type 1 and the latter class Type 2[42].

Examination of temporal behavior of spectra during intensity variations may be useful, for different emission mechanisms predict different behavior[24].

High quality spectral measurements in the critical region of ~100–1000 keV may turn out to be fruitful, for it may help distinguish between thermal models and non-thermal models. When electrons are non-thermal, the flat spectra can continue beyond ~1000 keV without breaks, but if thermal electrons of transrelativistic order ($kT_e \simeq m_e c^2$) are responsible for Compton scatterings, a break is expected in this energy range. In reality, things may be complicated by coexistence of both thermal and non-thermal electrons. Also, distinction between thermal and non-thermal particles becomes blurred, and cyclotron and synchrotron models may become indistinguishable, when the mean particle energy becomes relativistic.

Studies of soft photon absorption will be also important since it will give useful information on clouds in the immediate vicinity[25].

It is desirable that these points are taken into account in the observing programs of ASTRO-C, ESLAB, and other future space programs. For theorists, detailed work on physics of plasma around a supermassive black hole is still in a pioneering stage. Some of hypothetical models are shown in Fig. 7. In any case, for all concerned it appears that the problem of quasars and related objects remain as an exciting and rewarding field for some years to come.

We thank many of our colleagues for stimulating and useful discussion. We are especially grateful to professors M. Oda, Y. Tanaka, and the other members of ISAS (the Institute of Space and Astronautical Science), for their hospitality and help. The manuscript was prepared in ISAS.

REFERENCES

[1] For instance, see Burbidge, G., Burbidge, M., 1967, "Quasi Stellar Objects", W.H. Freeman Co.

[2] For instance, see Rees, M.J., 1978, Physica Scripta, Vol. 17, 193.

[3] Tsuruta, S., 1977, Astron. Astrophys., Vol. 61, 647.

[4] For instance, see Tananbaum, H. et al., 1978, Ap. J., Vol. 223, 74, etc.

[5] Tananbaum, H., 1980,"X-Ray Astronomy", Giacconi, R., Setti, G. (eds.), Reidel, Dordrecht, 291.

[6] Shleier, E. et al., 1979, Ap. J., Vol. 234, L39.

[7] Pounds, K.A., 1980, "X-Ray Astronomy ", Giacconi, R., Setti, G. (eds.), Reidel, Dordrecht, 273, etc.

[8] Baity, W.A. et al., 1981, Vol 244, 429.

[9] Urray, C.M., Mushotzky, R.F., 1981, NASA preprint 82101.

[10] Tennant, A.F. et al., 1982, Ap. J., in press.

[11] Delvaille, J.P., Epsrein, A., Shnopper, H.W., 1978, Ap. J., Vol. 219, L81.

[12] Ricketts, M.J., Cook, B.A., Pounds, K.A., 1976, Nature, Vol. 259, 546.

[13] Marshall, N., Warwick, R.S., Pounds, K.A., 1981, M.N.R.A.S., Vol. 194, 987.
[14] Holt., S.S., 1980, "X-Ray Astronomy", Giacconi, R., Setti, G. (eds.), Reidel, Dordrecht, 327.
[15] Boldt, E., 1979, "HEAO Science Symposium", Dailey, C. et al. (eds.), NASA CP-2112, 82.
[16] Mushotzky, R.F. et al., 1980, Ap. J., Vol. 235, 377.
[17] Worral, D.M. et al., 1981, Ap. J., Vol. 243, 53, etc.
[18] Zamorani, G. et al., 1981, Ap. J., Vol. 245, 357.
[19] Worrall, D.M. et al., 1980, NASA preprint 80645, etc.
[20] Pravdo, S.H. et al., 1982, Ap. J., in press.
[21] Pietsch, W. et al., 1981, Astron. Astrophys., Vol. 94, 234.
[22] Primini, F.A. et al., 1979, Nature, Vol. 278, 234.
[23] Coe, M.J. et al., 1981, M.N.R.A.S., Vol. 195, 241.
[24] Bignami, C.F. et al., 1979, Ap. J., Vol. 232, 649.
[25] Hayes, M.J.C. et al., 1982, "the 15th ESLAB Symp. on X-Ray Astr.", in press.
[26] Fabian, A.C., Ross, R. R., 1981, M.N.R.A.S., Vol. 194, 29.
[27] Rees, M.J., Begelman, M.C., Blandford, R.D., 1982, Ann. N.Y. Acad. Sci., in press, etc.
[28] For instance, see Kellerman, K., 1978, Physica Scripta, Vol. 17, for review; Readhead, A.C.S., Cohen, M.H., Blandford, R.D., 1978, Nature, Vol. 272, 131, etc.
[29] Gunn, J.E., 1979, "Active Galactic Nuclei", Hazard Mitton (eds.), Cambridge Press, etc.
[30] Blandford, R.D., Konigl, A., 1979, Ap. J., Vol. 232, 34, etc.
[31] For instance, see Rees, M.J. et al., 1981, preprint; Blandford, R.D., Znajek, R.L., 1977, M.N.R.A.S., Vol. 179, 433, etc.
[32] Katz, J.I., 1976, Ap. J., Vol. 206, 910.
[33] Sunyaev, R.A., Titarchuk, L.G., 1980, Astron. Astrophys. Vol. 86, 121.

[34] Lightman, A.P., Rybicki, G.B., 1980, Ap. J., Vol. 236, 928.

[35] Takahara, F., 1981, Prog. Theor. Phys., Vol. 65, 883, etc.

[36] Lightman, A.P., 1981, Ap. J., Vol. 244, 392.
Lightman, A.P., 1981, CFA preprint No. 1542.

[37] Petschek, A.G., Colgate, S.A., Colvin, J.D., 1976, Ap. J., Vol. 209, 356.

[38] Takahara, F., Tsuruta, S., 1982, Prog. Theor. PhYs., in press; Takahara, F., Tsuruta, S., Ichimaru, S., 1981, Ap. J., Vol. 251, 26; Tsuruta, S., Takahara, F., Ichimaru, S., 1982," the 15th ESLAB Symp. on X-Ray Astr.", in press. (see also Eardley, D.M., Lightman, A. P., 1976, Nature, Vol. 262, 196; Maraschi, L. et al., 1981, preprint.)

[39] Guibert, P.W., Ross. R.R., Fabian, A.C., 1982, M.N.R.A. S., in press.

[40] Kriss, G.A. et al., 1980, Ap. J., Vol. 242, 492.

[41] Angel, R., Stockman, P., 1980, Ann. Rev. Astron. Astrophys.

[42] Ku, W. H. -M., Helfand, D.J., Lucy, L.B., 1980, Nature, Vol. 288, 323.

LUMINOSITY EVOLUTION OF QSOs

GONG Shumo (龚树模), XIA Changli (夏昌立)
(Purple Mountain Observatory,
Academia Sinica, Nanjing)

ABSTRACT

Based on the result of four articles on Hubble diagram with standard candle of QSOs, we first obtain the improved deceleration factor q_0 by means of the exact formula of luminosity distance of Friedmann universe model. Then from the QSOs' Catalogue of Hewitt and Burbidge, we derive the statistical relation of m(z). Finally the luminosity of QSOs as a function of redshift z or time t and their respective derivatives are calculated.

a) The improved value of q_0 is found as follows:

i) The general formula of the apparent magnitude-redshift relation is

$$\overline{m}(z) = \overline{M}(z) - 5 + 5\log d_L, \tag{1}$$

where $\overline{m}(z)$, $\overline{M}(z)$ are respectively the mean apparent and absolute magnitude of QSOs at Z and d_L is the luminosity distance of QSOs, in unit of parsec.

According to Friedmann model, in the case of matter

dominated universe, q_0 must be positive and so the exact formula of d_L [1] is

$$d_L = \frac{c}{H_0 q_0^2}\left\{ zq_0 + (q_0-1)(-1+\sqrt{1-2q_0 z})\right\}, \qquad (2)$$

where c is the velocity of light and H_0, the Hubble constant.

ii) The special formula of the apparent magnitude-redshift relation of QSOs with the standard candle is

$$\overline{m}(z) = M - 5 + 5\log \frac{c}{H_0} f(z) = a + 5\log f(z), \qquad (3)$$

where M, the absolute magnitude of QSOs, the so-called standard candle, is constant for a subgroup of QSOs. Thus, in Eq. (3), a = M−5+5log c/H_0 is also a constant. With the method of standard candle to fit Hubble relation, it is in general unable to express d_L with exact formula (2), but is often expressed as a second degree polynomial approximation, i.e.

$$d_L \frac{H_0}{c} \simeq z + \frac{1}{2}(1-q_0)z^2 = f(z). \qquad (4)$$

The formula (4) is only valid for small values of z, say, $z \leq 0.2$. However, we can use the result of the Hubble relation for the case of standard candle to obtain an improved value of q_0. The reason why the subgroup of QSOs could have an approximate Hubble relation (3) is that its luminosity function $\phi_z(M)$ might have a more or less similar form at various values of z, such as [2], in spite of values different for M(z), and the variation of $\overline{M}(z)$ with z might be monotonous and smooth. In such case the absolute magnitude M, the standard candle, could be considered as some average value which is taken, as a reasonable approximation, to be $\overline{M}(\overline{z})$ at the average value $\overline{z}$ of this subgroup.

Putting $\overline{M}(\overline{z})$ into Eqs. (1) and (3), with Eq. (2) we can determine the improved value of q_0 by the following equation,

$$\log\frac{H_0}{c}\, d_L = \log\frac{1}{q_0^2}\left\{\overline{z}q_0 + (q_0-1)(-1+\sqrt{1+2q_0\overline{z}})\right\} = \log f(\overline{z}). \quad (5)$$

Fig. 1 shows a series of curves of $\log H_0/cd_L$ with values of q_0 indicated. Were there really standard candle of QSOs, i.e. QSOs with no luminosity evolution, then the log f(z) curves derived from observation would coincide with the curves with appropriate q_0. Actually three curves of log f(z) intersect with the latter at four points corresponding to four subgroups[3-6] of QSOs. Two subgroup[3, 4] have the same f(z), but different intersecting points as shown by arrows in Fig. 1. The result obtained by this method

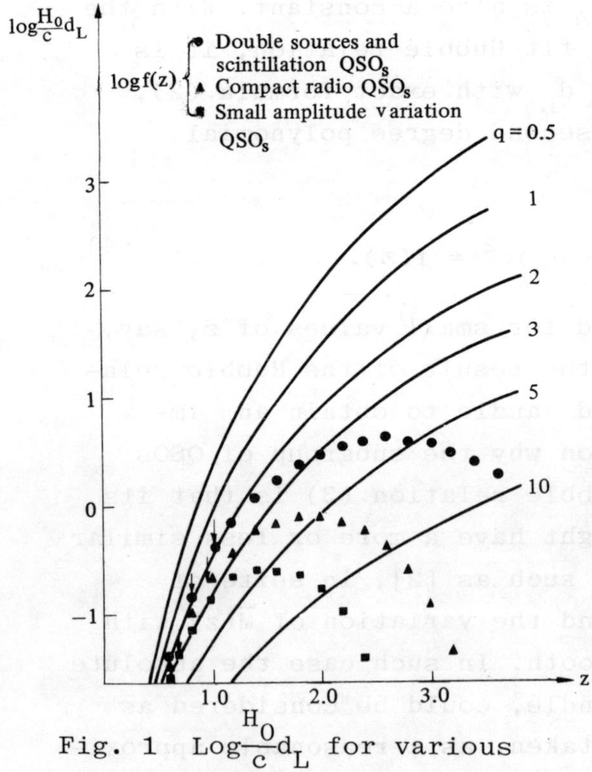

Fig. 1 $\mathrm{Log}\frac{H_0}{c}d_L$ for various q_0 and logf(z) at various z.

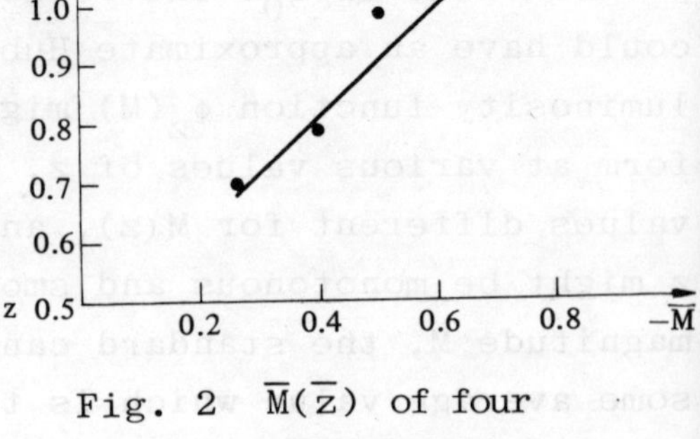

Fig. 2 $\overline{M}(\overline{z})$ of four subgroups of QSOs.

is tabulated in Table 1, in which n is the number of samples in each subgroup and values of q_0 in parentheses, the corresponding values obtained by approximate formula (4), are shown here for comparison. The values of $\overline{M}(\overline{z})$ are derived with H_0 = 75 km s^{-1} Mpc^{-1}.

Table 1. Improved values of q_0 obtained from four subgroups

QSOs Subgroups	n	$\overline{z}$	a	$\overline{M}(\overline{z})$	f(z)	q_0	References
(1)	(2)	(3)	(4)	(5)	(6)	(7)	
Double sources	11	0.79	18.64	-24.40	$z-0.19z^2$	1.65(1.38)	[3]
Scintillation	25	1.05	18.29	-24.72	$z-0.19z^2$	2.00(1.38)	[4]
Compact radio	8	0.98	18.50	-24.51	$z-0.26z^2$	2.45(1.52)	[5]
Small amplitude variation	24	0.69	18.74	-24.27	$z-0.325z^2$	2.18(1.65)	[6]

The references [3-6] refer respectively to QSOs of double radio sources, strong scintillation sources, compact radio sources and small amplitude variation ($\leq 0^m.5$) sources. The distribution of $M(\overline{z})$ for the four subgroups is shown in Fig. 2. The four points lie nearly on a straight line. Thus the member of each subgroup could be considered as that of the whole group of all four. The final value of q_0, obtained as the average of the four, weighted by n, is equal to +2.07±0.12.

b) We derive statistically the relation $\overline{m}(z)$ from Hewitt and Burbidge's new catalogue of QSOs[7].

After the correction of interstellar absorption[8] and K-correction[9], more than 1500 QSOs are divided into 36 intervals with Δz=0.1 and their median values of z=0.05, ..., 3.55. In each interval the average values of $\overline{m}(z)$ are obtained.

To get less scattered points in $\overline{m}(z)$-z diagram, we exclude all QSOs with their galactic latitude $\leqslant 20°$ and those, having uncertain value of z, i.e. with (z) noted in the catalogue[7].

By means of the least square solution, the fitted curve for the $\overline{m}(z)$ is

$$\overline{m}(z)=17.32+1.350z-0.371z^2. \quad (6)$$

Applying the method of reduced linear correlation analysis, we obtain the correlation coefficient r=-0.549 and residue standard error $s=0^m.29$. For n=33, and confidence of 99%, the corresponding r=0.449, so the relation (6) is significant. Fig. 3 shows the values of $\overline{m}(z)$ in

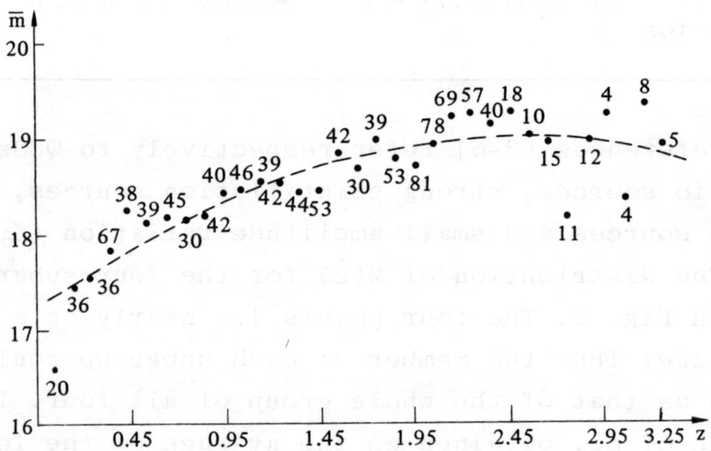

Fig. 3 The average apparent magnitude of QSOs at various z.

each interval and the fitting curve. The number beside each $\overline{m}(z)$ point is number of QSOs to be averaged.

c) With q_0=+2.07 and $\overline{m}(z)$ of [7], we can calculate $\overline{M}(z)$, and $d\overline{M}(z)/dz$. By means of the relation between z and time t[10], the values of $\overline{M}(t)$ and $d\overline{M}(z)/dt$ are computed. The calculated values are shown in Figs. 4 and 5.

d) The Malmquist effect[11] of our sample is examined. In

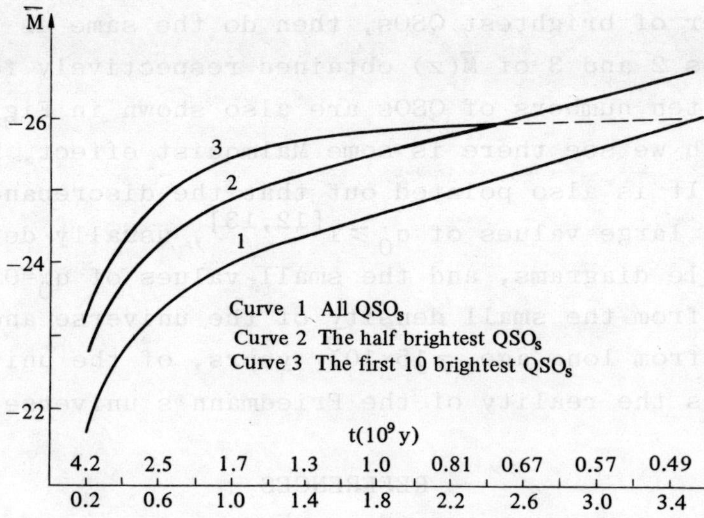

Fig. 4 The average absolute magnitude of QSOs at various z.

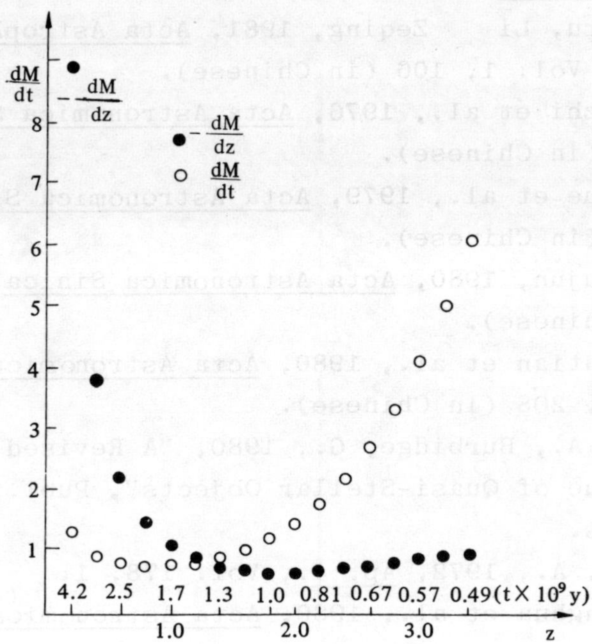

Fig. 5 $-\frac{dM}{dz}$ and $\frac{dM}{dt}$ of QSOs

each interval we take both the half number and the first ten number of brightest QSOs, then do the same as above. The curves 2 and 3 of $\overline{M}(z)$ obtained respectively for the half and ten numbers of QSOs are also shown in Fig. 4, from which we see there is some Malmquist effect, but not serious. It is also pointed out that the discrepancy between the large values of $q_0 \gtrsim 1$[12,13], usually derived from Hubble diagrams, and the small values of q_0=0.03-0.2[14], resulted from the small density of the universe and especially from long age, $\gtrsim 15\times10^9$ years, of the universe, challenges the reality of the Friedmann's universe model.

REFERENCES

[1] Mattig, M., 1958, Astron. Nachr., Vol. 284, 109; 1959, Astron. Nachr., Vol. 285, 1.

[2] Zhu Baoru, Li Zeqing, 1981, Acta Astrophysica Sinica, Vol. 1, 106 (in Chinese).

[3] Fang Lizhi et al., 1976, Acta Astronomica Sinica, Vol. 17, 134(in Chinese).

[4] Qu Qinyue et al., 1979, Acta Astronomica Sinica, Vol. 20, 98 (in Chinese).

[5] Chang Fujun, 1980, Acta Astronomica Sinica, Vol. 21, 7 (in Chinese).

[6] Yang Lantian et al., 1980, Acta Astronomica Sinica, Vol. 21, 208 (in Chinese).

[7] Hewitt, A., Burbidge, G., 1980, "A Revised Optical Catalogue of Quasi-Stellar Objects", Publ. Kitt Peak Nat. Obs.

[8] Sandage, A., 1972, Ap. J., Vol. 178, 1.

[9] Xiao Xinghua et al., 1980, Acta Astronomica Sinica, Vol. 21, 368 (in Chinese).

[10] Weinberg, S., 1972, "Gravitation and Cosmology", J. Wiley, New York, 482.

[11] Malmquist, K., 1921, Arkiv. Mat. Ast. Och. Fysik, Vol. 16, No. 23.

[12] Kristian, J., Sandage, A., Westphal, L., 1978, Ap. J., Vol. 221, 383.

[13] Meinel, R., 1981, Astron. Nachr., Vol. 302, 170.

[14] Sandage, A., Tammann, G., 1975, Ap. J., Vol. 197, 265.

THE STATISTICAL ANALYSIS OF Lα ABSORPTION LINES FROM A UNIFORM SAMPLE OF NINE HIGH REDSHIFT QSOs

ZOU Zhenlong(邹振隆), CHEN Jiansheng(陈建生),
BIAN Yulin(卞毓麟), TANG Xiaoying(唐小英),
CUI Zhenxing(崔振兴)
(Beijing Observatory, Academia Sinica, Beijing)

The high resolution spectra of the large redshift QSOs have revealed that there are more absorption lines on the short wavelength side of the emission line than on the long-wavelength side. According to a recent study[1-8], it is reasonable to believe that the unidentified absorption lines shortward of Lα emission are mainly caused by the hydrogen clouds.

A large number of well established Lα absorption lines will provide the sample for the study of the properties of the clouds. Using the data of five QSOs, Sargent et al.[9] (hereafter SYBT) concluded that the Lα clouds are cosmologically distributed entities (unassociated with QSOs or with intervening galaxies).

In this paper, we almost double the sample size to examine if the SYBT's conclusions still hold for the larger sample. Four new QSOs we added are PKS 0528-250[2], PKS 0805+046[3], PKS 1448-232[10], and PKS 1442+101[8]. All of these objects were observed at the Cassegrain focus of AAT by using the RGO spectrogragh, 1200/mm grating and 25cm camera and having nearly the same resolution and the ratio of signal to noise as those of SYBT. The data are believed to match the five ones of SYBT, so we obtained a larger uniform sample. We selected, following the SYBT's criteria, the Lα absorption lines with rest equivalent width $W_0 \geq 0.32Å$. In addition, we found the Lα sample satisfying the condition

$$1025.72(1+Z_{em}) \leq \lambda \leq 1215.67(1+Z_{em}).$$

Therefore, there are only L_α lines in the window and no L_β lines. This slightly modified the SYBT's sample by taking Off the lines outside the window and left 350 lines altogether for nine QSOs (Table 1).

Table 1. The L_α sample

QSO	λ_{min}(Å)	λ_{max}(Å)	z[a]	z[b]	No.
1 PKS 2126-158	4390	5203	2.611	3.280	47
2 Q 0002-422	3860	4575	2.175	2.763	35
3 PHL 957	3784	4486	2.113	2.690	20
4 Q 0453-423	3750	4445	2.085	2.656	37
5 B2 1225+31.7	3282	3890	1.700	2.200	20
6 PKS 0805+046	3976	4713	2.271	2.877	46
7 PKS 0528-250	3862	4577	2.177	2.765	41
8 PKS 1448-232	3301	3912	1.715	2.218	28
9 PKS 1442+101	4660	5524	2.833	3.544	76

In order to test whether a significant difference in the L_α line density occurs among the different QSOs and redshift intervals, we used the two-factor variance analysis and the comparison test[11] instead of the χ^2-test used by SYBT. In our case, the factor A represents the redshift level (we take m_A = 3 to make enough L_α lines in each interval), and the factor B represents QSOs, here m_B = 9

Table 2(1). Two factor variance analysis

	N(z)	$No(q_0=0)$	$No(q_0=0.5)$	$No(q_0=1)$
F_A	0.81	0.6	0.7	0.7
F_B	3.1	1.4	2.1	2.1

The theoretical values: $F_{0.05}^{(A)}=3.6$ $F_{0.05}^{(B)}=2.6$

Table 2(2). Comparison test

	N(z)	$No(q_0=0)$	$No(q_0=0.5)$	$No(q_0=1)$
$\overline{K}_{I}^{(A)}$	58.94	17.28	31.74	41.42
$\overline{K}_{II}^{(A)}$	70.84	20.19	37.76	49.42
$\overline{K}_{III}^{(A)}$	63.23	17.08	32.81	43.14
$\overline{K}_{1}^{(B)}$	70.29	17.91	35.47	46.86
$\overline{K}_{2}^{(B)}$	59.52	17.28	32.06	41.92
$\overline{K}_{3}^{(B)}$	34.69	10.10	18.71	24.46
$\overline{K}_{4}^{(B)}$	64.78	19.05	35.12	45.87
$\overline{K}_{5}^{(B)}$	39.99	13.70	23.40	30.13
$\overline{K}_{6}^{(B)}$	75.95	21.48	40.38	52.91
$\overline{K}_{7}^{(B)}$	69.69	19.87	37.20	48.72
$\overline{K}_{8}^{(B)}$	55.69	18.54	32.12	41.47
$\overline{K}_{9}^{(B)}$	107.07	25.72	52.46	69.59
T_A	25.26	7.93	14.14	18.36
T_B	60.30	18.92	33.74	43.82

means that nine QSOs are contained in our sample. The results are shown in Table 2, where the symbols have the same meaning as in [11]. From this we can conclude that the variation of the L_α line density with redshift is compatible with the Friedmann cosmological models, in which the number density of the absorption clouds per unit comoving volume is constant.

In order to test if the rest equivalent width W_0 of the L_α absorption line is related to redshift z (or the ejection velocity β), we used the nonparametric tests for the data points (z, Wo) and (β, W_0). (see, for example, [9,12]). The results are given in Table 3. We can see that there is no significant correlation between W_0 and z, and the correlation between W_0 and β is at the marginal level for our sample. The results above place strong limitations on the ejection hypothesis. They are entirely consistent with the intervening hypothesis.

Table 3(1). Rank correlation test

	n	ρ_s	$\mu(\rho_s)$	$\sigma(\rho_s)$	$(\rho_s-\mu)/\sigma$	ρ_k	$\mu(\rho_k)$	$\sigma(\rho_k)$	$(\rho_k-\mu)/\sigma$
(W,Z)	350	0.0001	0	0.0535	0.002	0.0097	0	0.0358	0.27
(W,β)	350	0.1198	0	0.0535	2.24	0.0938	0	0.0358	2.62

Table 3(2). Wilcoxon and chain test

i	j	L_i	L_j	L_i+L_j	μ_Q	Q	σ_Q	$\frac{Q-\mu}{\sigma}$	P	μ_P	σ_P	$\frac{P-\mu}{\sigma}$
1	2	47	35	82	1950	2094	106.7	1.34	38	41.1	4.40	-0.70
1	3	47	20	67	1598	1652	73.0	0.75	32	29.1	3.39	0.85
1	4	47	37	84	1997	2011	111.0	0.13	41	42.4	4.49	-0.31
1	5	47	20	67	1598	1746	73.0	2.03	30	29.1	3.39	0.27
1	6	47	46	93	2209	2113	130.1	-0.74	48	47.5	4.79	0.10
1	7	47	41	88	2091	2158	119.5	0.56	46	44.8	4.64	0.26
1	8	47	28	75	1786	1783	91.3	-0.03	36	36.1	4.02	-0.02
1	9	47	76	123	2914	3021	192.1	0.56	53	59.1	5.21	-1.17
2	3	35	20	55	980	949	57.2	-0.54	27	26.5	3.40	0.15
2	4	35	37	72	1277	1180	88.8	-1.09	35	37.0	4.21	-0.47
2	5	35	20	55	980	1056	57.2	1.34	23	26.5	3.40	-0.88
2	6	35	46	81	1435	1188	104.9	-2.35	38	40.8	4.39	-0.64
2	7	35	41	76	1347	1292	96.0	-0.58	32	38.8	4.30	-1.58
2	8	35	28	63	1120	1012	72.3	-1.49	33	32.1	3.89	0.23
2	9	35	76	111	1960	1808	157.6	-0.96	47	48.9	4.52	-0.42
3	4	20	37	57	580	548	207.2	-0.15	32	27.0	3.40	1.47
3	5	20	20	40	410	457	128.1	0.37	22	21.0	3.12	0.48
3	6	20	46	66	670	564	71.7	-1.47	31	28.9	3.40	0.62
3	7	20	41	61	620	623	65.1	0.05	27	27.9	3.41	-0.26
3	8	20	28	48	490	454	47.8	-0.75	25	24.3	3.33	0.21
3	9	20	76	96	970	937	110.8	-0.30	33	32.7	3.20	0.09
4	5	37	20	57	1073	1187	59.8	1.91	25	27.0	3.40	-0.59
4	6	37	46	83	1554	1444	109.1	-1.01	39	42.0	4.47	-0.67
4	7	37	41	78	1461	1510	99.9	0.49	38	39.9	4.38	-0.43
4	8	37	28	65	1221	1204	75.5	-0.23	32	32.9	3.92	-0.10
4	9	37	76	113	2109	2163	163.4	0.33	50	50.8	4.66	-0.17
5	6	20	46	66	670	482	71.7	-2.62	27	28.9	3.40	-0.41
5	7	20	41	61	620	531	65.1	-1.37	27	27.9	3.41	-0.26
5	8	20	28	48	490	390	47.8	-2.08	22	24.3	3.33	-0.69
5	9	20	76	96	970	750	110.8	-1.98	30	32.7	3.20	-0.84
6	7	46	41	87	2024	2187	117.6	1.39	47	44.4	4.62	0.67
6	8	46	28	74	1725	1790	89.7	0.73	41	35.8	4.02	1.29

Table 3(2) (Continued)

i	j	L_i	L_j	L_i+L_j	μ_Q	Q	σ_Q	$\frac{Q-\mu}{\sigma}$	P	μ_p	σ_p	$\frac{p-\mu}{\sigma}$
6	9	46	76	122	2829	3112	189.3	1.50	58	58.3	5.16	-0.06
7	8	41	28	69	1435	1385	81.8	-0.61	37	34.3	3.97	0.68
7	9	41	76	117	2419	2380	175.0	-0.22	50	54.3	4.90	-0.88
8	9	28	76	104	1470	1539	136.5	0.51	43	41.9	3.98	0.28
Δz<0.1	$<\Delta Z>$ 0.1	49	301	350	8599	7832	656.8	-1.77	93	85.3	4.48	1.72

We also computed the two-point correlation functions[13] for the L_α samples in four new QSOs. The results are given in Fig. 1. In Fig. 2, we have displayed the fine scale structure of the two-point correlation function for the L_α lines near the origin. The data for all nine QSOs were added together and shown on expanded scale with a finer bin interval ΔS_o. It is clear that the observed correlation function is flat, unlike the two-point correlation function for galaxies. A χ^2-test has been done for 4 new objects, and the results are shown in Table 4.

We also compared the two-point correlation function of the L_α clouds with that of C IV doublets. The QSOs involved are the three SYBT's objects (PHL 957, B 2 1225+31.7, and PKS 0237-233 and four new objects. Fig. 3 shows that a peak in the correlation function of the C IV doublets still exists, while the correlation function of the L_α clouds is quite flat, indicating that they may be a new kind of objects in the universe. The properties of the clouds will be studied in the next paper.

We are gratefull to Dr. Xie Zhongjie of Beijing University for many discussions about the statistic test. The authors are indebted to Drs. Morton D.C., Peterson B.A., Wright A.E., and Jauncey D.L., for allowing us to use a part of unpublished data of the cooperative work.

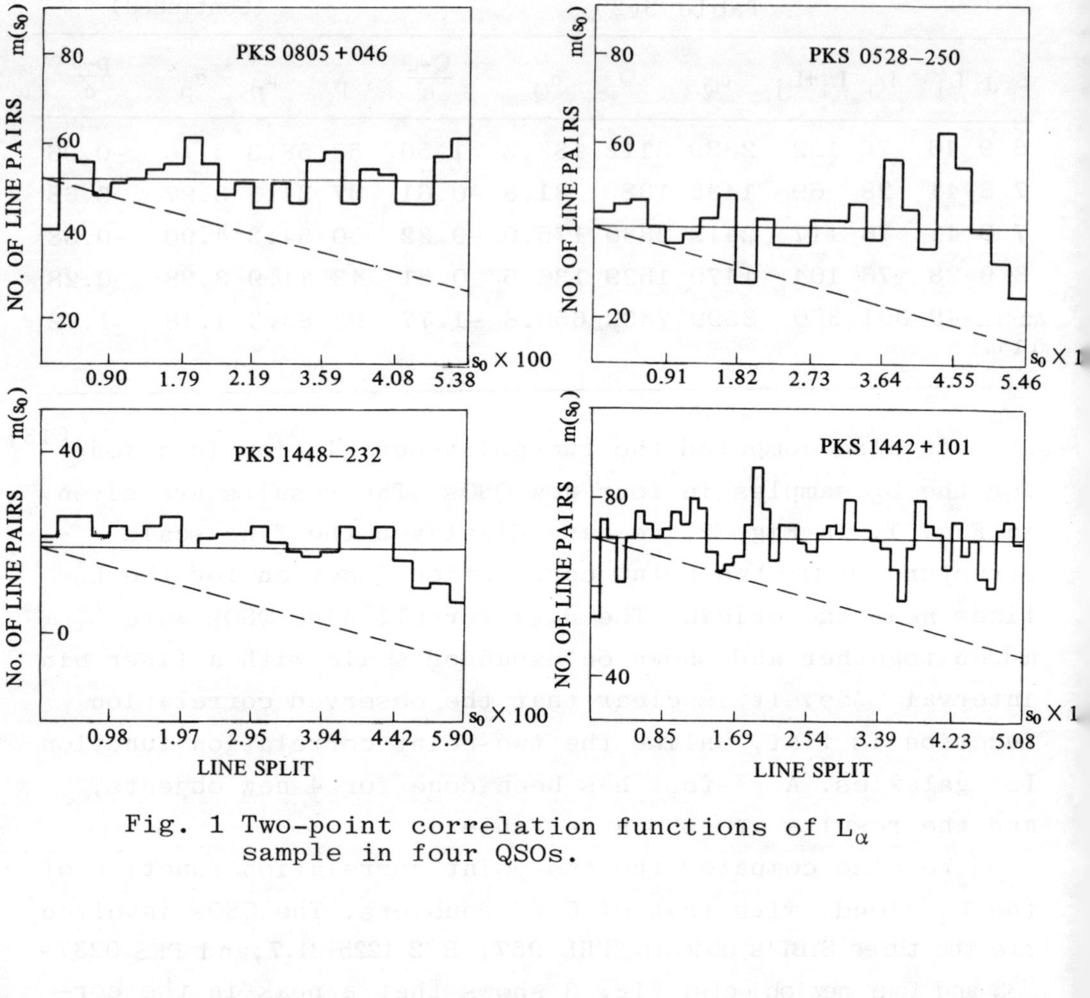

Fig. 1 Two-point correlation functions of L_α sample in four QSOs.

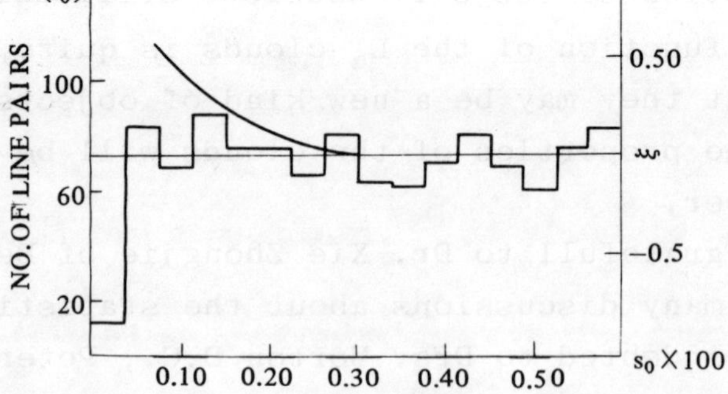

Fig. 2 Comparison between the two-point correlation function of L_α sample and that of galaxies (smooth curve).

Table 4. χ^2 test on two-point correlation function

QSO	n−2	χ^2_{n-2}	σ/σ_p	χ^2_{n-2} (Corrected)	$\chi^2_{n-2}(0.05)$
PKS 0805+046	22	9.44	0.67	20.95	33.9
PKS 0528−250	22	22.77	0.98	23.72	33.9
PKS 1448−232	22	21.03	0.97	22.50	33.9
PKS 1442+101	46	20.19	0.84	28.89	62.5

n = number of bins in splitting space

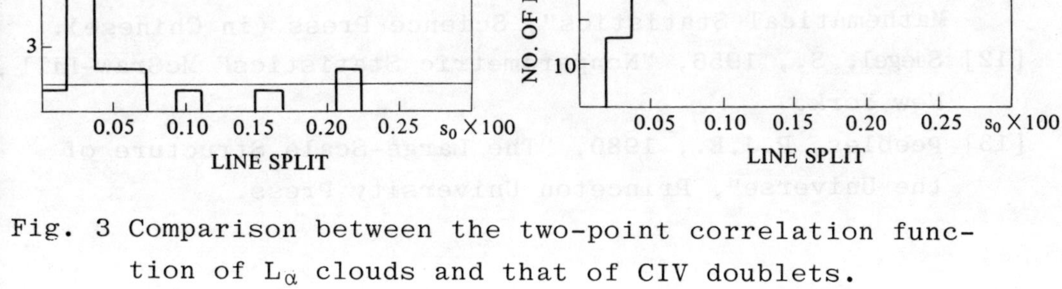

Fig. 3 Comparison between the two-point correlation function of L_α clouds and that of CIV doublets.

REFERENCES

[1] Chen Jiansheng, Morton, D.C., Peterson, B.A., Wright, A.E., Jauncey, D.L., 1981, Acta Astrophysica sinica, Vol. 1, 89.

[2] Morton, D.C., Chen, J.S., Wright, A.E., Peterson, B.A., Jauncey, D.L., 1980, M.N.R.A.S., Vol. 193, 399.

[3] Chen,J.S., Morton, D.C., Peterson, B.A., Wright, A.E., Jauncey, D.L., 1981, M.N.R.A.S., Vol. 196, 715.

[4] Cui Zhenxing, Chen Jiansheng, Tang Xiaoying, Bian Yulin, Zou Zhenlong, 1982, Acta Astrophysica Sinica (to be published).

[5] Lynds, C.R., 1971, Ap. J., Vol. 164, L73.

[6] Lynds, C.R., Oemler, A., 1975, Ap. J., Vol. 199, 558.

[7] Young, P.J., Sargent, W.L.W., Boksenberg, A., Caswell, R.F., Whelan, J.A.J., 1979, Ap.J., Vol. 229, 891.

[8] Chen, J.S., Morton, D.C., Peterson, B.A., Wright, A.E., Jauncey, D.L., (in preparation).

[9] Sargent, W.L.W., Young, P.J., Boksenberg, A., Tytler, D., 1980, Ap. J. Suppl. Ser., Vol. 42, 41.

[10] Chen, J.S., Morton, D.C., Peterson, B.A., Wright, A.E., Jauncey, D.L., (in preparation).

[11] Division of Probability and Statistics, Institute for Mathematics, Academia Sinica, 1973,"Usual Method of Mathematical Statistics", Science Press (in Chinese).

[12] Siegel, S., 1956, "Nonparametric Statistics" McGraw-Hill, New York.

[13] Peebles, P.J.E., 1980, "The Large-Scale Structure of the Universe", Princeton University Press.

THE NUCLEAR MASS OF TYPE I SEYFERT GALAXIES

LIU Ruliang (刘汝良)
(Purple Mountain Observatory, Academia Sinica, Nanjing)

I. THE NUCLEI OF SEYFERT I GALAXIES

The mass of central object in an active galactic nucleus is important in connection with the question of ultimate energy source of radiation. Woltjer[1] first obtained the nuclear masses of three Seyfert galaxies on the assumption that the observed widths of broad emission lines in active nuclei correspond to the velocity of matter (projectiles) freely falling in the gravitation field of the central compact object. In a recent paper[2], Dibay determined the nuclear mass and luminosity for 42 Seyfert I galaxies and 14 QSO's. However, his results led to a somewhat peculiar feature in the histogram of the nuclear mass, that is, QSO-Seyfert II-Seyfert I sequence. This is inconsistent with the current view that the Seyfert I may be interpreted as the mini-quasar. The peculiarity of Dibay's result may have originated in his use of the full width at half maximum (FWHM) of the Balmer lines as the measure of the velocity in the broad line region (BLR) of Seyfert I.

The hydrogen emission line in Seyfert I generally shows a broad component of the order of $\Delta v \cdot 10^3 - 10^4$ km s^{-1}. The broad component is formed by a system of moving dense clouds or filaments immersed in a much larger system of less dense, more homogeneous, and more slowly moving hot plasma in the central compact source which is less than 0.1pc in size. Outside this dense region there is a more rarefied extended region with size of the order of kpc in which the narrow components of the hydrogen line and narrow forbidden lines of other ions are formed. Many models of the BLR of Seyfert I have been suggested, which may be classified as: a) a spherical system of clouds in turbulent motion; b) a rotating, cylindrical system of clouds in turbulent motion[3] and c) a spherically expanding or collapsing system of discrete clouds[4].

In any case, if quasi-stationary state of such a system surrounding the central nuclei of Seyfert I galaxies is governed by the gravitational field of the central massive object, then the mass of central object may be written as

$$M = \frac{v^2 R}{G} = \frac{k v_{eff}^2 R}{G} , \qquad (1)$$

where G is the gravitational constant, R is the size of BLR, v is the characteristic velocity of the discrete gas clouds in the BLR, v_{eff} is the projection of v on the line of sight, and k is a transformation factor from v into v_{eff}. The value of k depends on the model of BLR.

II. THE SPECTRA OF MASS AND L_{bol} FOR SEYFERT GALAXIES

For the sake of simplicity, we take $k \sim 1$, a good choice for the case of spherically symmetric gas flow. The data of $v_{eff} = \frac{FWZI}{2}$ are taken mainly from Osterbrock[5], (FWZI means full width zero intensity) and the luminosities of Balmer line, $L(H_\beta)$, from Kriss et al.[6] and Weedman[7].

The Hubble constant is assumed to be $H_0 = 55$. The size of BLR, R, may be obtained from L (H_β) and so the nuclear mass M can be obtained by formula (1).

The bolometric luminosity L_{bol} is determined by Weedman's empirical formula[7]

$$\log L_{bol} = \log H_\beta + 3.2. \tag{2}$$

The Eddington luminosity is given by $L_{Edd} \simeq 1.3 \times 10^{38}\ M/M_\odot$ erg s^{-1}. Thus we can construct the mass spectra for 50 samples of Seyfert I, 17 samples of Seyfert II and 14 samples of low red-shift QSOs, as shown in Fig. 1. In Fig. 2 We also draw the historgram of the ratio L_{bol} / L_{Edd} for two types of Seyfert galaxies, in comparison with Dibay's data. It is noted that our result shows a larger overlap than Dibay's one.

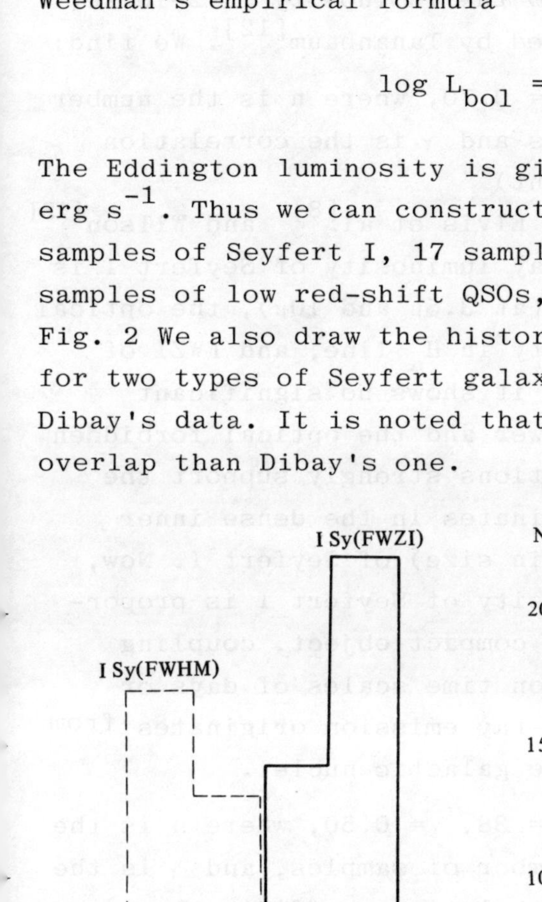

Fig. 1

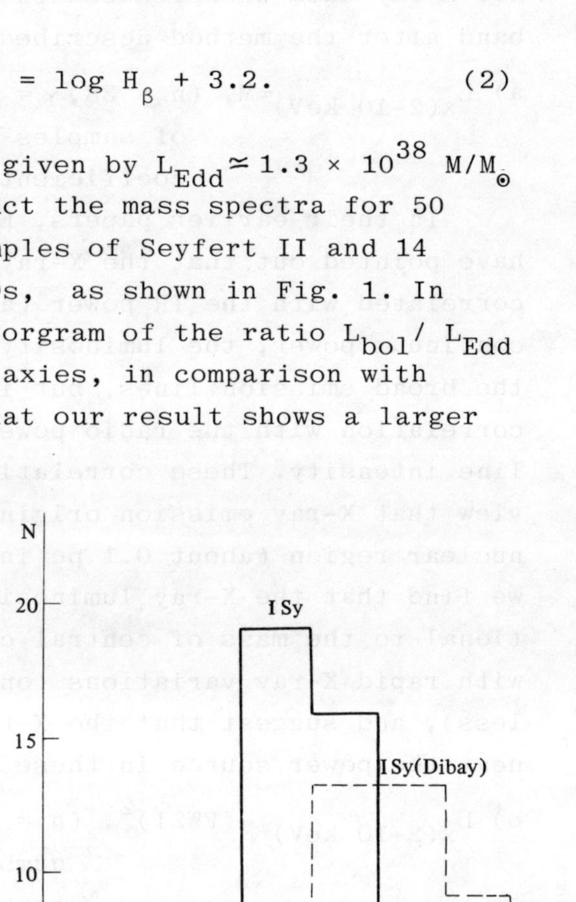

Fig. 2

III. L_x AND M OF SEYFERT I GALAXIES

The X-ray luminosities for 38 Seyfert I are taken from Kriss[6], Elvis[8], Tananbaum[9], Marshall[10], and Ward[11]. All X-ray data were reduced to the luminosity in 2-10 keV band after the method described by Tananbaum[12]. We find:

a) $L_{x(2\text{-}10\ keV)} \propto M$, (n = 38, γ = 0.70, where n is the number of samples and γ is the correlation coefficient)

In their earlier papers, Elvis et al.[8] and Wilson[13] have pointed out that the X-ray luminosity of Seyfert I is correlated with the IR power (at 3.5μ and 10μ), the optical continuum power, the luminosity in H_α line, and FWZI of the broad emission lines, but it shows no significant correlation with the radio power and the optical forbidden line intensity. These correlations strongly support the view that X-ray emission originates in the dense inner nuclear region (about 0.1 pc in size) of Seyfert I. Now, we find that the X-ray luminosity of Seyfert I is proportional to the mass of central compact object, coupling with rapid X-ray variations (on time scales of days or less), and suggest that the X-ray emission originates from near the power source in these galactic nuclei.

b) $L_{x(2\text{-}10\ keV)} \propto (FWZI)^2$, (n = 38, γ= 0.50, where n is the number of samples, and γ is the correlation coefficient).

This result could be compared with those of Комберг[14], and the L_x is connected with the dynamics of the clouds, i.e. with the falling matter accretion rate $\dot{M}$, if the accretion velocity should be given by the FWZI of the broad line.

Optical broad lines seem to arise from clouds moving at speeds of about ten thousands km s^{-1}. Shock wave could be produced by collision of such high velocity clouds.

Charged particles could be accelerated via shocks, or by some electromagnetic processes occurring near the massive central object. These energetic particles can emit X-rays as the synchrotron radiation or inverse Compton scattering. Therefore it is expected that L_x depends on the mass of central compact object and the physical, dynamic condition surrouding the nucleus of Seyfert I.

If the radiant energy ultimately comes from the gravitational energy, then it may be expected that

$$L_x = \alpha \frac{GM}{R} \dot{M} ,$$

where $\dot{M}$ is the mass accretion rate, and $\alpha(<1)$ is a numerical factor.

Hence we conclude: Our statistical result $L_x \propto M$ implies that, if the spherically symmetric accretion model is assumed, the ratio $\dot{M}/R$ is required to be nearly the same for most Seyfert I galaxies.

REFERENCES

[1] Woltjer, L., 1959, Ap. J., Vol. 130, 38.

[2] Dibay, Eh, A., 1981, AZ Lett., Vol. 7, 451.

[3] Osterbrock, D.E., 1979, A.J., Vol. 84, 901.

[4] Su Hongjun, Simkin, S.M., 1980, Ap. J., Vol. 238, L1.

[5] Osterbrock, D.E., 1977, Ap. J., Vol. 215, 733.

[6] Kriss, G.A. et al., 1981, Ap. J., Vol. 242, 492.

[7] Weedman, D.W., 1976, Ap.J., Vol. 208, 30.

[8] Elvis, M. et al., 1978, M.N.R.A.S., Vol. 183, 74.

[9] Tananbaum, H. et al., 1978, Ap. J., Vol. 223, 74.

[10] Marshall, F.E. et al., 1979, Ap. J.Suppl., Vol. 40, 657.

[11] Ward, M.J. et al., 1978, Ap. J., Vol. 223, 788.

[12] Tananbaum, H. et al., 1979, Ap. J., Vol. 234, L9.

[13] Wilson, A.S., Penston, M.V., 1979, Ap. J., Vol. 232, 389.

[14] Комберг, Б.В., 1981, ИН-Т Космич Исслед. АНСССР. Премр, Vol. 627, 38.

THE THEORY OF AN OBJECT WITH A COMPACT CORE AND ITS APPLICATIONS

FANG Lizhi (方励之),XIANG Shouping (向守平)

(Astrophysics Research Division, University of Science and Technology of China, Hefei)

I. INTRODUCTION

The configuration of a compact core located in the centre of an object is often related to high energy astrophysical phenomena. For example, the X-ray bursts of globular clusters have been considered as evidence of the existence of a compact core in such clusters. A bright core with a more bright "jet" coming off it are found in the elliptical galaxy M87. Moreover, neutrino halo models suggested by several authors[1] also have the core-halo structure.

In this paper we propose a theory for two classes of objects with the core-halo structure, namely, the loaded polytropes and the loaded neutrino objects. The theory is applied to analyze some properties of galactic nucleus and quasars.

II. THE RELATIVISTIC LOADED POLYTROPES

The classical loaded polytropes have been discussed by Huntley and Saslaw[2]. The equation of polytropic state is given by

$$P = \kappa\rho^{1+\frac{1}{n}}. \tag{1}$$

In the case of relativistic polytropes, the Emden equation is[3]

$$\xi^2 \frac{d\theta}{d\xi} \cdot \frac{1-2\sigma(n+1)[(\lambda\xi_0^3/3)+m(\xi)]/\xi}{1+\sigma\theta} + \frac{1}{3}\lambda\xi_0^3 + m + \sigma\xi\theta \frac{dm}{d\xi} = 0, \tag{2}$$

$$\frac{dm}{d\xi} = \xi^2\theta^n, \tag{3}$$

where

$$\left.\begin{array}{ll} \theta = (\rho/\rho_c)^{1/n}, & \sigma = \kappa\rho_c^{\frac{1}{n}}/c^2, \\ \xi_0 = r_0/a, & a = [\frac{(n+1)\sigma}{4\pi G\rho_c}]^{1/2}, \\ \lambda = \rho_0/\rho_c, & \rho_0 = 3M_0/4\pi r_0^3, \end{array}\right\} \tag{4}$$

r_0 and M_0 are the radius and mass of the core, respectively. The characters of the relativistic loaded polytropes can be summarized as follows:

a) The central peak of the density distribution is more remarkable than that of the classical cases, and the larger the σ, relativistic parameter, or the density of the core, the sharper is the central peak (Fig. 1).

b) The stable finite equilibrium configurations can exist only if the central core satisfies the following condition

$$2\sigma(n+1)\lambda\xi_0^2/3<1. \tag{5}$$

c) The stable equilibrium configurations can exist for all values of polytropic index in the range of n>0.

d) For n=3 loaded polytropes, besides the stable range with smaller σ, there is another metastable range at larger σ (Fig.2). The energy released during the possible transition between the metastable and the stable states is about 10^{60} erg.

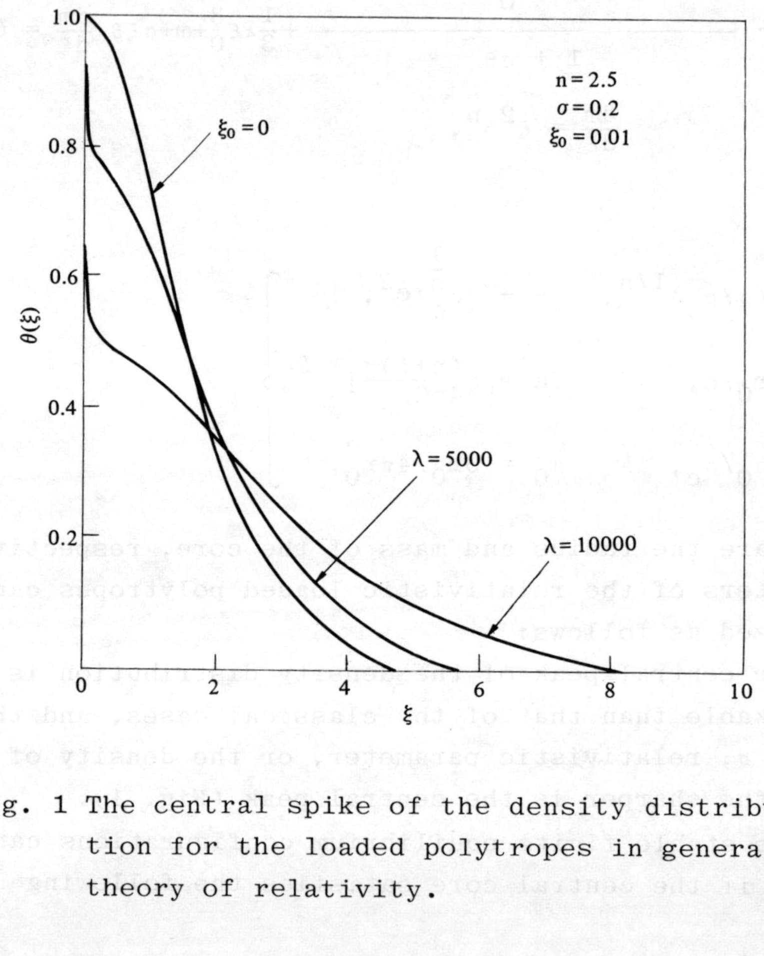

Fig. 1 The central spike of the density distribution for the loaded polytropes in general theory of relativity.

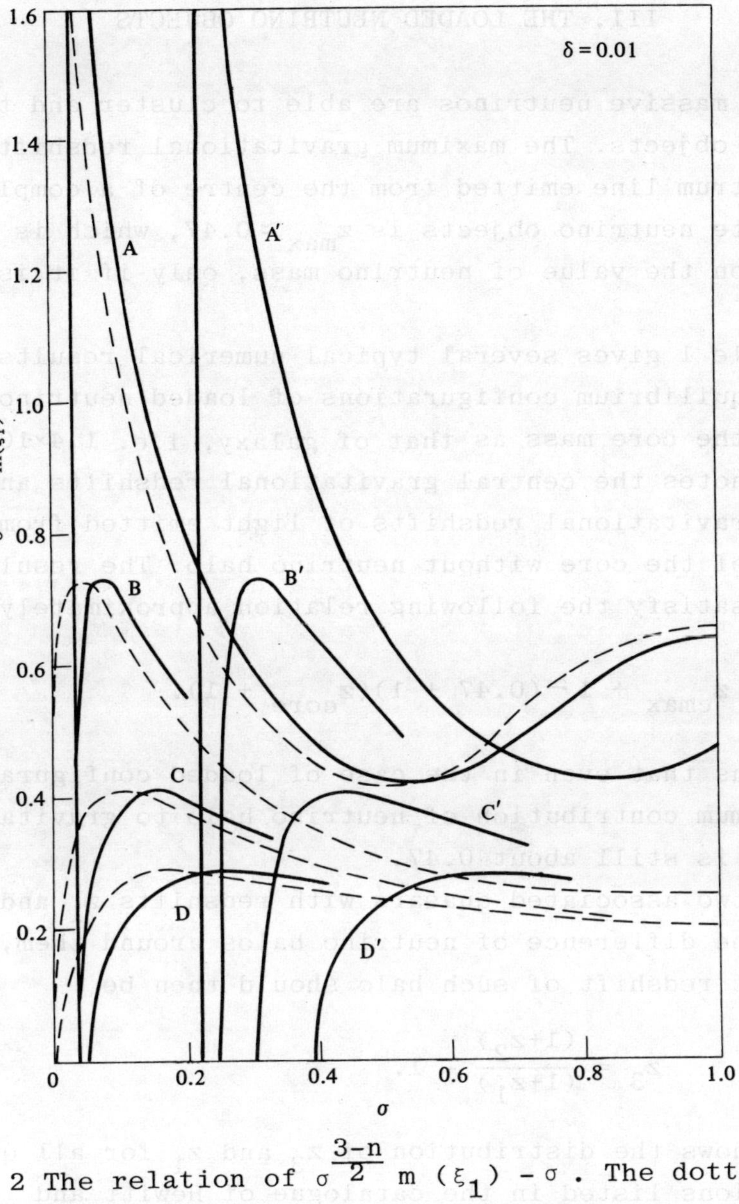

Fig. 2 The relation of $\sigma^{\frac{3-n}{2}}\ m\ (\xi_1) - \sigma$. The dotted lines show the situations without core. The letters (A,B,C,D) and (A', B', C', D') represent $\gamma = 0.2$ and $\gamma = 0.8$, respectively. The polytropic indices are, n=3(A,A'), n=2.5(B,B'), n=2(C, C'), n=1.5(D,D').

III. THE LOADED NEUTRINO OBJECTS

The massive neutrinos are able to cluster and to form neutrino objects. The maximum gravitational redshift of the spectrum line emitted from the centre of a completely degenerate neutrino objects is $z_{max} \simeq 0.47$, which is independent on the value of neutrino mass, only if it is nonzero[4].

Table 1 gives several typical numerical results of stable equilibrium configurations of loaded neutrino objects. We take the core mass as that of galaxy, i.e. $1.4\times10^{11}\ M_{\odot}$. z_{cmax} denotes the central gravitational redshifts and z_{core} is the gravitational redshifts of light emitted from the surface of the core without neutrino halo. The results in Table 1 satisfy the following relation approximately

$$z_{cmax} + 1 \simeq (0.47 + 1)(z_{core} + 1). \tag{6}$$

This means that even in the case of loaded configurations, the maximum contribution of neutrino halo to gravitational redshift is still about 0.47.

If two associated quasars with redshifts z_1 and z_2 are due to the difference of neutrino halos around them, the intrinsic redshift of such halo should then be

$$z_3 = \frac{(1+z_2)}{(1+z_1)} - 1. \tag{7}$$

Fig. 3 shows the distribution of z_3 and z_1 for all quasar associations listed in the catalogue of Hewitt and Burbidge[5], where z_1 is the smallest redshift in the association. No statistical difference can be found in between $z_3<0.47$ and >0.47. In both ranges the distributions under the upper limitline are rather uniform. This may imply that either there are only a few neutrino objects in the universe, or no massive halo of neutrino appears during the formation of quasars.

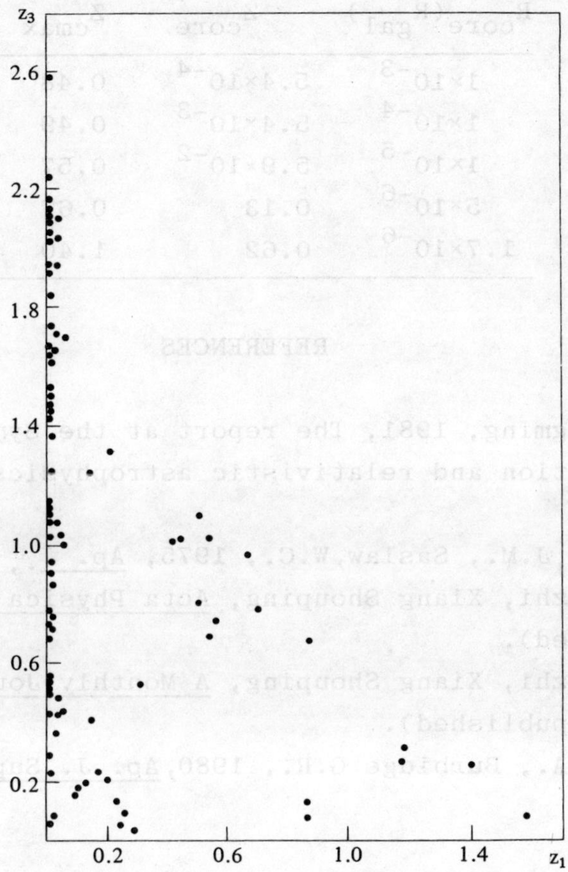

Fig. 3 The distribution of Z_1, Z_3 for all the associated quasars listed in the catalogue of Hewitt and Burbidge[5].

Table 1. The stable degenerate neutrino objects with a core

$R_{core}(R_{gal})$	Z_{core}	Z_{cmax}
1×10^{-3}	5.4×10^{-4}	0.48
1×10^{-4}	5.4×10^{-3}	0.49
1×10^{-5}	5.9×10^{-2}	0.57
5×10^{-6}	0.13	0.67
1.7×10^{-6}	0.62	1.40

REFERENCES

[1] Xu Chongming, 1981, The report at the Symposium on gravitation and relativistic astrophysics (Hangzhou, China).

[2] Huntley, J.M., Saslaw, W.C., 1975, Ap. J., Vol. 199, 328.

[3] Fang Lizhi, Xiang Shouping, Acta Physica Sinica (to be published).

[4] Fang Lizhi, Xiang Shouping, A Monthly Journal of Science (to be published).

[5] Hewitt A., Burbidge G.R., 1980, Ap. J. Suppl., Vol. 43, 57.

VIII. MASSIVE NEUTRINO AND COSMOLOGY

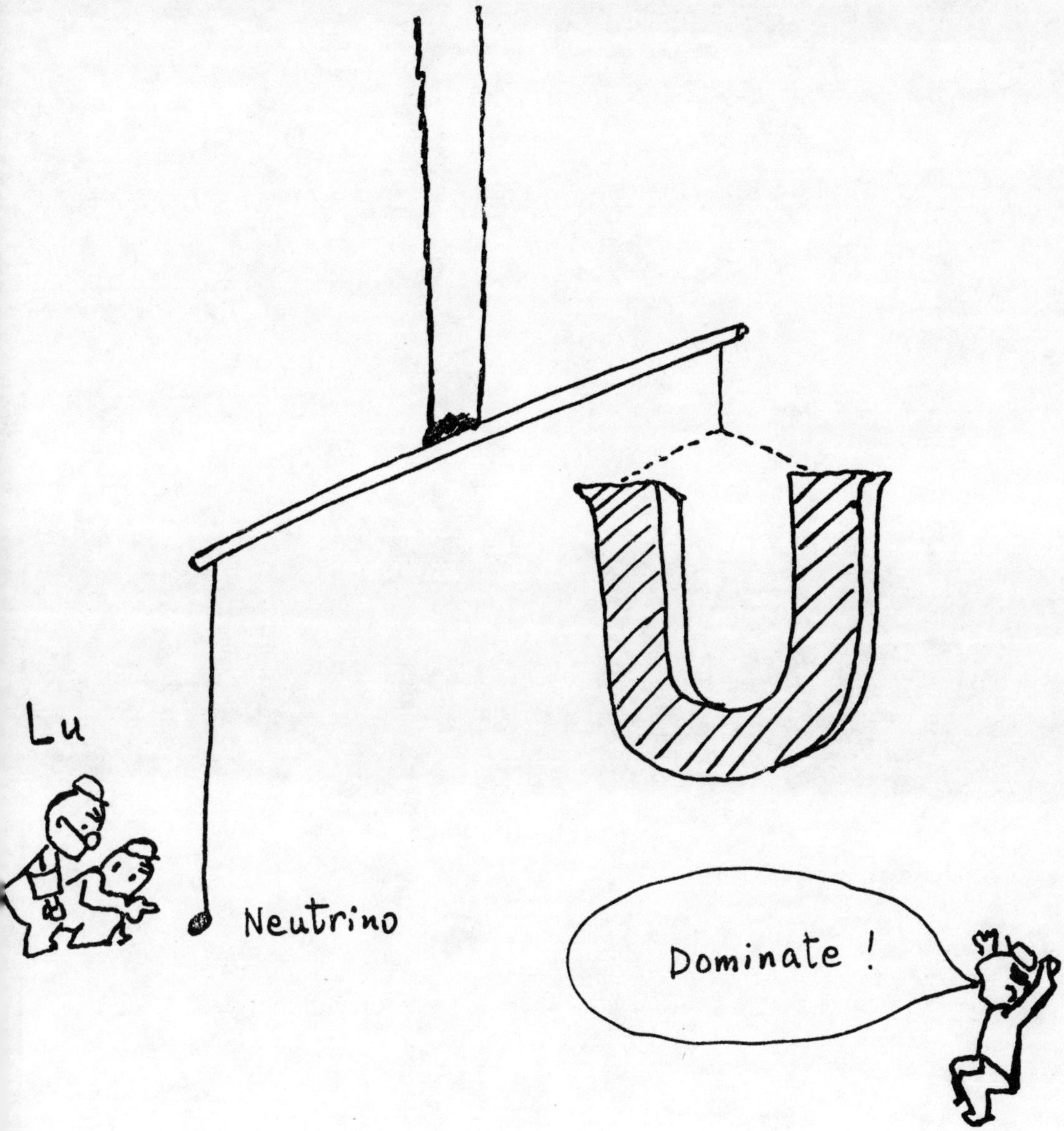
Lu
Neutrino
Dominate !

THE CLUSTERING OF MATTER IN
A NEUTRINO-DOMINATED UNIVERSE

LU Tan (陆 埮)

(Department of Astronomy, Nanjing University, Nanjing)

FANG Lizhi (方励之)

(Astrophysics Research Division, University of Science and Technology of China, Hefei)

I. INTRODUCTION

When he first received the news of the experimental proof of nonconservation of parity in beta decay, Prof. W. Pauli wrote in his letter that "this particle neutrino, of the existence of which I am not innocent, still persecutes me."[1] Indeed, the neutrino has been and is still being the most mysterious particle. It has persecuted many physicists, but it may open new ways especially for astrophysicists to solve many astrophysical, especially cosmological problems.

Recently, it was claimed that neutrino's rest-mass might be nonvanishing[2–3]

$$14 \text{ eV} < m_{\nu_e} < 46 \text{ eV}, \tag{1}$$

$$m_{\nu_e} \simeq m_{\nu_\mu} \; (\simeq m_{\nu_\tau}). \tag{2}$$

This fact is very important for cosmology. Neutrinos are rather abundant than baryons by a factor of about 10^9. Therefore, the mass density of neutrinos may be 100 times higher than that of baryons even if neutrino mass is as small as 10 eV.

So the nonvanishing rest mass of neutrinos should certainly induce many interesting astrophysical effects[4–5]. Within recent two years[6–21], there have been many achievements on this topic. In this paper, we will pick up some problems to review them.

The influences of neutrino rest mass on energy density in universe can be divided into two stages. When neutrinos are nonrelativistic, their rest mass itself would have a remarkable contribution to energy density. When neutrinos are relativistic, their rest mass can be neglected with respect to their kinetic energy. However, clue to the nonvanishing rest mass might double the species of neutrino, the energy density also increase in the latter case.

As is well known, all observed neutrinos are left-handed (ν_L). If neutrino rest mass is not zero, there should also be right-handed ones (ν_R). Note that the description of right-handed neutrinos is model dependent. If neutrino is of the Majorana type, $\nu_R \equiv \bar{\nu}_L$, namely neutrino and antineutrino are identical but with different chirality, then no doubling of species is needed in considering $m_\nu \neq 0$. In this case the neutrino is still of 2-component. However, it is more likely that the neutrino is of the Dirac type. This means that the neutrino with non-zero mass should be of 4-component. Hence, the right-handed neutrinos result in doubling the number of spin states with respect to the massless case. Up to now, no right-handed neutrinos have been detected in experiments of particle physics. This implies that even though ν_R exists, its interaction should be much weaker than the ordinary weak interaction. Among the related models of right-handed neutrinos, those with left-right symmetry are most natural, for example, $SU(2)_L \times SU(2)_R \times U(1)$[22,23] and $SU(2)_L \times SU(2)_R \times U(1)_L \times U(1)_R$[24,25]. In these models the interaction of right-handed neutrinos would be suppressed by the spontaneously breaking mechanism of vacuum.

Since the interaction of right-handed neutrinos should be very weak, they will decouple at an earlier time. We know, the decoupling of left-handed neutrinos took place after the annihilations of the hadrons and muons. Thus, if the decoupling of right-handed neutrinos had happened before these annihilations, all these annihilations should lead to the heating of left-handed ones, and therefore to the enhancement of them over right-handed ones.

For the minimal revision to the Weinberg-Salam-Glashow standard model of weak-electromagnetic unification, assuming the right-handed neutrinos to be a singlet of $SU(2) \times U(1)$, they would only interact with Higgs particles but not with gauge particles, right-handed neutrino would never be in thermal equilibrium in the early universe, and its abundance would be much more uncertain.

After all, we can say that the right-handed neutrinos are far less abundant than left-handed ones, and could at least in the first approximation be neglected.

II. JEANS INSTABILITY

Most important achievement made after consideration of nonzero mass of neutrino is about the formation mechanism of galaxies, clusters of galaxies and superclusters. As Dolgov and Zeldovich recently said "On the blue sky of modern cosmology there is a cloud: the observed structure of the universe shows a considerable inhomogeneity at the scale of less than or about 100 Mpc."[26] The existence of nonzero rest mass of neutrinos might open new ways to dispel this cloud.

Most people believe that the formation of galaxies, clusters of galaxies and superclusters could be regarded as due to the clustering of matter from the uniform background.[27] The first stage of clustering is the Jeans instability stage. This means that if there appears a small perturbation with scale λ larger than λ_J upon a uniform background with density ρ, the perturbation would then be instable and growing. Here

$$\lambda_J = (\pi v_s^2 / G\rho)^{1/2} \tag{3}$$

is known as the Jeans wave length, v_s denotes the speed of sound in the background medium. In other words, if the mass M in the perturbed region is larger than the critical mass M_J, the matter in this region would contract and cluster. Here

$$M_J = \frac{4\pi}{3} \rho \lambda_J^3 \tag{4}$$

is known as the Jeans mass. The physical meaning is obvious. The perturbation in the background medium may be regarded as the sound wave in a self-gravitating medium. There is a battle between pressure and gravity in that medium. When $\lambda > \lambda_J$ or equivalently $M > M_J$, the gravity overcomes the pressure, then the perturbations would be amplified. When $\lambda < \lambda_J$ or $M < M_J$, the pressure overcomes the gravity, then the perturbations would propagate in the medium as a sound, and it would be attenuated by viscosity and radiative diffusion.

In the case of $m_\nu = 0$, we can find:

a) Jeans wave length λ_N:

Before the time of the recombination, $T > 4\times10^3$ K, the Jeans wave length of matter coupling with radiation is

$$\lambda_N = 1.6\times10^{11} \rho_N^{-1/3} \left(1 + \frac{\sigma k T}{m_N c^2}\right)^{-1} \sigma^{2/3} \text{ cm}, \tag{5}$$

where ρ_N is the baryon density, and

$$\sigma = (4aT^3/3\ n_N k) = 3.6\ \eta^{-1}, \tag{6}$$

$$\eta = n_N/n_\nu . \tag{7}$$

After the recombination, $T<4\times10^3$ K, matter being decoupled from radiation, we have

$$\lambda_N = 1.6\times10^{20}(T/4\times10^3\ \text{K})^{-1/2}\quad \text{cm}, \tag{8}$$

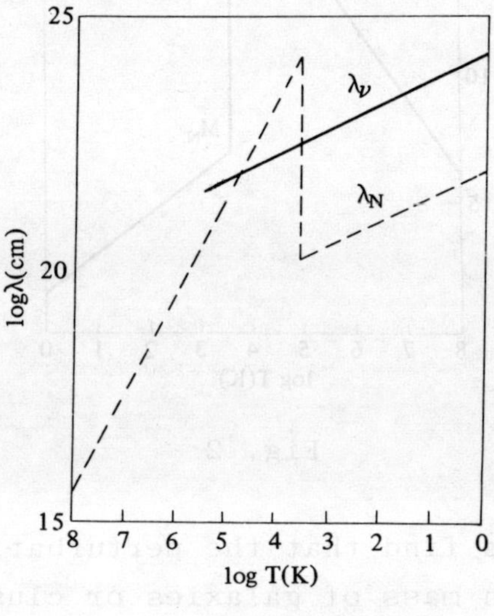

Fig. 1

Fig. 1 shows the Jeans wave length λ_N as a function of the cosmic temperature T, in which we have taken the present values of the Hubble constant $h_0 = 0.75$ in unit of 100 km $s^{-1}Mpc^{-1}$, T = 2.7 K and $\rho_{no} \simeq 3.1\times10^{-31}$ g cm^{-3}.

b) Jeans mass M_N:

Before the recombination, we have the Jeans mass

$$M_N = \frac{4\pi}{3}\ \rho_N \lambda_N^3 = 8.9\left(1 + \frac{\sigma kT}{m_N c^2}\right)^{-3}\ \sigma^2\ \ M_\odot ; \tag{9}$$

while after the recombination, we have

$$M_N = 8.7\times10^6 \ (T/4\times10^3)^{3/2} \ M_\odot . \tag{10}$$

Fig. 2 shows the Jeans mass as a function of T.

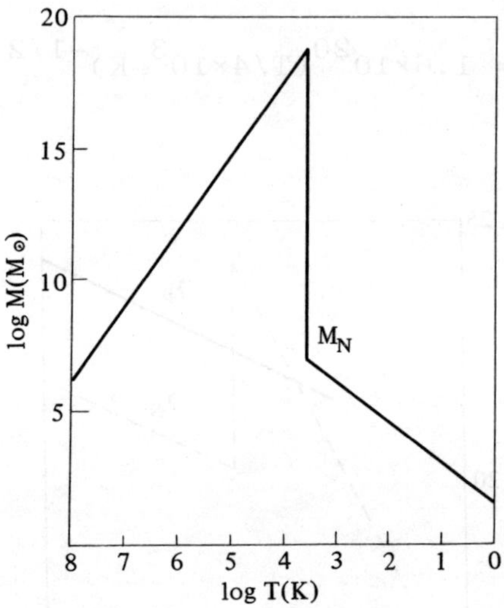

Fig. 2

From Fig. 2 we find that the perturbations in the early universe with mass of galaxies or clusters of galaxies must go through a unamplified and perhaps damping regions. After the recombination, the perturbation with mass larger than about $10^6 \ M_\odot$ could be amplified. No preferential clustering mass exists. This is in conflict with the fact that the mass of galaxy and the mass of cluster of galaxies are relatively definite, about $10^{11} \ M_\odot$ and $10^{14} \ M_\odot$, respectively, within a factor of about 100.

III. THE EFFECTS OF MASSIVE NEUTRINOS ON THE CLUSTERING OF MATTER

If $m_\nu \neq 0$, we must consider the effects of the cosmic neutrinos on the Jeans instability. Because the Jeans clustering of neutrino occurred only after its decoupling, so in the period of the Jeans clustering the matter in the universe could be described as a two-component fluid: one is the neutrino gas; one is the matter (plasma of ionized H and He, before the recombination) plus the background radiation. This two components interact with each other only through the gravitational force.

Now, let us first discuss the Jeans problem in such a gravitationally coupled two component fluid. The fundamental equations for the non-relativistic two-component fluid are[6–7]:

Component N:

$$\frac{\partial \rho_N}{\partial t} + \nabla\cdot(\rho_N \vec{v}_N) = 0,$$
$$\frac{\partial \vec{v}_N}{\partial t} + (\vec{v}_N\cdot\nabla)\vec{v}_N = -\frac{1}{\rho_N}\nabla P_N + \vec{g}; \tag{11}$$

Component ν:

$$\frac{\partial \rho_\nu}{\partial t} + \nabla\cdot(\rho_\nu \vec{v}_\nu) = 0,$$
$$\frac{\partial \vec{v}_\nu}{\partial t} + (\vec{v}_\nu\cdot\nabla)\vec{v}_\nu = -\frac{1}{\rho_\nu}\nabla P_\nu + \vec{g}; \tag{12}$$

Gravitational field $\vec{g}$:

$$\nabla\times\vec{g} = 0,$$
$$\nabla\cdot\vec{g} = -4\pi G(\rho_N + \rho_\nu). \tag{13}$$

From the above equations, one can easily derive two coupled dispersion equations for density perturbation upon a static and homogeneous background

$$\ddot{\rho}_{1k} = \overline{v}^2_{ks}\nabla^2\rho_{1k} + 4\pi G\rho_k(\rho_{1\nu} + \rho_{1N}), \tag{14}$$

where subscript 1 denotes the quantities of perturbation and $k = \nu$, N. We can then show that the necessary and sufficient condition for the onset of Jeans instability of this gravitationally coupled two-component system is

$$\lambda > \lambda_J, \tag{15}$$

$$\lambda_J^{-2} = \lambda_\nu^{-2} + \lambda_N^{-2}, \tag{16}$$

where λ_J is the Jeans length of the coupled system;

$$\lambda_\nu = (\pi v_{\nu s}^2 / G\rho_\nu)^{1/2}$$

and

$$\lambda_N = (\pi v_{Ns}^2 / G\rho_N)^{1/2}$$

are the Jeans length of the components ν, N respectively.

In order to estimate λ_ν, we should calculate the characteristic speed in neutrino gas $v_{\nu s}$. After decoupling and e^+e^- annihilation, and as neutrino still being relativistic, neutrino distribution would be

$$dn_\nu = \frac{4\pi}{(2\pi\hbar)^3} p^2 dp\{e^{pc/kT'_\nu} + 1\}, \tag{17}$$

where T'_ν is the effective temperature of neutrino gas, which is related to the radiation temperature by

$$T'_\nu = 0.71\ T. \tag{18}$$

However, as neutrino gas has become non-relativistic, the distribution is still correct, but its effective temperature T_ν should instead be[7,10]

$$T_\nu = 6.3\times 10^{-6}\ (30\ \text{eV}/m_\nu) T^2\ \text{K}. \tag{19}$$

Although neutrino gas has become collisionless, due to its dispersion in motion, we can still define the pressure of the neutrino gas as

$$P_\nu = \int v_z p_z \, dn_\nu . \tag{20}$$

Therefore, the characteristic speed $v_{\nu s}$ could be found as

$$\begin{aligned} v_{\nu s}^{2} &= (\partial P_\nu / \partial \rho_\nu)_{adiabatic} \\ &= 2.28\times10^{-10} c^2 (30 \text{ eV}/m_\nu)^2 (T/2.7\text{K})^2 . \end{aligned} \tag{21}$$

With this in mind, we can readily obtain λ_ν. Fig. 1 also shows λ_ν as a function of cosmic temperature T, in which we have taken m_ν = 30 eV and g_ν = 2, g_ν is the number of spin states of neutrino.

As can be seen by comparing with λ_N, before the recombination, $\lambda_\nu \ll \lambda_N$, so $\lambda_J \simeq \lambda_\nu$; after the recombination, $\lambda_\nu \gg \lambda_N$, so $\lambda_J \simeq \lambda_N$. This means the Jeans clustering of neutrinos could occur before the recombination, and it could carry ordinary matter to cluster together. Therefore, the mass of ordinary matter carried by neutrinos during their contracting and clustering should be

$$\begin{aligned} M_J &= \frac{4}{3}\pi \rho_N \lambda_\nu^3 \\ &= 3.8\times10^{12} (m_\nu/30 \text{ eV})^{-9/2} (g_\nu/6)^{-1} (T/4\times10^3)^{3/2} M_\odot . \end{aligned} \tag{22}$$

And after recombination, we have

$$M_J \simeq M_N , \tag{23}$$

the same as Eq. (10). Fig. 3 shows M_J as a function of T.

On the other hand, there is another important mass limit known as the horizon mass M_{hor}, which is the mass of ordinary matter within the particle horizon

$$M_{hor} \simeq \frac{4}{3}\pi\rho_N(ct)^3$$

$$\simeq 1.1\times10^{17}(T/4\times10^3)^{-3}\ M_\odot. \qquad (24)$$

M_{hor} as a function of T is also shown in Fig. 3.

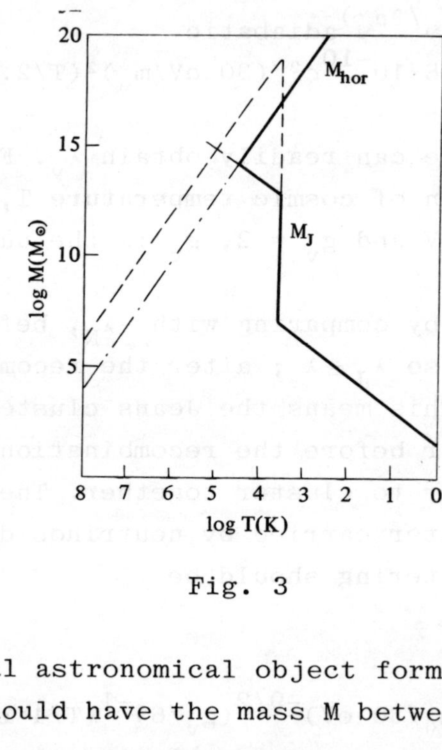

Fig. 3

The actual astronomical object formed through Jeans instability should have the mass M between M_J and M_{hor}, namely

$$M_J < M < M_{hor}. \qquad (25)$$

From the cross point of the M_J line and the M_{hor} line in Fig. 3, one can find a mass called preferential clustering mass to be

$$M_{pre}=8.4\times10^{13}(m_\nu/30\text{eV})^{-3}(g_\nu/6)^{-2/3}(T_0/2.7\text{K})^{-2}\ M_\odot, \qquad (26)$$

where T_0 is the cosmic background temperature of radiation. The value of formula (26) gives the earliest mass scale the Jeans instability to take place in the early universe. It is important to note that this preferential mass is just of the order of clusters of galaxies. This result may mean that the formation of clusters of galaxies depends mainly on purely gravitational clustering process, while galaxies themselves might be due to the fragmentation of cluster of galaxies in the successive time.

Strictly speaking, we must deal with the problem of cosmological Jeans clustering by the relativistic and expanding model. One can however adopt safely the Newtonian and non-expanding model, if our interests cover only the Jeans length and Jeans mass, but not the growing rate of small perturbations. The relativistic model changes only the growing rate obtained from the Newtonian model, but not the Jeans scales.

In the expanding universe, instead of Eq. (14), the dispersion equations should be[9–10]

$$\ddot{\eta}_k + 2\left(\frac{\dot{R}}{R}\right)\dot{\eta}_k + \frac{v_{ks}^2 q^2}{R^2}\eta_k = 4\pi G(\rho_\nu \eta_\nu + \rho_N \eta_N), \quad (27)$$

where

$$\rho_{1k}(t) = \delta\rho_k(t) e^{i\vec{q}\cdot\vec{r}/R(t)}, \quad (28)$$

$$\eta_k(t) = \delta\rho_k(t)/\rho_k(t). \quad (29)$$

With these equations, the above results could also be derived.

However, the clustering of ordinary matter before the recombination could make the microwave background radiation non-isotropic, so the observational fact that this radiation is highly isotropic has usually been regarded as a strong disproof for any clustering before the recombination.

But before the recombination the ordinary matter is ionized, there is strong coupling between ordinary matter and radiation. The ability of neutrino to carry ordinary matter during its contraction is rather weak. It can be shown[9,10]* from Eq. (27) in a neutrino dominated universe that the degree of non-uniformity of the ordinary matter reached at the era of the recombination is only in the order of about $\eta_N \simeq 10^{-4}$, which is no conflict with observational isotropic facts of microwave background radiation. The rapid growing of perturbation in ordinary matter could occur only after the recombination, while the mass of clustering matter would still keep the preferred value in the early universe.

IV. THE COLLAPSING STAGE OF CLUSTERING OF MATTER

In the later stage of the Jeans clustering, the pressure could be neglected relative to the gravity, we could then use the dust approximation[14-17]. Suppose that at the initial time t_i the perturbation in a region with mass in the range of (25) forms a overdense region with $\delta\rho_i = \bar{\rho}_i - \rho_{ci} > 0$, and begins to grows up, where $\bar{\rho}_i$ is the density in this region, and ρ_{ci} is the critical density at t_i.

For simplicity in quantitative analysis, we discuss an overdense region with spherical symmetry configuration.

As is well known in cosmology that the expanding dynamics of the universe with spherical symmetry is described by the equation

$$\dot{R}^2/c^2 + k = (8\pi G/3c^2)\rho R^2, \tag{30}$$

where R(t) is the famous Robertson-Walker scale factor and k is a constant with value + 1, 0, or -1 according as

---

* See also the paper by Lu and Zhu in this Proceedings.

$$\Omega = \rho/\rho_c \tag{31}$$

being positive, zero or negative, where

$$\rho_c \equiv 3H^2/8\pi G, \tag{32}$$

$$H \equiv \dot{R}/R \quad \text{(Hubble Constant)}. \tag{33}$$

For the spherical overdense region, the dynamics also satisfies the Eq. (30). In order to distinguish these two problems, we will instead use a(t) to denote the scale factor of the local overdense sphere, of which the radius can be denoted as

$$r(t) = r(t_i)\, a(t), \tag{34}$$

$r(t_i)$ being its initial radius. Obviously we have the initial conditions

$$a(t_i) = 1 \tag{35}$$

and

$$\dot{a}(t_i)/a(t_i) = \dot{R}(t_i)/R(t_i) = H_i \tag{36}$$

in the assumption that the expansion rates of the local sphere is the same as the cosmic background at the initial time t_i.

According to the explanation of Newton's theory, the -k term in Eq. (30) is proportional to the total energy of the system, so the system will expand forever if k<0, and it will turn into contraction at some time if k>0. Hence for our problem we can write the k as

$$k = \frac{8\pi G}{3c^2}\,\delta\rho_i > 0, \tag{37}$$

with

$$\delta\rho_i = \bar{\rho}_i - \rho_{ci}, \tag{38}$$

$$\rho_{ci} = 3H_i^2/8\pi G. \tag{39}$$

For non-relativistic case, the density ρ in the perturbed region is related to the initial density $\bar{\rho}_i$ by

$$\rho = \bar{\rho}_i/a^3. \tag{40}$$

Thus Eq. (30) for local sphere can be written as

$$\dot{a}^2 = \frac{8\pi G}{3a}\bar{\rho}_i - \frac{8\pi G}{3}\delta\rho_i. \tag{41}$$

Its solution is

$$a = \left(\frac{\bar{\rho}_i}{\delta\rho_i}\right)\frac{1-\cos\theta}{2}, \tag{42}$$

$$t = \frac{\bar{\rho}_i\rho_{ci}^{1/2}}{2H_i(\delta\rho_i)^{3/2}}(\theta - \sin\theta) + \xi, \tag{43}$$

where ξ is a small constant determined by the initial condition (35).

As can be seen from formula (42), a(t) reaches the maximum at $\theta = \pi$

$$a_{max} = \bar{\rho}_i/\delta\rho_i, \tag{44}$$

and the corresponding time, namely the time of the expansion to be stopped, is

$$t_s \simeq \pi\rho_{ci}^{3/2}/2H_i(\delta\rho_i)^{3/2}. \tag{45}$$

As $\theta = 2\pi$, a(t) decreases to zero, namely the end of the collapse stage, the corresponding time is

$$t_c \simeq \pi \rho_{ci}^{3/2} / H_i (\delta\rho_i)^{3/2} \tag{46}$$

Therefore, the condition that a clustering process could have been completed at present is given by

$$t_c < t_0, \tag{47}$$

where t_0 is the present age of the universe. Taking $q_0 \simeq 1/2$ (deceleration factor at present) and $h_0 \simeq 0.5$ (Hubble constant at present in unit of 100 km s^{-1} Mpc^{-1}), condition (47) can be written as

$$m_\nu > 7.5\ (10^{-3}/(\delta\rho_i/\rho_{ci}))^{3/4}\ \text{eV}. \tag{48}$$

This is a constraint on m_ν from the formation of the dynamical equilibrium objects.

Equilibrium objects obeys the virial theorem. Although no hydrodynamical viscosity are working in the neutrino component, this component still will undergo a strong relaxation process in the clustering. In fact, every neutrino moves in the gravitational field determined by other neutrinos. Such a field is variational with time and not a conservative field. In other words, this is a kind of very-many-body collision through gravitation. This process will bring relaxation. After the relaxation process accompanied with the clustering, the resulting objects could be in dynamical equilibrium states.[28]

If the radius of the clustering sphere decreases by a factor of β ($\simeq 2$, from the estimation[15]) in the period from t_s to the time of contraction stop, then the density of the equilibrated objects is given by

$$\langle\rho\rangle = \beta^3 a_{max}^{-3}\ \bar{\rho}_i$$

$$= \beta^3(\delta\rho_i / \bar{\rho}_i)^3(T_i/T_0)^3\rho_0, \tag{49}$$

so the radius of the object is

$$r = \frac{1}{\beta}\left(\frac{3M}{4\pi\rho_0}\right)^{1/3} \frac{(\bar{\rho}_i/\delta\rho_i)}{(T_i/T_0)} . \tag{50}$$

This could help us to estimate the scale of the neutrino object.

V. FURTHER REMARKS

In above discussions, we have only considered the spherical case. In reality, the problem is however more complicated. It is rather implausible to suppose contractions in all three directions being identical. The more probable case is that the contraction or clustering would have proceeded faster in some direction than in other two perpendicular directions. Thus, the density enhancements should be asymmetric, and so the highly flattened pancake-like distribution of matter should result.

Several detailed calculations about these problems have been done[5,20,29,30]. When the baryons contract gravitationally with neutrinos, their kinetic energy should be dissipated by, for instance, a shock wave process, and forming the objects separated from neutrinos. As a result, the system of baryons and neutrinos would become a system selfbound about the central plane, leading to a sandwich-like structure. This means that the shocked baryons cooling in the central plane would be surrounded by neutrinos.

It should be noted that such formed pancake-like structure would be the clusters of galaxies or even superclusters. This baryon pancake could continue to develop a high density, heat up, and fragment into smaller parts, to form galaxies. Neutrinos with low enough velocities would be bound to them as a neutrino halos. The larger the fragments are, the higher the velocities should be for neutrinos to escape. Hence the mass fraction in bound neutrinos, behaving as the so called "missing mass", increases with the size of the fragment. This is in agreement with the observational fact of the mass-to-light ratio of astronomical objects rising with increasing size[8].

Moreover, as the perturbations of matter develop locally and collapse into pancake-like objects, there should leave behind some large empty regions or huge holes void of matter with scales of about 40—100 Mpc[5]. It is very interesting that these features seem to have been discovered observationally.

If pancake-like object intersects with their neighboring pancakes, there might appear filamentary density enhancements. This feature remains to be observed in the future.

For neutrino astronomical objects, there have been some detailed discussions and calculations in the degeneracy approximation. Because of the transparency and the strong gravitational field in these objects, one can observe the light even from its centre, and these light spectrum should exhibit large redshifts. As the result of gravitational lens effect, the light therefrom may even be amplified enormously.[31,32—37]

It is interesting to note that the density perturbations behaving as sound waves with large wave length before the recombination might become as the "seeds" of Jeans instability after the recombination. Recently, the periodicity in the distribution of quasar redshifts has been found relative to the argument

$$x(z,q_0) = \int_0^z \frac{dz}{(1+z)(1+2q_0 z)^{1/2}} \; . \qquad (51)$$

This periodicity may be explained as the remains of those

density wave-like perturbations.[38,39]

REFERENCES

[1] Wu, C.S., 1960, "Theoretical Physics in the Twentieth Century", Fierz, M., Weisskoph, V.F. (eds.).

[2] Reines, F., Sobel, S.W., Pasierb, E., 1980, Phys. Rev. Lett., Vol. 45, 1307.

[3] Lyubimov, V.A. et al., 1980, Phys. Lett., Vol. 94B, 266.

[4] Lu Tan, Fang Lizhi, 1982, Prog. Phys., Vol. 2, No. 1, (in Chinese); Lu Tan, Fang Lizhi, 1982, "Physics and Contemporary Needs", Riazuddin (ed.), Vol. 6, Plenum Publishers.

[5] Bond, J.R., Szalay, A.S., Proc. Neutrino-81.

[6] Lu Tan, Luo Liaofu, Yang Guochen, 1980, Symposium on Cosmic Ray and High Energy Astrophysics (China); 1981, Acta Astronomica Sinica, Vol. 22, 251.

[7] Fang Lizhi, Liu Yongzhen, 1981, Nuovo Cimento Lett., Vol. 32, 129; 1981, Acta Astrophysica Sinica, Vol. 1, 291.

[8] Fang Lizhi, Kiang T., Hu Fuxing, Cheng Fuhua, 1982, Prog. Phys., Vol. 2, No. 3.

[9] Lu Tan, Zhu Peichen, 1981, Nature Journal, Vol. 4, 952.

[10] Lu Tan, Zhu Peichen, 1982, Kexue Tongbao, Vol. 27, 484.

[11] Fang Lizhi, Xiang Shouping, (to be published).

[12] Gao Jiangong, Ruffini, R., 1981, Acta Astrophysica Sinica, Vol. 1, 19.

[13] Qing Chengrui et al., 1981, Acta Astrophysica Sinica, Vol. 1, 9.

[14] Sato, H., Takahara, F., 1980, Prog. Theor. Phys., Vol. 64, 2029.

[15] Sato, H., 1980, preprint, RIFP-410; 1981, preprint,

RIFP-423.
[16] Sato, H., Takahara, F., 1981, Prog. Theor. Phys., Vol. 66, 508.
[17] Wasserman, I., 1981, Ap. J., Vol. 248, 1.
[18] Olive, K.A. et al., 1980, Enrico Fermi Institute, Preprint No. 80–42.
[19] Vishniac, E.T., 1981, APJMAR 1582.
[20] Bond, J.R., Efstathiou, G., Silk, J., 1980, Phys. Rev. Lett., Vol. 45, 1980.
[21] Klinkhamer, F.R., Norman, C.A., 1981, Ap. J., Vol. 243, L1.
[22] Pati, J., Salam, A., 1974, Phys. Rev., Vol. D10, 275; 1975, ibid, Vol. D11, 1137.
[23] Mohapatra, R. et al., 1975, Phys. Rev., Vol. D11, 566; 1977, ibid. Vol. D16, 2843.
[24] Shafi, Q., Watterich, Ch., 1978, Phys. Lett., Vol. 73B, 65.
[25] Luo Liaofu, Lu Tan, Yang Guochen, 1980, Phys. Lett., Vol. 95B, 257.
[26] Dolgov, A.D., Zeldovich, Ya. B., 1981, Rev. Mod. Phys., Vol. 53, 1.
[27] Weinberg, S., 1972, "Gravitation and Cosmology", J. Wiley, New York.
[28] Lynden-Bell, D., 1967, M.N.R.A.S., Vol. 136, 101.
[29] Zeldovich, Ya. B., 1970, Astron. Astrophys., Vol. 5, 84.
[30] Doroshkevich, A.G. et al., 1980, M.N.R.A.S., Vol. 192, 321.
[31] Xu Chongming, 1981, Nature Journal, Vol. 4, 878.
[32] Markov, M.A., 1964, Phys. Lett., Vol. 10, 122.
[33] Schramm, D.N., Steigman, G., 1981, Ap. J., Vol. 243, 1.
[34] Cowsik, R., McClelland, J., 1972, Phys. Rev. Lett., Vol. 29, 669.
[35] Gunn, J.E. et al., 1978, Ap. J., Vol. 223, 1015.

[36] Tremaine, S., Gunn, J.E., 1979, Phys. Rev. Lett., Vol. 42, 407.
[37] Jones, B.J.T., 1976, Rev. Mod. Phys., Vol. 48, 107.
[38] Fang Lizhi, Chu Yaoquan, Liu Yongzhen, Cao Shenglin, Astron. Astrophys. (to be published).
[39] Fang Lizhi, 1982, Preprint, RIFP-467.

DECELERATION OF THE EXPANDING UNIVERSE

FANG Lizhi(方励之)

(University of Science and Technology of China, Hefei)

HU Fuxing(胡佛兴)

(Purple Mountain Observatory, Academia Sinica, Nanjing)

ABSTRACT

The various methods of determining the deceleration parameter q_0 are reviewed. The Hubble diagram method for galaxies and various subsets of quasars consistently gives $q_0>1/2$, the mean mass density method always gives $q_0<1/2$, while the θ-z method more often than not follows the "Euclidean" relation $\theta \propto z^{-1}$. Thus, the ultimate question of whether the universe is open or closed now hinges on the question as to which factor is more important: luminosity decrease, or the existence of a quasi-uniform, non-luminous, non-baryonic mass distribution. We are inclined to the second alternative because it makes use of, rather than ignores, the clue from the mean density results. The recent discovery of a non-zero rest-mass of the electron neutrino offers a possible identification of the postulated mass distribution.

I. INTRODUCTION

The Hubble constant H_0 and the deceleration parameter q_0 are two most important parameters in standard cosmology. The Hubble constant H_0, indicating the expanding rate of the present universe, has now being narrowed down to the range about from 50 to 100km s^{-1} Mpc^{-1}[1–3]. The deceleration parameter q_0, indicating the deceleration of the expanding universe, however, still remains far unsolved.

The deceleration parameter q_0 is directly related to the basic question whether the universe is finite or infinite. Is the universe finite or infinite? It was a very interesting problem for philosophers even early before modern science rose into prominence. The discussions then, however, were largely philosophical and speculative. A lot of astronomical problems have been researched by comparing observational data with theories, since Galilean and Newton developed the classical dynamics, but except the finite-infinite question. This is because the observed astronomical phenomena are always local, yet the finite-infinite question is related to the universe as a whole.

The situation did not change until modern cosmology developed. The finite-infinite question now is no longer isolated. It is closely related to a lot of local observational phenomena. We can judge which cosmological model is more reasonable through the observations of local phenomena, and then infer whether the universe is open or closed.

This sort of local-whole connection is especially simple in standard cosmology. It depends on the value of deceleration parameter q_0.

For a pressureless universe, we have:

$$q_0 = 4\pi G\rho_0/3H_0^2, \tag{1}$$

$$\rho_{crit} = 3H_0^2/8\pi G = 1.9\times10^{-29}\ h^2\ g\ cm^{-3}, \tag{2}$$

$$\Omega_0 = \rho_0/\rho_{crit}, \qquad (3)$$

and

$$q_0 = \Omega_0/2, \qquad (4)$$

where h is the Hubble constant H_0 in units of 100 km s^{-1} Mpc^{-1} and ρ_{crit} means the critical density, the close or open of the universe depends on whether the present mean mass density ρ_0 is larger than ρ_{crit}

Therefore the universe is closed if $q_0>1/2$ or $\Omega_0>1$, and it is open if $q_0<1/2$ or $\Omega_0<1$.

Here we will review the present status of the determination of q_0 and outline some preliminary conclusions (cf.[4]).

II. THE DETERMINATIONS OF DECELERATION PARAMETER q_0

(1) Mean Mass Density in the Universe

Perhaps the most direct method is to determine the mass density ρ_0 in the universe. As early as in 1934, Hubble already estimated the contribution to ρ_0 from galaxies by $\rho_G = N\ M_G$, where N is the number of galaxies per unit volume and M the mean mass of galaxies.

A great many improvements have been made in recent decades. Instead of $\rho_G = N\ M_G$, we use $\rho_G = L_T\langle M/L\rangle$, where L_T is the luminosity of galaxies and $\langle M/L\rangle$ denotes mass-to-luminosity ratio. Some results are listed in Table 1. one find all the results suggested $\rho_G<<\rho_c$, (i.e. $\Omega<<1$). This means if the matter in the universe were mainly provided by galaxies, then $q_0<1/2$, namely, the universe should be infinite.

Table 1. Some results from the method of the mass-to-luminosity ratios of galaxies

Author	Samples of galaxies	$\Omega_G=\rho_G/\rho_c$
Shapiro, 1971[5]	brighter than magnitude 12(m_{pq})	0.01
Gott and Turner, 1976[6]	brighter than magnitude 12(m_{pq})	0.08
Davis et al., 1978[7]	all sky, complete to magnitude 13(m_{pq})	0.01—0.2

(2) Deuterium Abundance Method

The matter in the universe, however, may not all be presented in the form of galaxies. For example, intergalactic and intercluster gas or dust are not in the form of galaxies. We have to look for such a way that can be used to detect the matter not in the form of galaxies.

One such way are deuterium (D) abundance. As we know, D is very active, and it cannot be produced through stellar thermonuclear reaction. The fission reaction raised by the collision between cosmic ray and interstellar matter can lead to the production of deuterium. However, the quantities are very few[8]. Shock wave in supernova explosion probably produce more D, but, a lot of other lighter nucleus will be produced at the same time, e.g. ^{10}Be, ^{10}B etc[9,10]. Therefore if we detected alone higher abundance of D, this sort of D most probably be produced in early universe, rather than during the stellar evolutional processes. Since 1970s, the abundance of D has been detected. The most reliable measurement of the present value of abundance of D comes from the Lyman $\beta,\gamma,\delta,\varepsilon$ lines of atomic deuterium on board the Copernicus satellite[11—14]. The abundance from 10 stars' spectra are summarized in the first column of

Table 2.

Table 2. Deuterium abundance from observations of atomic lines of 10 stars and deduced upper bounds to baryon density

Value	Observed D/H (10^{-5})	Mass fraction X_D (10^{-5})	Upper bound to ρ_B (10^{-31} g cm^{-3})	Upper bound to Ω_B (h^{-2})
Max (Min)	2.5	3.5	4.3	0.023
Median	1.4	2.0	6.1	0.032
Min (Max)	0.63	0.88	7.8	0.041

For deuterium is depleted more easily through nuclear reaction, the higher the density of nuclide (or, exactly say, the baryon density) the lower is the deuterium abundance. According to Wagoner et al. 1967[15], Wagoner 1973[16], 1974[17], the fraction of deuterium by mass, X_D, decreases sharply with increasing the present baryon density ρ_B. In general, stellar processing can only lead to a depletion of D[18] (Lang 1974[19], p. 438), and the depletion in stars probably outweighs any production by SN or comic rays, so we should regard the present observed D abundance as a lower bound to the amount in the big-bang and the deduced value of ρ_B as an upper bound to the true one. Other corresponding quantities are also listed in Table 2. In the light of the numbers given in Table 2, we would prefer to conclude that

$$0.02\ h^{-2} \leqslant \Omega_B \leqslant 0.04\ h^{-2}. \qquad (5)$$

The result shows that if mass in the universe were mainly contributed by baryons, still, we should have $q_0 < 1/2$, i.e. it leads to the same conclusion: the universe should be infinite.

The main matter content in the universe, however, might not be baryons. For example, the number of neutrinos in the universe is 10^8–10^9 times more than the baryon number. We are now forced to look for some way to get the total mass.

(3) Dynamical Method

Dynamical measurement may be the oldest and most efficent method to estimated masses of systems in astronomy. The masses of astronomical systems, such as, galaxies, binary galaxies, galactic groups and clusters of galaxies etc. have been dynamically estimated. A most striking result here is that M/L increases with the linear size of system considered[20]. This tendency is shown clearly in Table 3, which is taken mainly from[21]. M/L appears to increase about 50-fold as we go from the kpc-sized single galaxies to the Mpc-sized Abell clusters. This means that only a small fraction of the mass of a cluster is in the form of luminous galaxies. The investigation of systems even larger than Abell cluster is desirable.

Table 3 Galactic mass-to-luminosity ratio

Objects	⟨M/L⟩ ($h\,M_\odot/L_\odot$)	mean ⟨M/L⟩
Single S galaxies	8 – 12	10
Single E+SO galaxies	10 – 20	15
Double galaxies	12 – 70	40
Groups of galaxies	60 – 180	100
Clusters of galaxies	200 – 900	500

Such a system is the Local Supergalaxy. On this scale we can no longer ignore the Hubble expansion. What we have to do is to link the deviation from the Hubble flow with any density enhancement over and above the mean background. We can obtain Ω_G as in the following:

$$\Omega_G = 2(v/V)\bar{\delta} \ , \qquad (6)$$

where v is the peculiar velocity of the Local Group in the direction of Virgo cluster and has been determined by a number of authors[22–25,3]. V is the Hubble velocity of the Local Group with respect to the Virgo cluster and can be taken to be 1050 km s^{-1}[26,27]. We find $v/V \simeq 0.3$. Yahil et al. (1980)[26] found a mean density contrast of $\bar{\delta}$ = 3.5 in the distribution of galaxies inward of the Local Group, then we have $\Omega_G \sim 0.2$. Davis et al. (1980)[28] obtained similar result:

$$0.3 < \Omega_G < 0.5. \qquad (7)$$

This shows that the contribution to the total density by invisible matter continues to rise right up to a scale of 10 h^{-1} Mpc. It also suggests that the invisible matter is not as clustered as the galaxies are.

A more recently developed dynamical method is by the cosmic energy theorem. Here the entire galactic population is regarded as a field of fluctuating density in random motion with respect to the Hubble flow. We seek the relation between the potential energy W in the density fluctuation and the kinetic energy of the peculiar velocities. The relation between T and W in an evolving universe is given by Layzer-Irvine equation[29]:

$$\frac{d}{dt}(T + W) + \frac{\dot{R}}{R}(2T + W) = 0. \qquad (8)$$

Solutions of the form $T = -\beta W$ are known in some particular cases, with $\beta = \frac{1}{2}$ and $\frac{2}{3}$[30] and $\beta \simeq \frac{4}{7}$[31]. Geller and Davis (1978)[32] used the cosmic energy theorem to a sample complete to magnitude 13 and obtained

$$0.2 \leqslant \Omega_G \leqslant 0.3. \tag{9}$$

One difficulty in cosmic energy theorem is that the quantity $\langle v^2 \rangle$, namely the mean square value of the peculiar velocities v of galaxies, is not directly observed. What is observed is Δv, the radial velocity difference between two galaxies. This motivates Peebles to work out the second cosmic virial theorem[33]. Peebles (1979)[34] has applied the second theorem to the data of 166 galaxies of Kirshner et al. (1978)[35] and obtained:

$$0.2 \leqslant \Omega_G \leqslant 0.6. \tag{10}$$

Thus, the results (9) and (10) from the statistical approach are in agreement with those from the Local Supergalaxy(7). It appears that there exists non-luminous and uniform material background.

All these results obtained by dynamical method, however, belong to some astronomical systems, and one cannot consider them as the total mean mass density in the universe, because all the considered regions are the mass concentrated ones. Sandage, Yahil et al. (1980)[36] have worked out a method to infer the mean mass density through the dynamical measurement of some special system. One key assumption here is: the ratio of the distribution of the total mass density to that of luminous matter is the same everywhere inside the system. By using this assumption they have obtained dynamically the result for the Local Supercluster as:

$$q_0 = 0.02 \pm 0.01, \tag{11}$$

i.e. the universe is infinite again.

The assumption mentioned above, however, is conflicted with the facts that the mass-to-luminosity. ratio increases with the size of system considered. If the ratio is always the same everywhere (i.e. so the ratio of non-luminous matter to luminous one does), then the M/L cannot increases with the linear size of systems. So, it is sure that the distribution of non-luminous matter is not as concentred as galaxies, it must spread in a much more uniform way.

(4) The Hubble Diagrams

A completely different way to determine the deceleration parameter q_0 is so-called apparent-magnitude redshift method. In standard cosmological model, we have the well-known relation[37,38]:

$$m - M = 5 \log f(z,q_0),$$

$$f(z,q_0) = q_0^{-2} \left\{ q_0 z + (q_0-1)(\sqrt{1-2q_0 z} - 1) \right\}, \quad (12)$$

where m and M are, respectively, the apparent and absolute magnitude of the objects, z is the redshift. If a group of objects have the same absolute magnitude, then we can find q_0 by fitting m-z relation (12) to their m-z plot. Here the key problem is to identify objects with the same luminosity.

The m-z relation of galaxies dates back to the beginning of the modern observational cosmology when Hubble established his famous velocity-distance elation. Since Humason et al. (1956)[39] did whose pioneer work, researchers have tended to concentrate on the first ranked galaxies of a cluster as a standard candle, whose scatter in m-z diagram is about 0.3 mag[40], and a great number of improvements have been made in sample size and quality. Again,

Table 4 Determinations of q_0 from the m-z relation of Galaxies

Author	Objects	N	z_{max}	$m_{v,max}$	q_0
Humason et al. (1956)[39] and Sandage(1961)[42]	Mean of 1st, 3rd, 5th and 10th ranked cluster galaxies	18	0.20	16.7	2.5±1.0
Baum (1961)[43]	Mean of 2–4 brightest E galaxies in clusters	8	0.46	19.5	1.0±0.5
Sandage (1972b)[44]	Brightest cluster galaxies	39	0.46	20.1	0.96±0.6
Sandage and Hardy (1973)[45]	Brightest cluster galaxies	97	0.46	19.0	0.95±0.5
Gunn and Oke (1975)[46]	Brightest cluster galaxies	60	0.46	19.2	0.3±0.7
Sandage et al. (1976)[47]	Brightest cluster galaxies	66	0.46	19.0	1.0±0.3
Smith (1977)[48]	3CR radio galaxies	52	0.75	21.	>1.0*
Kristian et al. (1978)[49]	Brightest cluster galaxies	50	0.38	19.0	1.6±0.4
Meinei (1981)[50]	Brightest cluster galaxies+	66	0.38	19.0	2.3

* $q_0 \sim 3$ according to Jaakola et al. (1979)[51]. + With allowance for intergalactic absorption.

Table 5 Determinations of q_0 by the m-z relation of various quasar subsets

Author	Classifier/Indicator	N	s(ΔM) (mag)	published q_0	Re-calculated log q_0	median q_0
Setti and Woltjer (1973)[54]	Steep radio spectrum	81	0.9	~1	...	...
Stannard (1973)[55]	Flat radio spectrum	28	0.9	~1	...	...
Bahcall and Hills (1973)[56]	Brightest in z-bin	7	0.5	~1	...	...
Davidsen et al. (1977)[57]	La line width	6	0.4	~1	...	...
Netzer et al. (1978)[58]	Flat optical continuum	47	0.9	~1	...	...
Qu et al. (1979)[59]	Strong scintillation	27	0.9	~1.4	0.35±0.59	2.2
Yang et al. (1980)[60]	Small optical variation	24	0.7	...	0.90±0.43	7.9
Cheng (1981)[61]	Common to preceding 2	15	0.8	...	0.30±0.78	2.0
Fang et al. (1979)[62]	Radio component separation	17	0.6	0.9-1.5	0.40	2.5
Baldwin et al. (1978)[63]	C IV line width	33	0.6	2.4±1.4	0.04±0.26	1.1
Kiang and Cheng (1981)[53]	CIV, MgII, L_α line widths	53	0.6	...	0.27±0.14	1.9

radio galaxies also have about the same luminosity[41]. The various determinations of q_0 by this method are summarized in Table 4.

The difficult here is very hard to judge if all the objects on the plot have the same luminosity. Furthermore, when we consider a group of objects with a wide redshift range, it is very difficulty to get rid of the influence of their evolution effects in principle.

One way out may be to select the subsets with similar morphologies. It has been widely used in quasars. The selected criterion have radio structures, radio spectrum types, the widths of spectral lines, etc. The main results are summarized in Table 5, in which S(ΔM) gives the dispersion in luminosity. One can find that all the published q_0 listed in column 5 are greater than 1/2. This is also the case, apart from 1 exception, in Table 4, but here the fact is even more significant because the samples are selected in many more different ways. The last two columns are the recalculated results published by Cheng et al. (1981)[52], according to Kiang and Cheng's (1981)[53] procedure that can be applied to all cases and not only give a best estimate but also a standard error.

Though we do not think the evolution effects have been overcomed, for all different subsets. However, all the results nearly consistently lead to the same conclusion: $q_0>1/2$. This conclusion does not depend on any special subset. In a word, the m-z method for quasars, and also galaxies, almost always suggest: the universe is finite.

(5) Apparent Angular Diameter-Redshift Method

For an object of linear size ℓ, we have the relation between its apparent angular size θ and redshift z:

$$\theta = (\ell H_0/c)q_0^2(1+z)^2/[q_0z+(q_0-1)(\sqrt{(1+2q_0z)}-1)]. \tag{13}$$

Instead of a group with the same luminosity in m-z method, here we require a group with the same linear size. This idea has been tried on a) extended radio sources with resolved components, b) compact radio sources, c) individual galaxies in cluster and d) galaxy clusters as units.

Miley (1968, 1971)[64,65] found an obvious correlation between the largest angular separation (LAS) among the components of an extended source and its redshift. Table 6 lists the results by fitting Eq.(13) to the upper envelopes of LAS-z plots. We note that the majority resulted in the simple inverse relation $\theta \propto z^{-1}$, which is unphysical within the standard model. The possible reasons may be a) evolution in the linear size, and b) observational selection effects. For compact radio sources, however, Hewish et al. (1974, 1976)[66,67] got a different result. They measured the angular sizes of compact sources or compact components (the "hot spots") by means of an interplanetary scintillation technique. They found $0.5<q_0<20$, but their sample is too small (n = 23) to give any firm conclusion.

This method has also been used in galaxies[74–76] and clusters of galaxies[77–81]. Main results can be found in Table 7. Bahcall (1977)[80] has found that rich regular clusters have similar linear core radii, R_c. For 15 cluster, she derived:

$$R_c = 0.12 \pm 0.02 \ h^{-1} \text{Mpc}. \quad (14)$$

Baier (1976)[81] and Bruzual and Spinrad (1978)[79] obtained similar results. Thus rich regular cluster promises to be a standard yardstick. At present , however, the data with high redshift are rather scanty, and we must keep in mind when we stretch our samples to higher redshift, that the evolution effects and selection effects in the θ-z method will certainly become more serious[82–86,73].

Table 6 The angular diameter-redshift relation of extended radio sources

Author	Type of Sources	Sample Size	Largest z	Result
Legg (1970)[68]	QSO_s and galaxies	57	2.0	$\theta \propto z^{-1}$
Miley (1971)[65]	QSO_s and galaxies	85	2.0	$\theta \propto z^{-1}$
Kellermann (1972)[69]	QSO_s and galaxies	104	2.1	$\theta \propto z^{-1}$
Wardle and Miley (1974)[70]	QSO_s	166	2.9	$\theta \propto z^{-1}$
Brecher and Kieras (1977)[71]	QSO_s and galaxies	123	2.0	$\theta \propto z^{-1}$
Grueff et al. (1977)[72]	QSO_s and galaxies	233	2.1	$\theta \propto z^{-1}$
Wills (1979)[73]	QSO_s	68	2.8	$q_0=0$

Table 7 Various determination of q_0 using the θ-z relation

Author	Type of objects	Sample Size	Largest z	q_0
Hewish et al.(1974)[66]	Compact radio sources	23	1.6	0.5-2.0
Baum (1972)[74]	Cluster galaxies	~ 8 × 4	0.46	0.3±0.2
Sandage (1972)[75]	Brightest cluster galaxies	68	0.46	~+1.
Bahcall (1977)[80]	Rich clusters	16	0.38	0-2
Hickson (1977)[78]	Rich clusters	88	0.46	-0.9±0.2
Bruzual and Spinrad (1978)[79]	Rich clusters	54	0.95	+0.25±0.5

(6) Other Determination of q_0

The age of the universe can also be used to find the value of q_0 [87,88,2]. Silk and White (1978)[89] offered the method using X-ray and microwave observations of hot gas in a rich cluster of galaxies. Using rank statistic method to 119 3CR and 4C quasars, Turner (1979)[90] obtained a high and admittedly uncertain value of q_0. A similar result has been found by Hu (1982)[91] for four subsets of quasars. Fang et al. (1981)[92] suggested that the periodicity in the distribution of emission line redshifts of quasars might be the remnant of a density wave in the early universe. By using this idea, they found $q_0 \sim 3$.

III. CONCLUSION AND DISCUSSION

Fig. 1 summarizes the various determinations of q_0 made so far. Obviously there is a very large dispersion among the results. The problem of the value of q_0 remain far unsolved, even if q_0 is larger or smaller than 1/2, the basic qualitative conclusion, is still uncertain. However, from Fig. 1, we can find out that the differences are systematic and depend strongly on the method used. The mean mass density method always gives $q_0<1/2$, while the θ-z method more often than not follows the "Euclidean" relation $\theta \propto z^{-1}$. The Hubble diagram (i.e. the m-z method) consistently gives $q_0>1/2$.

In such case, we are forced to face three alternatives:

a) The standard cosmological model is correct, but the special difficulties of each method lead to their own different errors. After the measurements have been improved, the results from various methods should tend to be consistent.

b) The standard model is incorrect. This lead to the fact, various methods obtain qualitatively different and even opposite conclusions.

c) The standard model is correct and various results are reasonable; the inconsistence among them is apparent, not substantial. We can find a way to make them in consistence with each other.

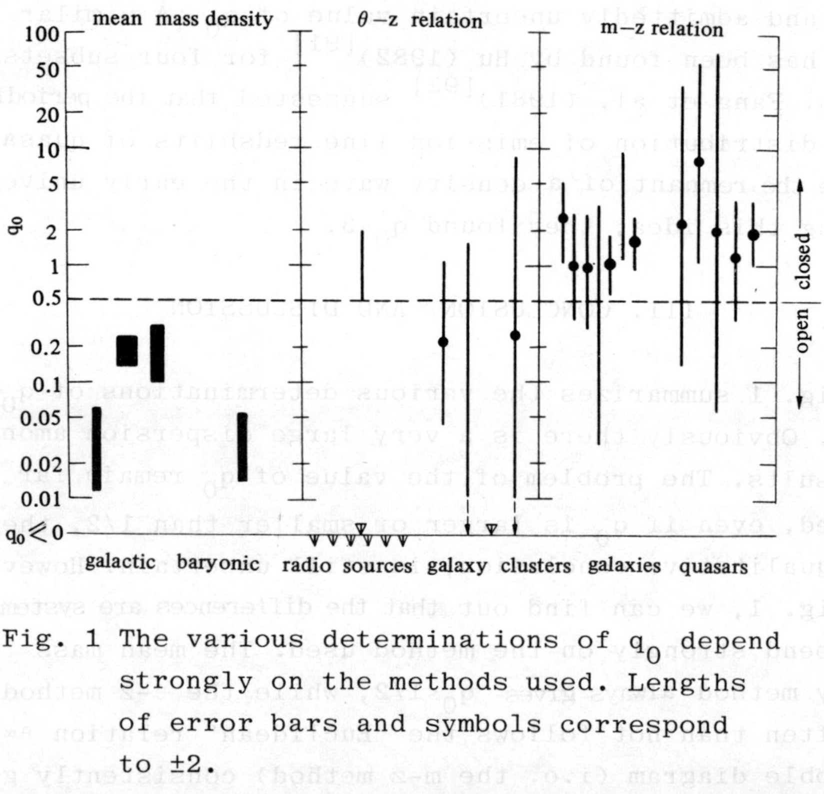

Fig. 1 The various determinations of q_0 depend strongly on the methods used. Lengths of error bars and symbols correspond to ±2.

We prefer to the third possibility. In fact, all the results would be able to be consistently put in the framework of standard model if matter has properties like these:

a) It is non-luminous, therefore the mass-to-luminosity ratio method of galaxies cannot reveals its contribution to ρ_0.

b) It is of non-baryon, therefore, the deuterium abundance

method should not also be able to detect its existence.

c) It spreads much more uniformly than galaxies, clusters of galaxies and superclusters therefore its contribution to ρ_0 cannot be measured dynamically.

d) It is the main contributer to mass density in the universe, therefore, the ρ_0 and/or q_0 obtained from m-z method will certainly be greater than the others.

It has now preliminarily been verified that there is such a sort of matter with the four characteristics mentioned above. It might be neutrino with non-zero rest mass[93,94]. If it is true, neutrinos would have the least rest mass among all known kinds of particles. However, it is probable that neutrinos have been governing the expansion of the universe.

We hope the proper allowance of a quasi-uniform material background, such as neutrino or massive photon, and evolutionary effects will eventually bring the various determination of q_0 into agreement. If not, then the validity of the standard cosmological model will be cast in doubt.

REFERENCES

[1] De Vaucouleurs G., 1972, IAU Symp. No. 44., 353.

[2] Sandage, A., Tammann, G.A., 1975, Ap. J., Vol. 197, 265, 277.

[3] Aaronson, M., Mould, J., Huchra, J., Sullivan III W., Schommer, R., Bothun, G., 1980, Ap. J., Vol. 239, 12.

[4] Fang Lizhi, Kiang, T., Cheng Fuhua, Hu Fuxing, 1982, Quart. J.R.A.S., Vol. 22, No. 3, in press.

[5] Shapiro, S. L., 1971, A. J., Vol. 76, 291.

[6] Gott, J.R., Turner, E.L., 1976, Ap. J., Vol. 209, 1.

[7] Davis, M., Geller, M.J., Huchra, J., 1978, Ap. J., Vol. 221, 1.

[8] Reeves, H., Fowler, W.A., Hoyle, F., 1970, Nature, Vol. 226, 727.

[9] Reeves, H., Audouze, J., Fowler, W.A., Schramm, D.N., 1973, Ap. J., Vol. 179, 909.

[10] Audouze, J., Tinsley, B., 1974, Ap. J., Vol. 192, 487.

[11] Rogerson, J., York, D.G., 1973, Ap. J., Vol. 186, L95.

[12] York, D.G., Rogerson, J., 1976, Ap. J., Vol. 203, 378.

[13] Vidal-Madjar, A., Laurent, C., Bruston, P., Audouze, J., 1978, Ap. J., Vol. 223, 589.

[14] Laurent, C., Vidal-Madjar, A., York, D.G., 1979, Ap. J. Vol. 229, 923.

[15] Wagoner, R.V., Fowler, W.A., Hoyle, F., 1967, Ap. J., Vol. 148, 3.

[16] Wagoner, R.V., 1973, Ap. J., Vol. 179, 343.

[17] Wagoner, R.V., 1974, IAU Symp. No. 63, 195.

[18] Grevesse, N., 1970, Geochim. Cosmochim. Acta, Vol. 34, 1129.

[19] Lang, K.R., 1974, "Astrophysical Formulae", Springer, Berlin.

[20] Ostriker, J.P., Peebles, P.J.E., Yahil, A., 1974, Ap. J., Vol. 193, L1.

[21] Faber, S.M., Gallegher, J.S., 1979, Ann. Rev. Astron. Astrophys., Vol. 17, 135.

[22] Yahil, A., 1980, private communication referenced in Davis et al., (1980).

[23] De Vaucouleurs, G., Bollinger, G., 1979, Ap. J., Vol. 233, 433.

[24] Schechter, P.L., 1980, A. J., Vol. 85, 801.

[25] Smoot, G.F., Lubin, P.M., 1979, Ap. J., Vol. 234, L83.

[26] Yahil, A., Sandage, A., Tammann, G.A., 1980, Ap. J., Vol. 242, 448.

[27] Mould, J., Asronson, M., Huchra, J., 1980, Ap. J., Vol. 238, 458.

[28] Davis, M., Tonry, J., Huchra, J., Latham, D.W., 1980, Ap. J., Vol. 238, L113.

[29] Irvine, W.M., 1965, Ann. Phys., Vol. 32, 322.
[30] Fall, S.M., 1975, M.N.R.A.S., Vol. 172, 23p.
[31] Davis, M., Peebles, P.J.E., 1977, Ap. J. Suppl., Vol. 34, 425.
[32] Geller, M.J., Davis, M., 1978, Ap.J., Vol. 225, 1.
[33] Peebles, P.J.E., 1976, Astrophys. Sp. Sci., Vol. 45, 3.
[34] Peebles, P.J.E., 1979, A. J., Vol. 84, 730.
[35] Kirshner, R.P., Oemler, A., Schechter, P.L., 1978, A. J., Vol. 83, 1549.
[36] Yahil, A., Sandage, A., Tammann, G.A., 1980, "Cosmologie Physique/Physical Cosmology", North-Holland Publishing Company.
[37] Mattig, W., 1958, Astron. Nachr., Vol. 284, 109.
[38] Mattig, W., 1959, Astron. Nachr., Vol. 285, 1.
[39] Humason, M.L., Mayall, N.U., Sandage, A., 1956, A. J., Vol. 61, 97.
[40] Sandage, A., 1976, Ap. J., Vol. 205, 688.
[41] Sandage, A., 1972, Ap. J., Vol. 178, 25.
[42] Sandage, A., 1961, Ap. J., Vol. 133, 355.
[43] Baum, W.A., 1961, IAU Symp. No. 15 "Problems of Extragalactic Research", 390.
[44] Sandage, A., 1972, Ap. J., Vol. 178, 1.
[45] Sandage, A., Hardy, E., 1973, Ap. J., Vol. 183, 743.
[46] Gunn, J.E., Oke, J.B., 1975, Ap. J., Vol. 195, 255.
[47] Sandage, A., Kristian, J., Westphal, J.A., 1976, Ap. J., Vol. 205, 688.
[48] Smith, H.E., 1977, IAU Symp. No. 74, 279.
[49] Kristian, J., Sandage, A., Westphal, J.A., 1978, Ap. J., Vol. 221, 383.
[50] Meinei, R., 1981, Astron. Nachr., Vol. 302, 177.
[51] Jaakola, T., Moles, M., Vigier, J.P., 1979, Astron. Nachr., Vol. 300, 229.
[52] Cheng Fuhua, Kiang, T., Fang Lizhi, 1981, Acta Astron. Sinica, Vol. 22, 357 (in Chinese).
[53] Kiang, T., Cheng Fuhua, 1981, Journal of the University

of Science and Technology of China (submitted) (in Chinese).
[54] Setti, G., Woltjer, L., 1973, Ap. J., Vol. 181, L61.
[55] Stannard, D., 1973, Nature, Vol. 246, 295.
[56] Bahcall, N.A., Hills, R.E., 1973, Ap. J., Vol. 179, 699.
[57] Davidsen, A.F., Hartig, G.F., Fastie, W.G., 1977, Nature, Vol. 269, 203.
[58] Netzer, H., Yahil, A., Yaniv, R., 1978, M.N.R.A.S., Vol. 184, 21p.
[59] Qu Qinyue, Qin Zhihai, Han Chuanshou, Zhou Fengan, 1979, Acta Astron. Sinica, Vol. 20, 98 (in Chinese). English Translation in Chinese Astron. (1980) Vol. 4, 97.
[60] Yang Lantian et al., 1980, Acta Astron. Sinica, Vol. 21, 208 (in Chinese).
[61] Cheng Fuhua, 1982, Acta Astrophys. Sinica (in press) (in Chinese).
[62] Fang Lizhi, Zhou Youyuan, Cheng Fuzhen, Chu Yaoquan, Cheng Fuhua, 1979, Scientia Sinica, Vol. 22, 1292.
[63] Baldwin, J.A., Burke, W.L., Gaskell, M.G., Wampler, E. J., 1978, Nature, Vol. 273, 431.
[64] Miley, G.K., 1968, Nature, Vol. 218, 933.
[65] Miley, G.K., 1971, M.N.R.A.S., Vol. 152, 477.
[66] Hewish, A., Readhead A.C.S., Duffett-Smith, P.J., 1974, Nature, Vol. 252, 657.
[67] Hewish, A., Readhead, A.C.S., 1976, Ap. Lett. Vol. 18, 1.
[68] Legg, T.H., 1970, Nature, Vol. 226, 65.
[69] Kellerman, K.I., 1972, Ap. J.,Vol. 77, 531.
[70] Wardle, J.F., Miley, G.K., 1974, Astron. Astrophys., Vol. 30, 305.
[71] Brecher, K., Kieras, C., 1977, Astrophys. Space Sci., Vol. 47, 25.

[72] Grueff, G., Schiavo-Campo, P., Vigotti, M., Zanni, M., 1977, Astron. Astrophys., Vol. 60, 321.
[73] Wills, D., 1979, Ap. J. Suppl., Vol. 39, 291.
[74] Baum, W.A., 1972, IAU Symp. No. 44, 393.
[75] Sandage, A., 1972, Ap. J., Vol. 173, 485.
[76] Tinsley, B. M., 1972, Ap. J., Vol. 178, L39.
[77] Noonan, T., 1972, A. J., Vol. 77, 134.
[78] Hickson, P., 1977, Ap. J., Vol. 217, 964.
[79] Bruzual, A.G., Spinrad, H., 1978, Ap. J., Vol. 220, 1.
[80] Bahcall, N.A., 1977, Ann. Rev. Astron. Astrophys., Vol. 15, 505.
[81] Baier, F.W., 1976, Astron. Nachr., Vol. 297, 295.
[82] Rees, M.J., Setti, G., 1968, Nature, Vol. 219, 127.
[83] Van der Laan, Perola, G.C., 1969, Astron. Astrophys., Vol. 3, 468.
[84] Christiansen, W., 1969, M.N.R.A.S., Vol. 145, 327.
[85] Miley, G.K., 1976, "The Physics of Non-Thermal Radio Sources", 1.
[86] Longair, M.S., 1978, "Observational Cosmology", 220, Saas Fee, Switzerland.
[87] Weinberg, S., 1972, "Gravitation and Cosmology", Wiley, New York.
[88] Gott, J. R., Gunn, J. E., Schramm, D.N., Tinsley, B. M., 1974, Ap. J., Vol. 194, 543.
[89] Silk, J., White, S.D.M., 1978, Ap. J., Vol. 226, L103.
[90] Turner, E.L., 1979, Ap. J., Vol. 230, 291.
[91] Hu Fuxing, 1982, Acta Astrophys. Sinica., Vol. 2, No. 2, 1 (in Chinese).
[92] Fang Lizhi, Chu Yaoquan, Liu Yongzhen, Cao Shenglin, 1981, Astron. Astrophys. (submitted).
[93] Lyubimov, V.A., Norikov, E.G., Nozik, V.Z., Tretyakov, E. F., Kozik, V.S., 1980, Phys. Lett., Vol. 94B, 266.
[94] Reines, F., Sovel, H.W., Pasierb, E., 1980, Phys. Rev. Lett., Vol. 45, 1307.

THE LARGE-SCALE INHOMOGENEITIES IN THE UNIVERSE

FANG Lizhi（方励之）

(Astrophysics Research Division,
University of Science and Technology of China, Hefei)

I. INTRODUCTION

Our universe seems to be homogeneous on the very large scales and considerably inhomogeneous on the scale below or around a few 100 Mpc.

The inhomogeneities can be described mainly by two parameters: the length scale and the "amplitude" of the density perturbations. The "amplitude" of the inhomogeneities is defined by

$$\delta^2 = \overline{(\rho(x) - \bar{\rho})^2/\overline{\rho^2}}, \qquad (1)$$

where $\rho(x)$ is the density distribution and $\bar{\rho}$ its mean value. The length scale is given by $\lambda = 2\pi/\bar{k}$, and

$$\bar{k} = \int_0^\infty k|\delta_k|^2 dk / \int_0^\infty |\delta_k|^2 dk, \qquad (2)$$

where δ_k denotes the Fourier component of the perturbed density, namely

$$\delta_k = \int(\rho(x) - \bar{\rho})e^{ik\lambda}dx/\bar{\rho}V. \qquad (3)$$

In a recent review Dolgov and Zeldovich call the problem of the inhomogeneity a cloud on the blue sky of modern cosmology[1]. They mean that there is still no theory to explain completely why the observed structure of the universe shows such inhomogeneity. We want, however, to emphasize that the "cloud" shrouds not only the explanation of forming the "observed structure", but also the "observed structure" itself. The observed picture of the inhomogeneity may not be the dominative component in the density perturbations of our universe. It cannot be the observational base to develop the clustering theory. In this connection, we need to know the length scale and the amplitude of the dominative component in the inhomogeneities. This is the main topic of my review.

II. THE DENSITY INHOMOGENEITIES OF VISIBLE MATTER

The sketch of cosmological inhomogeneities given by visible objects is shown in Table 1.

Table 1. Length scales and amplitudes of the inhomogeneities of various objects

objects	length scale	amplitude
galaxies	10 kpc	10^3
clusters of galaxies	5 Mpc	1–10
superclusters or voids	100 Mpc	1
periodic structures of quasars and intervening absorbing objects	300 Mpc	10^{-2}–10^{-3}

The clustering of cosmic matter on the scales of galaxy and cluster of galaxies has been studied in great detail. Recently, the progress is to find the preliminary evidence of the existence of density inhomogeneities on larger scale.

(1) Void

The deep redshift surveys of three small angular regions in the constellation Boötes by Kirsher et al. show identical $\Delta z \simeq 0.02$ gaps in the redshift distribution of galaxies down to 16th magnitude[2]. The three sampled regions, each 1°.4 square, form a roughly equilateral triangle on the celestial sphere, about 35° on a side. Assuming that the entire triangular region will ultimately prove to be empty in this same $\Delta z \simeq 0.02$ redshift interval, centered at $z \simeq 0.05$, one would deduce the existence of a void with length scale of

$$60\ h^{-1}\ \mathrm{Mpc}$$

at about 150 h^{-1} Mpc distance from us, where $h = H_0/100$ and the Hubble constant H_o is in the unit of km sec^{-1} Mpc^{-1}.

(2) Clustering of Quasars in 100 Mpc

The nearest neighbor test has been made by Chu et al.[3] for the quasars from a survey by Savage and Bolton in a field of 5° × 5° centred on (02^h, -50).[4] The completeness of this sample is considered more than 90 percent.

Figs. 1 and 2 show the distributions of nearest neighbor separation D for the observational sample and for a mean of 10 Monte Carlo samples. In Fig. 1 the Monte Carlo simulations are generated by replacing the observed red-shifts with random numbers, and in Fig. 2 by the same approach for the observed RA and DEC.

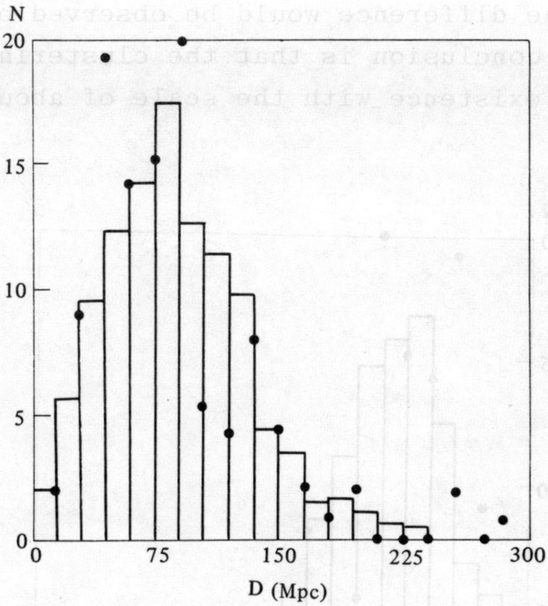

Fig. 1 Nearest neighbor test for quasar distribution. Circular points are from the observed results of Savage and Bolton on the field 02^h, -50°. Histogram is the mean of 10 Monte Carlo simulations by randomizing z.

The observed distribution is different significantly from the Monte Carlo simulations in the range of less than about 100 h^{-1} Mpc. The statistical significances η of such difference are given in Table 2, the definition of η is

$$\eta = n^{1/2}(<D> - D^*)/\theta, \quad (4)$$

where $< D >$ is the sample's mean, D^* and θ are the mean and standard deviation for Monte Carlo samples, respectively, and n the number of objects. In Table 2, $P(\eta)$ is the probability that the difference would be observed by chance Therefore, the conclusion is that the clustering of quasars seems to be in existence with the scale of about 100 Mpc.

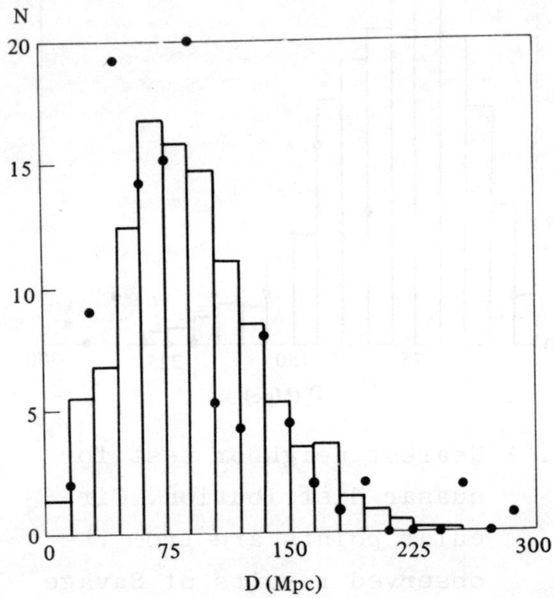

Fig. 2 Nearest neighbor test for quasar distribution. Circular points are from the observed results of Savage and Bolton on the field 02^h, $-50°$. Histogram is the mean of 10 Monte Carlo simulations by randomizing RA and DEC.

Table 2. Nearest neighbor test for quasar distribution

n	<D>(Mpc)	D^*(Mpc)	$\hat{\sigma}$(Mpc)	η	$P(\eta)$
109	166.7	191.5	94.1	-2.76	99.5%
109	166.7	193.2	91.1	-3.04	99.9%

(3) Periodic Inhomogeneity

The distribution of emission line redshifts for quasars has been studied by means of a statistical analysis of the power spectrum.[5] It has been demonstrated that the distribution have, at 99% confidence level, a periodic component with respect to the argument F defined as

$$F(z,q_0) = \int_0^z \frac{dz}{(1+z)(1+2q_0 z)^{1/2}} , \tag{5}$$

where q_0 is the deceleration parameter. Namely, there exists a set of peaks in the redshift distribution of quasars and the redshift values corresponding to these peaks are given by

$$F(z,q_0) = An + B , \tag{6}$$

where n equals zero or positive integer and A and B are parameters given by statistical result. In the case of $q_0=0$, Eq. (6) becomes[6]

$$\ln(1+z) = An + B . \tag{7}$$

This periodicity indicates the existence of wave-like perturbation in the quasar distribution[7]. The present value of "wave-length" for such wave-like perturbation is

$$\lambda_0 = cA/H_0 . \tag{8}$$

By using the statistical results, one finds

$$\lambda_0 \simeq 600 \text{ Mpc}. \tag{9}$$

The evidence for the periodic inhomogeneity may also be found from the redshift distribution of absorption line z_{ab} for quasars. In Fig. 3 we show the histogram of β for the absorption systems[8] listed in the catalog of Hewitt and Burbidge[9], β is defined by

$$\beta = \frac{(1+z_{em})^2-(1+z_{ab})^2}{(1+z_{em})^2+(1+z_{ab})^2} . \tag{10}$$

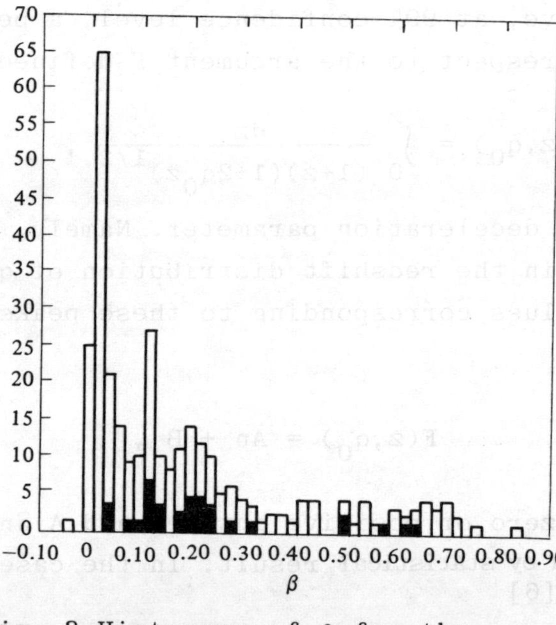

Fig. 3 Histogram of β for the absorption systems. The blank histogram corresponds to the data from catalog [9] and the black portions of the diagram correspond to the data from[18].

One can find two remarkable peaks located in $0.10<\beta<0.12$ and $0.18<\beta<0.20$, respectively. This means that, corresponding to the peaks, the redshifts are given by

$$\ln \frac{1+z_{em}}{1+z_{ab}} = A'\Delta n, \qquad (11)$$

where $\Delta n = 0, 1, 2$ and $A' = 0.103$. If we consider that z_{em} concentrate on the values of Eq. (7), we have from Eq. (11) then:

$$\ln(1+z_{ab}) = A'n + B. \qquad (12)$$

Thus if the absorption lines are due to intervening objects, Eq. (12) also indicates that a periodic component exists in the distribution of intervening absorbing objects[10]. The "wave-length" is now

$$\lambda_0 \simeq 300 \text{ Mpc}.$$

(4) Amplitude

If the scale for a kind of objects is much less than the average distance between them, the amplitude δ of the inhomogeneity due to these objects can then be estimated by

$$\delta = (\rho_{obj}/\bar{\rho})^{1/2} \qquad (13)$$

where ρ_{obj} and $\bar{\rho}$ are the mean densities in the objects and in our universe, respectively.

Using Eq. (13), one finds the amplitudes of inhomogeneities to be about

$$\delta \simeq 10^3$$

on galactic scale and

$$\delta \simeq 1 - 10$$

on the scale of cluster of galaxies.

The scale of void is the same as or larger than that of supercluster and the ratio of the galactic density in voids to that in normal regions is about or less than 0.1, therefore the amplitude on the scale of void or supercluster is still on the order, or even larger than, of unit.

The spectral power P(n) can be used as a measurement for the periodic component n in the distribution of quasars and the ratio P(n)/P(0) is the relative intensity of periodic component n superposed on the homogeneous background. The values of these relative intensities for various sets of quasars are in the range of 10^{-3}-10^{-2}.[7]

(5) Summary

The observed structure of visible objects in our universe shows the considerable inhomogeneity, i.e. $\delta \gtrsim 1$, at the scale of less than or about 100 Mpc.

If we consider that such structure were formed from the uniform distribution in the early universe, then the present density perturbation was already in a highly nonlinear regime. Thus the fundamental problem in the formation of cosmic inhomogeneity is: which is the mechanism to be responsible for the formation of inhomogeneities with so large amplitude at the length scales of galaxy, cluster of galaxies and supercluster or viod.

III. THE DENSITY INHOMOGENEITIES OF INVISIBLE MATTER

An assumption is implied in the above discussion, namely, with respect to the visible objects, the invisible matter in our universe can always be neglected. However it may not be correct.

(1) Evidence for the Existence of Invisible Matter

The rotation curves of spiral galaxies can not be explained by the visible matter distribution in the galaxies. It is possible to infer the existence of a invisible halo[11]

The mass-to-light ratio M/L for group and cluster are much higher than for single galaxies.[12] This means that only a small fraction of the mass of a cluster is in the form of luminous galaxies and the length scale of the inhomogeneities of invisible matter should be much larger than that of galaxies.

The results of various determinations of q_0 show a systematic difference, namely the results depend mainly on the method used in the determination[13] The Hubble diagram method consistently gives $q_0 > 1/2$, the mean mass density method always gives $q_0 < 1/2$, while the θ-z method more often than not follows the "Euclidean" relation $\theta \propto z^{-1}$. One of the possible reasons for this disagreement is the existence of invisible matter with a quasi-uniform distribution. (see Dr. Hu's review in this volume)

What can this invisible matter be? From the measurement of deuterium abundance, one concludes the main composition of the invisible matter may not be nucleons or bodies made up of nucleons. It is probably made up of massive neutrinos, mini black holes and gravitinos. Our universe will be neutrino-dominative, if, as the claim of Lyubimov et al., the mass of electron neutrino may be between 14 and 46 eV[14] Recently, it has been found that the measured ratio of the flux of antiprotons to protons($\bar{p}/p$ ratio) is significantly greater than the ratio expected from the commonly adopted model of cosmic ray propagation. This excessive antiprotons might be produced by Hawking evaporation of mini black holes.[15]

(2) An Upper Limit for the Inhomogeneity due to Invisible Matter

In the case of invisible-matter-dominated universe, the total density inhomogeneity in the universe should be determined mainly by the density distribution of invisible matter. However, the question is how to determine the length scale and the amplitude of the density inhomogeneity of invisible matter.

An upper limit for the amplitude of such inhomogeneity can be found by the small angular anisotropy of the microwave background[16] In the most theories concerning the origin of the anisotropy, they discuss only the effects of inhomogeneity of the temperature, velocity and gravitational potential at the recombination epoch, but not taking into account the gravitational deflection due to the density inhomogeneities after the recombination epoch. The propagation of radiation will be affected by the metric perturbation, so the intensity fluctuation can also be caused by the stochastic small gravitational deflection. It is analogous to the scintillation of light signal caused by a refractive medium with statistically distributed inhomogeneities.[17]

The characteristic time scale of the scintillation is the order of λ/v, where λ is the length scale of inhomogeneities of invisible matter and v is the average peculiar velocity associated to them. Since λ/v is so large, the temporal scintillation is not observable. Nevertheless, the scintillation effect for the microwave background may be observable as a directional intensity fluctuation if we observe it with an angular resolution of the order of

$$\theta_s \simeq \lambda/d, \tag{14}$$

where d is the distance of the source. Suppose the extended

source with uniform intensity. The light signals from the source have experienced a statistically independent stochastic field if their arriving directions are different. The intensity fluctuation are caused from these different histories of the inhomogeneities. Therefore, we can expect that the directional fluctuation is the same order with the mean square temporal fluctuation.

The mean square temporal fluctuation of intensity is given as

$$\left(\frac{\Delta I}{I}\right)^2 = \frac{2}{3} Qd^3 \tag{15}$$

and

$$Q = 2\overline{h^2} \int_0^\infty \left[\frac{1}{r^2}\frac{d^2}{dr^2} M(r) - \frac{1}{r^3}\frac{d}{dr} M(r)\right]dr. \tag{16}$$

$\overline{h^2}$ and M(r) can be expressed by the density perturbation spectrum δ_k (Eq. (3)) as follows:

$$\overline{h^2} = \left(\frac{6\pi G\bar{\rho}}{c^2}\right)^2 \overline{k^{-4}}\ \delta^2, \tag{17}$$

$$M(r) = \int_0^\infty dk\, k^{-3}|\delta_k|^2 \sin kr \Big/ \int_0^\infty dk\, k^{-2}|\delta_k|^2 r, \tag{18}$$

where $\overline{k^{-4}}$ is the average of k^{-4} over the $|\delta_k|^2$.

If the angular resolution of detection, $\Delta\theta$, is much larger than θ_s, the average fluctuation will be decreased to

$$\left(\frac{\Delta I}{I}\right)_{\Delta\theta} \simeq \left(\frac{\Delta I}{I}\right)\frac{\theta_s}{\Delta\theta} \simeq \Omega_0 \sqrt{\frac{\lambda}{c/H_0}}\ \delta\ \frac{\lambda}{\lambda(\Delta\theta)}\ , \tag{19}$$

where $\lambda(\Delta\theta) \sim \Delta\theta(c/H_0)$. If we put the observed limit for $\Delta(\Delta\theta)_{-4} \equiv (\Delta I/I)_{\Delta\theta}/10^{-4}$ as $\Delta(\Delta\theta)_{-4} \simeq 1$, the relation (19) gives

$$\delta < 2\times 10^{-3} \lambda_{10}^{-3/2} h_0^{-3/2} \Omega_0^{-1} (\Delta\theta)_1 \Delta(\Delta\theta)_{-4}, \tag{20}$$

where $\lambda_{10} \equiv \lambda/10\text{Mpc}$, $(\Delta\theta)_1 \equiv \Delta\theta/1^\circ$. This result suggests that the inhomogeneity of the total matter distribution is much smaller than that of the visible nucleon matter if $\Omega_0 \simeq 1$, i.e. the universe is dominated by the invisible matter (see Fig. 4).

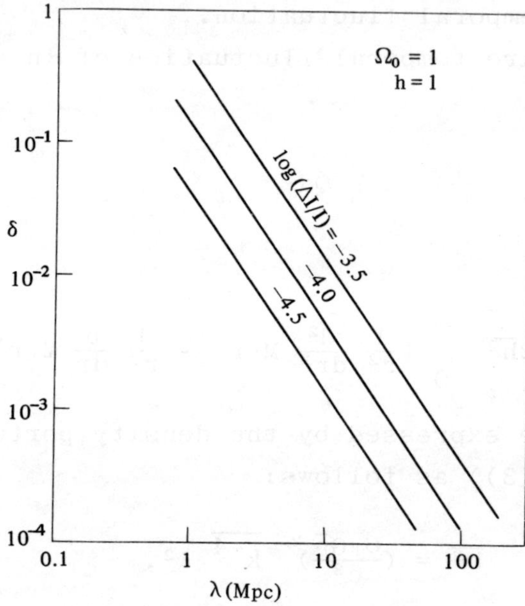

Fig. 4 Upper limit of the inhomogeneities given by gravitational "scintillation" of the microwave background.

IV. CONCLUSION

The pictures of the inhomogeneities are quite different for the visible objects and for the invisible matter in a closed universe. In opposition to the clustering of visible objects, even at present the total density perturbations with scale of about or larger than 10 Mpc are still very small, and it can be described in a linear regime of the growth of density perturbations.

Therefore, the problems and difficulties in explaining the clustering are also very different in the open ($\Omega_0 \ll 1$) universe and in the closed ($\Omega_0 \gtrsim 1$) universe. In the former one paradox is hard to get around. According to the standard argument, density fluctuations would have grown by a factor of about 10^3 in the interval from the recombination to the present. Thus, a 1% density fluctuation at the recombination epoch should exist as the origin of the present inhomogeneity. If such 1% density fluctuations were common at recombination, we should be seeing 1% fluctuation in the microwave background. However, the 3K background is observed to be smooth to one part in 10^4.

The cosmic invisible matter can extricate the cosmological clustering problem from the above mentioned predicament. In the case of closed universe, a much less than 10^{-4} fluctuation at the recombination is enough to evolve the total density inhomogeneity with the amplitude limited by the gravitational scintillation of background radiation. The amplitude of inhomogeneity due to visible matter is much more pronounced than the true density fluctuation given by visible plus invisible matter.

In a word, the fundamental problem in cosmic clustering is no longer to search the mechanism of forming a large density perturbation, but to explain the difference between the small total density perturbation and the large apparent amplitude in the inhomogenity of visible objects.

REFERENCES

[1] Dolgov, A.D., Zeldovich, Ya. B., 1981, Rev. Mod. Phys., Vol. 53, 1.

[2] Kirshner, R.P., Oemler, A. Jr, Schechter, P.L.,

Schectman, S.A., 1981, Ap. J., Vol. 248, L57.
[3] Chu Yaoquan, Zhu Xingfen, (to be published).
[4] Savage, A., Bolton, J.G., 1979, M.N.R.A.S., Vol. 188, 599.
[5] Fang Lizhi, Chu Yaoquan, 1981, Bulletin A.A.S., Vol. 13, 530.
[6] Karlsson, K.G., 1977, Astron. Astrophys., Vol. 58, 237.
[7] Fang Lizhi, Chu Yaoquan, Liu Yongzhen, Cao Shenglin, 1982, Astron. Astrophys., Vol. 106, 287.
[8] Fang Lizhi, Chu Yaoquan, 1982, Bulletin A.A.S., Vol. 14.
[9] Hewitt, A., Burbidge, G.R., 1981, Ap.J. Suppl., Vol. 43, No. 1.
[10] Chu Yaoquan, Fang Lizhi, Liu Yongzhen, (to be published).
[11] Bosma, A., 1978, "The Distribution and Kinematics of Neutral Hydrogen in Spiral Galaxies of Various Morphological Types", Ph. D. Thesis, Groningen.
[12] Ostriker, J.P., Peebles, P.J.E., Yahil, A., 1974, Ap. J. Lett., Vol. 193, 65.
[13] Fang Lizhi, Kiang, T., Cheng Fuhua, Hu Fuxing, Quart. J. R.A.S., (to be published).
[14] Lu Tan, Fang Lizhi, 1982, Prog. Phys, Vol. 2, No. 1 (in Chinese).
[15] Kiraly, P., Szabelski, J., Wdowczyk, J., Wolfendale, A.W., 1981, Nature, Vol. 293, 120.
[16] Fang Lizhi, Sato, H., Prog. Theor. Phys., (to be published).
[17] Fang Lizhi, Scientia Sinica, (to be published).
[18] Sargent, W.L.W., Young, P.J., Boksenberg, A., Tytler, D., 1980, Ap.J. Suppl., Vol. 42, 41.

THE REST-MASS OF NEUTRINO AND THE COSMIC CLUSTERING PHENOMENA

LU Tan (陆 埮)
(Department of Astronomy, Nanjing University, Nanjing)
ZHU Peichen (朱沛臣)
(Department of Physics, Nanjing University, Nanjing)

The formation mechanisms of galaxies, clusters of galaxies and superclusters are famous difficult problems in cosmology. However, the possibility of neutrino having non-vanishing rest-mass may open new ways to solve these problems. If we take $m_\nu \simeq 30$ eV as recently claimed in experiments[1,2], the universe would be dominated by neutrinos, and there might appear many interesting astrophysical effects[3].

One of the most important consequences is the existence of the preferential mass in neutrino clustering, and this clustering process would begin to appear prior to the era of the recombination[4]. During their contracting and clustering, massive neutrinos could carry the ordinary matter through their gravitational coupling, contracting to form the object with mass just at the right order of cluster of galaxies. Hence, the massive neutrinos may provide the clue to explain naturally the formation of clusters of galaxies. However, as is well known, the clustering of ordinary matter before the era of the recombination could make the microwave background radiation non-isotropic, so the observational fact that this radiation is highly isotropic has usually been regarded as the strong disproof for any clustering before the recombination. In this paper, we will discuss in some detail the contracting process from the starting point of preferential clustering ($T_1 \simeq 4\times10^4$ K) to the era of the recombination ($T_2 \simeq 4\times10^3$ K), and prove the nonuniformity reached at T_2 to be too small to conflict with the observational fact in microwave background radiation.

The hydrodynamic equations describing the neutrino-matter system coupled through gravitation in the expanding universe can be written as[5]

$$\ddot{\eta}_\nu + 2\frac{\dot{R}}{R}\dot{\eta}_\nu + \frac{v_{s\nu}^2 q^2}{R^2}\eta_\nu = 4\pi G(\rho_\nu \eta_\nu + \rho_m \eta_m) \tag{1}$$

$$\ddot{\eta}_m + 2\frac{\dot{R}}{R}\dot{\eta}_m + \frac{v_{sm}^2 q^2}{R^2}\eta_m = 4\pi G(\rho_\nu \eta_\nu + \rho_m \eta_m), \tag{2}$$

where

$$\eta_{\nu,m}(t) = (\delta\rho)_{\nu,m}/\rho_{\nu,m} \tag{3}$$

with $(\delta\rho)_{\nu,m}$ being the amplitudes of density disturbance, namely the solution for disturbance has been taken as

$$(\delta\rho)_{\nu,m}\exp\{i\vec{q}\cdot\vec{r}/R(t)\}, \tag{4}$$

and v_{sm} is the speed of sound in matter, $v_{s\nu}$ the similar characteristic speed in neutrino gas. Note $(\delta\rho)_{\nu,m}$, $\rho_{\nu,m}$ and R are all dependent on time t, while $\vec{q}$ is a constant vector. In the neutrino-dominated universe, Eqs.(1) and (2) can be simplified as

$$\ddot{\eta}_\nu + 2\frac{\dot{R}}{R}\dot{\eta}_\nu + \frac{v_{s\nu}^2 q^2}{R^2}\eta_\nu = 4\pi G\rho\eta_\nu \tag{5}$$

$$\ddot{\eta}_m + 2\frac{\dot{R}}{R}\dot{\eta}_m + \frac{v_{sm}^2 q^2}{R^2}\eta_m = 4\pi G\rho\eta_\nu. \tag{6}$$

Note

$$R \simeq R_1 (t/t_1)^{2/3} , \quad (7)$$

$$v_{s\nu}^2 \simeq v_{s\nu 1}^2 (t/t_1)^{-4/3} , \quad (8)$$

and we have approximately $v_{sm}^2 \simeq c^{2/3}$ during the period from T_1 to T_2. Here and in the following, the subscripts "1" and "2" refer to quantities at $T = T_1$ and T_2 respectively. Then, Eq.(5) can easily be solved[6]:

$$\eta_\nu \simeq t^{-1/6} J_{\mp 5/2}(3\Lambda t^{-1/3}) , \quad (9)$$

where J is the Bessel function, and

$$\Lambda = (v_{s\nu 1}^2 \, q^2 / R_1^2 \, t_1^{-8/3})^{1/2} . \quad (10)$$

At $t \gg \Lambda^3$, there are two asymptotic solutions. One is growing:

$$\eta_\nu \simeq \eta_{\nu 1} (t/t_1)^{2/3} ; \quad (11)$$

and the other is damping:

$$\eta_\nu \simeq t^{-1} . \quad (12)$$

We are only interested in the growing solution (11), which describes the contraction and clustering of cosmic neutrino gas.

Substituting Eq.(11) into Eq. (6), we can find its solution[6]

$$\eta_m \simeq \eta_{\nu 1} (3 v_{s\nu 1}^2 / c^2)$$

$$+\left(\frac{t_1}{t}\right)^{-1/3}\left(Ae^{i\sqrt{2}\left(\frac{c}{v_{s\nu 1}}\right)\left(\frac{t}{t_1}\right)^{1/3}}+Be^{-i\sqrt{2}\left(\frac{c}{v_{s\nu 1}}\right)\left(\frac{t}{t_1}\right)^{1/3}}\right) \tag{13}$$

where A and B are two integration constant. The second term in (13) appears to be damping oscillation, and after its damping out, η_m should approach to a constant value before recombination:

$$\eta_m \simeq \eta_{\nu 1}(3\, v_{s\nu 1}^2/c^2). \tag{14}$$

As can be seen from Eqs. (11) and (14), the ratio η_m/η_ν will be decreasing in the period from T_1 to T_2. This means that the strong coupling between matter and radiation before the recombination greatly weakens the ability for neutrinos to carry the ordinary matter to contract with themselves.

As the disturbance η_m of (14) appears before the era of the recombination, it would influence the background radiation through various mechanisms.

For example, η_m may influence the temperature fluctuation of background radiation through the adiabatic mechanism as

$$\begin{aligned}\delta T/T &\simeq (1/3)\,(\delta\rho/\rho)_m \\ &\simeq 0.3\,\eta_{\nu 1}.\end{aligned} \tag{15}$$

$\eta_{\nu 1}$ is a small quantity $\sim 10^{-3}-10^{-4}$ [6]. Therefore, the deviation from the isotropism of the microwave background radiation in radiation temperature is only about

$$\delta T/T \lesssim 10^{-4}. \tag{16}$$

Usually, the change with time of the density of matter

would induce some characteristic velocity $\vec{u}$ in disturbance region,

$$\frac{\partial}{\partial t}\left(\frac{\delta\rho}{\rho}\right) = \nabla\cdot\vec{u}. \tag{17}$$

The radiation scattered from this region would be red-shifted or blue-shifted through Doppler effect. However, in our case, $(\delta\rho/\rho)_m$ keeps constant after its oscillation term being sufficiently damped down. Hence, this mechanism makes no or only little influence on the background radiation.

Between neutrino and radiation there is no electromagnetic coupling, but gravitational coupling. In other words, the density fluctuation of neutrino will induce the fluctuation of gravitational field, and thus give rise to additional gravitational red-shifts to the background radiation. As the density contrast of neutrino η_ν was growing, it would be at the epoch of the recombination

$$\eta_\nu(T_2 = 4\times10^3\text{K}) = \eta_{\nu 1}\ T_1/T_2. \tag{18}$$

From this density fluctuation, $\delta\rho_\nu = \eta_\nu\rho_\nu$, one can easily estimate the additional red-shift δz as

$$\delta z \simeq G\delta\rho_\nu\lambda^2_{J\nu 1}(T_1/T_2)^2 \simeq 10^{-4}, \tag{19}$$

where $\lambda_{J\nu 1}$ denotes the Jeans wave length for neutrino at T_1. This effect will also give rise to the non-isotropism to the microwave background radiation in temperature in the order $(\delta T/T)\simeq 10^{-4}$.

In summary, we see that the clustering for massive neutrinos begun from T_1 could only induce the deviation from the isotropism in microwave background radiation at T_2 totally in the order of $(\delta T/T)\simeq 10^{-4}$, which is too small to conflict with the observations. The rapid growing of matter could occur only after the recombination, while the mass of the clustering matter would still nearlly keep the value ($\sim 10^{14} M_\odot$) carried by neutrinos at its preferential clustering.

REFERENCES

[1] Lyubimov, V.A. et al., 1980, Phys. Lett., Vol. 94B, 266.

[2] Reines, F., Sobel, S.W., Pasierb, E., 1980, Phys. Rev. Lett., Vol. 45, 1307.

[3] Lu Tan, Fang Lizhi, 1982, Prog. Phys., Vol. 2, No. 1 (in Chinese);
Lu Tan, Fang Lizhi, 1982, "Physics and Contemporary Needs", (ed. Riazuddin), Vol. 6, Plenum Publishers.

[4] Fang Lizhi, Liu Yongzhen, 1981, Lett. Nuovo Cimento, Vol. 32, 129; 1981, Acta Astrophysica Sinica, Vol. 1, 291.

[5] Lu Tan, Luo Liaofu, Yang Guochen, 1981, Acta Astronomica Sinica, Vol. 22, 251.

[6] Lu Tan, Zhu Peichen, 1981, Nature Journal, Vol. 4, 952; 1982, Kexue Tongbao, Vol. 27, 501.

[7] Sato, H., Takahara, F., 1980, Prog. Theor. Phys., Vol. 64, 2029; 1981, ibid., Vol. 66, 508.

ON THE NEUTRINO HALOES OF GALAXIES OR CLUSTERS OF GALAXIES AND THEIR RESTRICTION ON NEUTRINO MASS

HUANG Wuliang (黄无量), XU Chunxian (徐春娴)

(Institute of High Energy Physics,
Academia Sinica, Beijing)

QING Chengrui (庆承瑞), HE Zuoxiu (何祚庥)

(Institute of Theoretical Physics,
Academia Sinica, Beijing)

The non-zero rest mass of the neutrino is an interesting and comprehensive problem [1]. It may, perhaps, be related to the problem of galaxy formation [2-4]. In this paper we consider a simple and crude model of galaxies, consisting of baryons and neutrinos, (B + ν).

The baryon matter is rotating and balancing in the central region as a nucleus (i.e. the general galaxy region), but the static and degenerate neutrinos will remain in this region and surround it as a halo.

Using this model, we obtain the neutrino percent content in the galaxy region (δ_ν), the halo radius (R_2) and mass (M''_ν), the ratio of the total number of neutrinos to the total number of protons (σ), and the upper limit of the neutrino mass, etc. As a first approximation, we consider a system with spherical symmetry.

The Newton's gravitation theory may be a good approximation, thus,

$$\Delta\phi = 4\pi G\rho, \qquad \Delta = \frac{1}{r^2}\frac{d}{dr}(r^2\frac{d}{dr}), \tag{1}$$

$$\rho = \rho_B + \rho_\nu \qquad 0 \leqslant r \leqslant R_1, \tag{2}$$

$$\rho = \rho_\nu \qquad R_1 \leqslant r \leqslant R_2. \tag{3}$$

According to the equilibrium of the neutrinos in the gravitational field [5], we have

$$m_\nu\phi + \mu_\nu = \text{const}, \tag{4}$$

$$\rho_\nu = b\mu_\nu^{3/2}, \quad b = (\frac{g}{2})\frac{2^{3/2}\, m_\nu^{5/2}}{3\pi^2\hbar^3}, \tag{5}$$

where μ_ν is the chemical potential, and g is the number of neutrino helicity states. Since the forces are balanced in the equatorial plane [6], so

$$\frac{d\phi}{dr} = \frac{v_\phi^2(r)}{r}. \tag{6}$$

We introduce a velocity distribution as follows: [7]

$$v_\phi(r) = v_0(\frac{r}{R_1})^{1/n} \qquad n \geqslant 1. \tag{7}$$

Substituting Eqs. (6), (7) into Eq. (1), we obtain

$$\rho = \bar{\rho}[\frac{2+n}{3n}(\frac{r}{R_1})^{\frac{2}{n}-2}], \tag{8}$$

where

$$\bar{\rho} = \frac{3\Omega^2}{4\pi G}, \qquad \Omega = \frac{v_0}{R_1}. \tag{9}$$

On the other hand, from Eqs. (4) and (6) we have

$$\frac{d\mu_\nu}{dr} = -m_\nu\frac{v_\phi^2(r)}{r}. \tag{10}$$

Using the continuity condition $\rho(R_1) = \rho_\nu(R_1)$, and integrating Eq. (10) we obtain

$$\mu_\nu = [\frac{\bar{\rho}}{b}\cdot\frac{2+n}{n}]^{2/3}+\frac{n}{2}m_\nu\Omega^2(R_1^{2-\frac{2}{n}})[R_1^{2/n}-r^{2/n}]. \quad (11)$$

It is obvious that ρ_ν has a minimum value at $r = R_1$:

$$\rho_{\nu min} = \bar{\rho}\cdot\frac{2+n}{3n}. \quad (12)$$

Thus,

$$\delta_\nu = \frac{\text{The total mass of } \nu \text{ in nucleus region } M'_\nu}{\text{The total mass of nucleus region } M} > \frac{\rho_{\nu min}}{\bar{\rho}},$$

i.e.

$$\delta_\nu > \frac{2+n}{3n}. \quad (13)$$

When $n\to\infty$, $(\frac{2+n}{3n})$ has the minimum value, $\frac{1}{3}$, and so $\delta_\nu > \frac{1}{3}$. It is interesting to note that the boundary condition on the interface between the nucleus and halo should obey

$$\left.\frac{d\rho(r)}{dr}\right|_{r=R_1} \geqslant \left.\frac{d\rho_\nu(r)}{dr}\right|_{r=R_1}, \quad (14)$$

and then, from Eqs. (5),(8) and (11), it follows that

$$m_\nu \leqslant \left[\frac{\pi^2\hbar^6}{6\Omega^2G^2R_1^6}\cdot\frac{(2+n)^2(n-1)^3}{n^5}(\frac{2}{g})^2\right]^{1/8}. \quad (15)$$

Since the term $\frac{(2+n)^2(n-1)^3}{n^5}$ has the maximum value 1.050 at n = 10, so

$$m_\nu \leqslant 1.006\left[\frac{\pi^2\hbar^6}{6\Omega^2G^2R_1^6}\cdot(\frac{2}{g})^2\right]^{1/8} \text{ eV}. \quad (16)$$

Certainly, this is the upper limit of the neutrino mass. Adopting the parameters in [8]: $\Omega=5\times10^{-16}s^{-1}$, $R_1= 3\times10^{22}$cm, and taking g = 2, we have $m_\nu\leqslant 20$eV; if g = 6, then $m_\nu \leqslant 15$eV.

In order to study some characteristics of the neutrino halo, we rewrite Eq. (1) in the form

$$\frac{1}{r^2}\frac{d}{dr}(r^2\frac{d}{dr})\mu_\nu = -\lambda\mu_\nu^{3/2}, \tag{17}$$

where

$$\lambda = \frac{2^{7/2}m_\nu^{7/2}\,G}{3\pi\hbar^3}\left(\frac{g}{2}\right), \tag{18}$$

and the boundary conditions are

$$\mu_\nu\Big|_{r=R_1} = \left(\frac{\bar{\rho}}{b}\cdot\frac{2+n}{3n}\right)^{2/3}, \quad \frac{d\mu_\nu}{dr}\Big|_{r=R_1} = -m_\nu\Omega^2 R_1. \tag{19}$$

As for the outer boundary of the halo, it may be determined from

$$\mu_\nu(R_2) = 0. \tag{20}$$

In order to find the exact value about the neutrino halo ultimately, we must integrate Eq. (17) numerically. The final results are shown in Figs. 1–4.

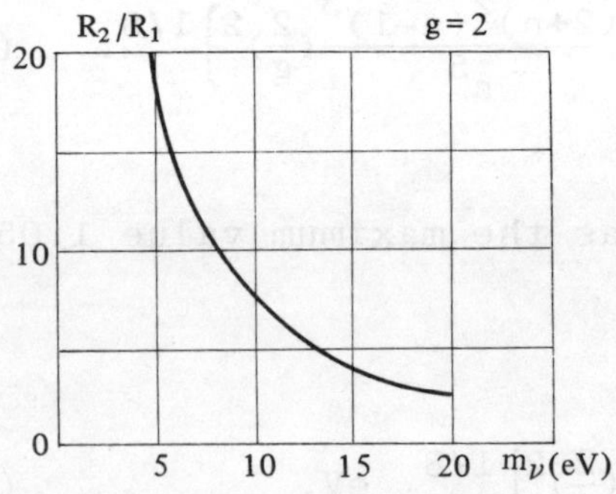

Fig. 1

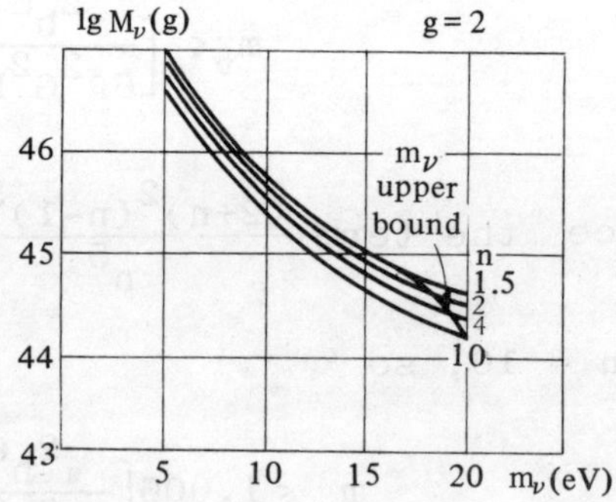

Fig. 2

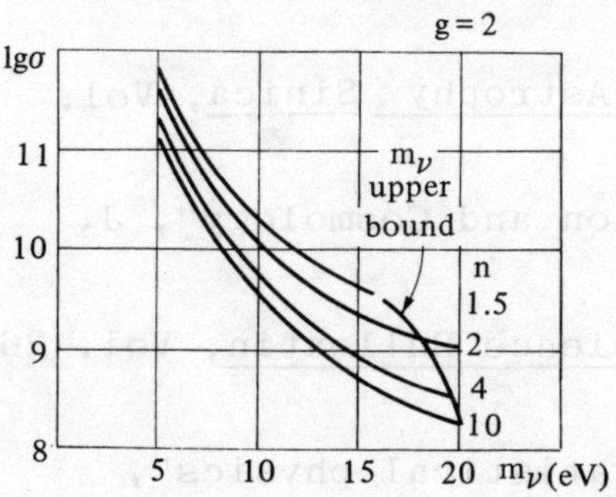

Fig. 3

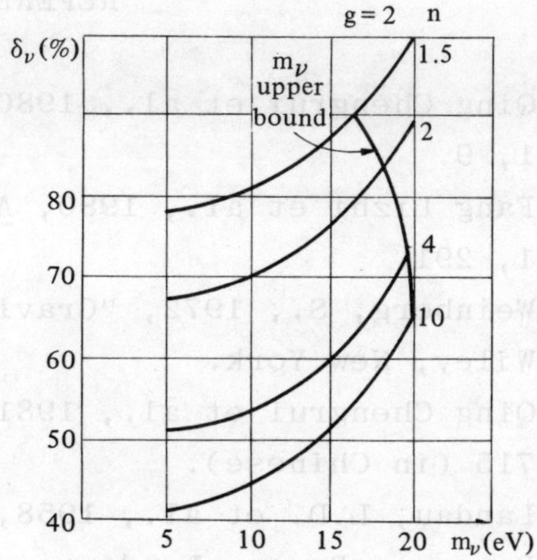

Fig. 4

In these Figures,

$$M_\nu = M'_\nu + M''_\nu \,, \tag{21}$$

and

$$\sigma = \frac{M_\nu}{M_B} \cdot \frac{m_p}{m_\nu} \,, \qquad M_B = M - M'_\nu \,. \tag{22}$$

As an example, by choosing $m_\nu = 15$ eV , n = 4, then from these Figures, we obtain $\delta_\nu = 60\%$. $\frac{R_2}{R_1} \simeq 4$, $M_\nu \simeq 0.6\times10^{45}$ g (i.e. $\frac{M''_\nu}{M} \simeq 5$, $\frac{M_\nu}{M_B} = 14$), $\sigma \simeq 0.9\times10^9$.

We would like to thank Xu Zongxing for completing the numerical calculation.

REFERENCES

[1] Qing Chengrui et al., 1980, Acta Astrophy. Sinica, Vol. 1, 9.

[2] Fang Lizhi et al., 1980, Acta Astrophy Sinica, Vol. 1, 291.

[3] Weinberg, S., 1972, "Gravitation and Cosmology", J. Wiley, New York.

[4] Qing Chengrui et al., 1981, Science Bullentin, Vol. 26, 715 (in Chinese).

[5] Landau, L.D. et al., 1958, "Statistical physics", Pergamon Press, London.

[6] Burbidge, E.M. et al., 1959, Ap. J., Vol. 130, 26; 1960, Ap. J., Vol. 131, 282.

[7] Sandage, A. et al., 1975, "Galaxies and the Universe", The University Press, Chicago.

[8] Liu Yongzhen et al., 1980, Science Bulletin, Vol. 25, 734 (in Chinese).

ON MASSIVE NEUTRINO HALOES AND GALACTIC STRUCTURE

A. Crollalanza

(University of Rome, Rome, Italy)

GAO Jian Gong (高建功)

(Xinjiang Engineering College, Wulumuqi, China)

Remo Ruffini

(University of Rome, Rome, Italy)

That self gravitating systems of massive degenerate neutrinos may play an important role in astrophysics was first pointed out by A. Markov[1]. Cowsik has been the first to point out the possible role of massive neutrinos in binding clusters of galaxies[2,3], while the main role of massive neutrinos on galactic structures has been pointed out in a series of recent works[4–8].

All these researchers have acquired particular momentum after the experimental work of Lubimov and Tretyakov on the determination of the neutrino rest mass[9].

The aim of this paper is to clarify, by using some simplified models, the interplay of "visible matter" (plasma, atoms, molecules, etc.) and massive neutrino halos in a galactic structure. It was pointed out by Gao and Ruffini[5] that for selected values of the neutrino mass and for selected values of the central densities of the equilibrium configurations of degenerate neutrinos the visible mass contribution can indeed be neglected. It is clear, however, that whenever the average density of visible galactic matter becomes comparable or larger than the neutrino density, due care should be taken in considering its effect on the neutrino distribution. In this paper we address ourselves to this issue and, more specifically, to configurations with a central region, where the visible matter dominates the neutrino density.

In order to simplify the problem we assume, for the "visible matter", two idealized distributions:

a) the density of visible matter is assumed to be constant and

$$\rho \simeq 2.2\times 10^{-22}\ \mathrm{g\ cm^{-3}}\ \text{for}\ 0 \leq r \lesssim 2\ \mathrm{kpc},$$
$$\rho \simeq 0\ \text{for}\ r \gtrsim 2\ \mathrm{kpc}$$

b) the density of visible matter is assumed to be constant and

$$\rho_1 \simeq 1.48\times 10^{-22}\ \mathrm{g\ cm^{-3}}\ \text{for}\ 0 \lesssim r \lesssim 2\ \mathrm{kpc},$$
$$\rho_2 \simeq 3.03\times 10^{-25}\ \mathrm{g\ cm^{-3}}\ \text{for}\ 2\ \mathrm{kpc} \lesssim r \lesssim 12.5\ \mathrm{kpc},$$
$$\rho \simeq 0\ \text{for}\ r \gtrsim 12.5\ \mathrm{kpc}.$$

The density distributions of the neutrino halos for both the above configurations are then compared and contrasted with the ones where the galactic matter density is neglected.

We have carried out the integration, as usual, by considering spherical symmetric configurations fulfilling the Newtonian equilibrium equations

$$\frac{dm(r)}{dr} = 4\pi r^2[\rho_\nu(r) + \rho_{vis}(r)] \tag{1}$$

$$\frac{dp_\nu(r)}{dr} = - Gm(r)\rho_\nu(r)/r^2, \tag{2}$$

where for P_ν and ρ_ν we have assumed the equation of state of a system of Fermi degenerate neutrinos. For every configuration we have, then, checked the validity of using the Newtonian versus the general relativistic approximation.

We have carried out the integration of the Eqs. (1) and (2) from the center to the border $R(P(R) = \rho(R) = 0)$, obtaining the radial distribution of matter, density and pressure. We have then computed the velocity curves due to the joint effect of the visible matter and the massive neutrinos. This quantity is of the greatest interest since it allows for observations directly related to the total mass distribution in the galaxy.

For the sake of example we have confined ourselves to one value of the neutrino rest mass, m = 8 eV and to configurations which have velocity curves with $v<400$ km s^{-1}. Numerical examples are summarized in Figs. 1a, 1b, 2a, 2b, 3a, and 3b.

The following general conclusions can be reached:

(1) The visible matter distribution can drastically modify the structure of the neutrino halo as long as the central density of visible matter is larger than the neutrino density $\rho_{vis}>\rho_\nu$ (see also Fig. 4).

(2) It is possible, by suitably choosing the neutrino rest mass and the central value of the density, to have a distribution of galactic matter dominated in the central region by "visible" matter and in the halo region by neutrinos (see Figs. 2a, 2b, and 3a, 3b).

(3) From the detailed knowledge of the velocity curves and of the visible matter distribution of a galaxy it is possible to infer the existence of a neutrino halo in that galaxy and, in principle, a sharp determination of the neutrino rest mass.

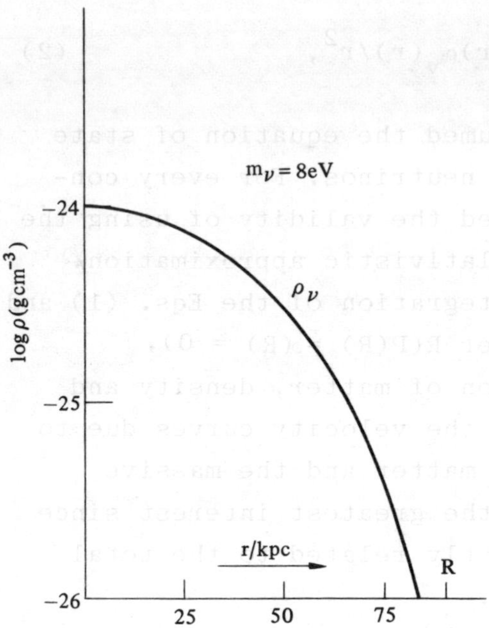

Fig. 1a Density distribution of degenerate massive neutrinos (m_ν=8 eV) corresponding to a central density of $\rho_c = 1.0 \times 10^{-24}$ g cm^{-3}. The radius of the configuration, R, is measured in kiloparsec. All effects of visible matter are here neglected.

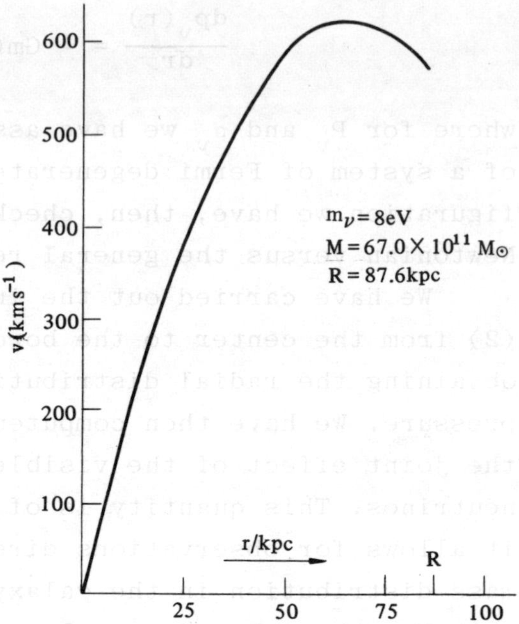

Fig. 1b Velocity curve, $v = [GM(r)/r]^{1/2}$ corresponding to the distribution given in Fig. 1a. M gives the total mass of the neutrino halo.

It is interesting to stress that, if neutrino halos do indeed exist in our galaxy, then the local particle density of neutrinos could be as high as $n_\nu \simeq 10^7$ cm^{-3}, namely larger by a factor of 10^5 from the one deduced within the framework of a strictly Friedmanian Universe (see e.g.[6]).

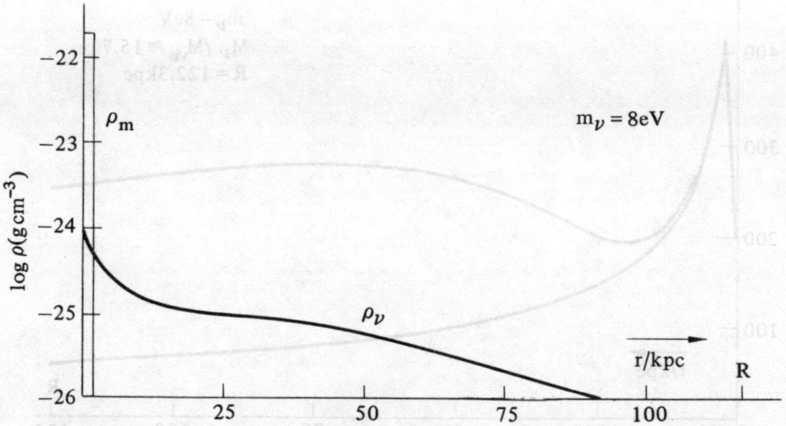

Fig. 2a Density distribution of massive neutrinos for the same parameters given in Fig. 1a but with a visible matter content given by $\rho=2.2\times10^{-22}$ g cm^{-3} for $0\leq r\leq2$ kpc and $\rho=0$ for $r>2$ kpc.

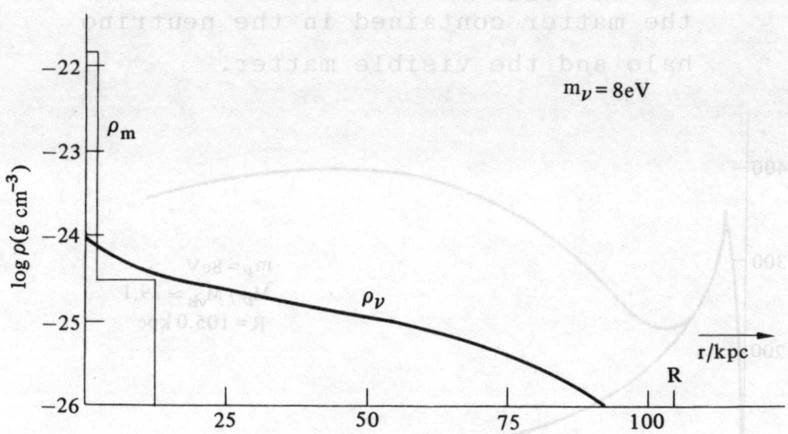

Fig. 2b Same as in Fig. 2a but with a visible matter content given by $\rho=1.48\times10^{-22}$ g cm^{-3} for $0\leq r\leq2$ kpc, $\rho=3.08\times10^{-25}$ g cm^{-3} for $2kpc\leq r\leq12.5$ kpc and $\rho=0$ for $r>12.5$ kpc.

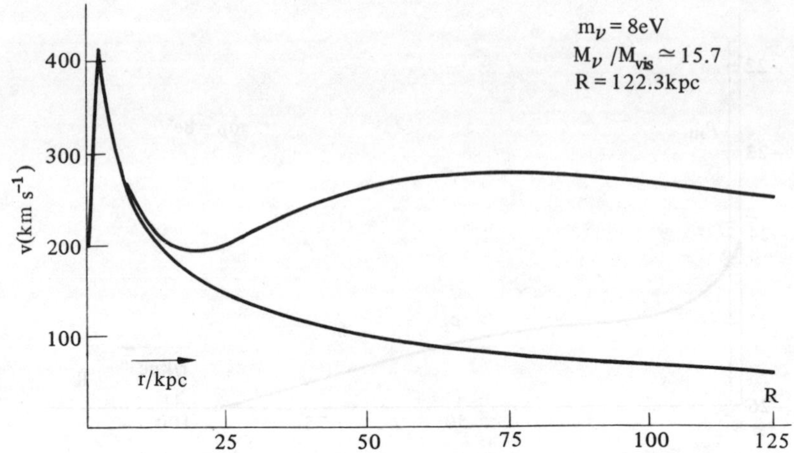

Fig. 3a Velocity curves corresponding to the matter distribution given in Fig. 2a. The lower curve corresponds only to visible matter, the upper curve to total matter (visible +neutrinos) and M_ν/M_{vis} gives the ratio between the matter contained in the neutrino halo and the visible matter.

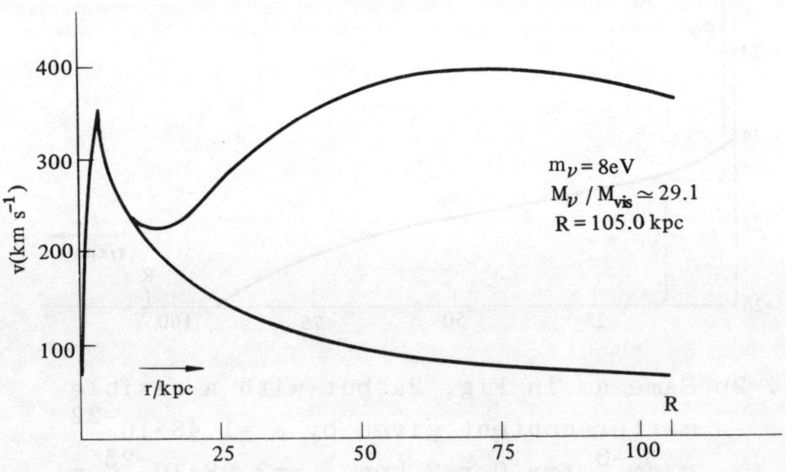

Fig. 3b Same as Fig. 3a, but corresponding to the matter distribution given in Fig.2b.

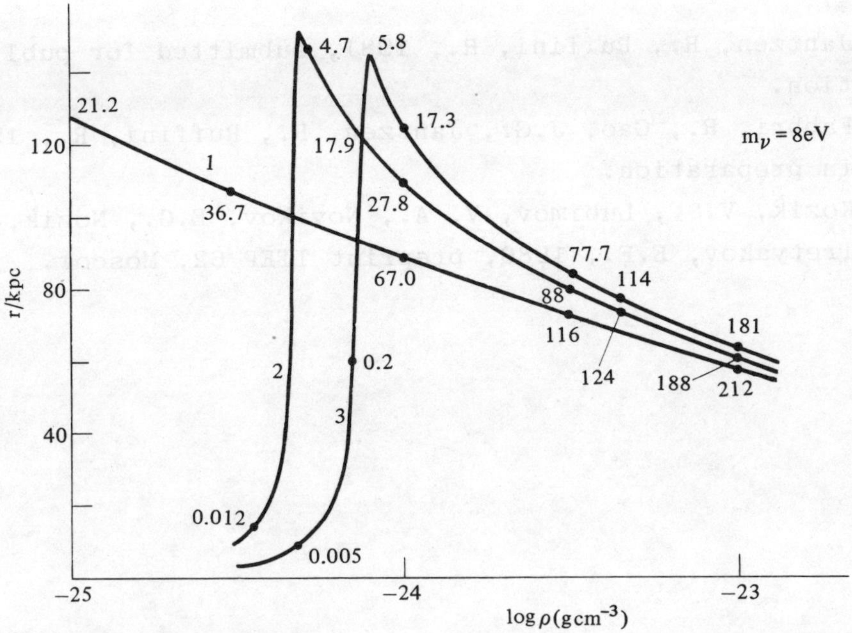

Fig. 4 This diagram emphasizes the effect of visible matter on the radius of the neutrino halos for selected values of the central densities. Curve I refers to absence of visible matter, and curve II (III) to the visible matter considered in Fig. 2a (2b).

REFERENCES

[1] Markov, M. A., 1964, Phys. Lett., Vol. 10, 122.

[2] Cowsik, R., McCleland, J.M., 1972, Phys. Rev. Lett., Vol. 29, 669.

[3] Cowsik, R., McCleland, J.M., 1973, Ap. J., Vol. 180, 7.

[4] Ruffini, R., 1980, Nuovo Cimento Lett., Vol. 29, 161.

[5] Gao, J.G., Ruffini, R., 1980, Phys. Lett., Vol. 97B, 388.

[6] Gao, J.G., Ruffini, R., 1981, Phys. Lett., Vol. 100B,

47.

[7] Jantzen, R., Ruffini, R., 1981, submitted for publication.

[8] Fabbri, R., Gao, J.G., Jantzen, R., Ruffini, R., 1981, in preparation.

[9] Kozik, V.S., Lubimov, V. A., Novikov, E.G., Nozik, V.Z., Tretyakov, E.F., 1980, preprint ITEP 62, Moscow.

GRAVITATIONAL LENS EFFECTS OF NEUTRINO CELESTIAL OBJECTS OF UNIFORM DENSITY

XU Chongming(须重明), WU Xuejun(吴雪君)

(Department of Physics, Fudan University, Shanghai)

The problems of quasars, such as the brightness of quasars[1] and the association of objects with different redshifts[2-4], are always explained by means of gravitational lens effects. Many authors tried to use gravitational focusing of light to interpret why quasars are brighter than usual objects and why quasars are frequently associated with foreground galaxies[5-7]. But up to now no one has discussed the situation of the gravitational focusing of light which is emitted by a source inside the transparent gravitational lens. According to our calculation, the brightness amplifications might be as large as 500 in some cases. It is also expected that the transparent object would cause a redshift. The idea comes from Hoyle and Fowler(1967)[8], who attempted to explain the quasar's high redshift Z by a gravitational redshift from a transparent object, since there is no limitation of Bondi (1964)[9] condition for the light emitted inside the transparent object. But these authors have not considered the brightness amplications caused by the gravitational lens effect.

Recently, it has been discovered from elementary particle experiments that the neutrino might have a non-zero rest mass m_ν[10,11]. Research on massive neutrino could greatly affect astrophysics[12]. The possible existence of stable objects of neutrinos was suggested by Markov (1964)[13]. Lately, neutrino celestial objects have been discussed by Gao and Ruffini (1981)[14]; Lu et al, (1981)[15] and Ching et al. (1981)[16]. As the Neutrino only joins weak interaction, almost everything is perfectly transparent with respect to it.

The purpose of this paper is to study the transparent gravitational lens effects of a neutrino celestial object. The solutions of the equations of null geodesics are obtained.

The metric of a transparent isotropic celestial object is given by

$$ds^2 = B(r)\, dt^2 - A(r)\, dr^2 - r^2\, d\theta^2 - r^2 \sin^2 \theta\, d\phi^2. \quad (1)$$

Suppose there is a source (star) S in the position with distance d from the centre of sphere O. OS is taken as X-axis. Since the problem has rotational symmetry for X-axis, we take $\theta=\frac{\pi}{2}$ for simplification (see Fig. 1, where R_0 is the radius of the transparent object).

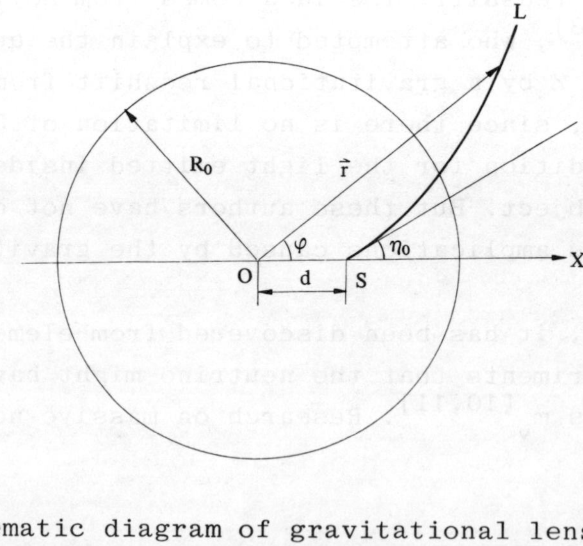

Fig. 1 Schematic diagram of gravitational lens effects.

In terms of solving the geodesic equations[17], we have

$$\frac{d\phi}{dr} = \pm \frac{d\sin\eta_0}{r}\left(\frac{B(r)A(r)}{r^2B(d)-d^2B(r)\sin^2\eta_0}\right)^{\frac{1}{2}}, \tag{2}$$

where η_0 is the angle between OS axis and ray at $r = d$ (initial condition). From Eq. (2), the solution of geodesic equation is found as

$$\phi(r) = \int_d^r \frac{d\sin\eta_0}{r}\left(\frac{B(r)\ A(r)}{r^2B(d)-d^2B(r)\sin^2\eta_0}\right)^{\frac{1}{2}} dr \quad (\eta_0 < \frac{\pi}{2}), \tag{3}$$

$$\phi(r) = \int_{r_0}^r \frac{d\sin\eta_0}{r}\left(\frac{B(r)\ A(r)}{r^2B(d)-d^2B(r)\sin^2\eta_0}\right)^{\frac{1}{2}} dr$$

$$-\int_d^{r_0} \frac{d\sin\eta_0}{r}\left(\frac{B(r)\ A(r)}{r^2B(d)-d^2B(r)\sin^2\eta_0}\right)^{\frac{1}{2}} dr \quad (\eta_0 > \frac{\pi}{2}), \tag{4}$$

where r_0 is the minimum distance from origin 0 to the geodesic line, which satisfies the equation $r_0^2\,B(d)-d^2B(r_0)\sin^2\eta_0 = 0$. Eqs. (3) and (4) is a general solution including the light trajectory from the source both inside and outside the transparent object. Moreover, if the object is not transparent and the source is outside, we can also use this formula to compute the deflection of light by the object. From Riemann geometry, the angle between ray and OS axis is

$$\chi = tg^{-1}\frac{\sqrt{A(r)}(tg\phi + r\frac{d\phi}{dr})}{A(r) - r tg\phi\frac{d\phi}{dr}}, \tag{5}$$

Obviously, as the curved space-time deflect the light rays

emitted from inside the transparent celestial object, the deflections are different for each η_0 (the deflected angle $\beta = |\chi-\eta_0|$), so that in some directions the light rays converge. In these directions, we would observe an increase in the source intensity, that is, the amplification of the brightness. Assume the luminous flames received at dA (located at r and ϕ) to be L_0 and L_N, corresponding to the cases when the gravitational lens exists and does not exist, respectively. The apparent amplification ratio K_v is defined as

$$K_v = L_0/L_N. \tag{6}$$

By means of geometrical method, and taking time dilation into account, we have

$$K_v = \frac{(r^2+d^2-2rd\cos\phi)^{3/2}\sin\eta_0 B(d)}{r^3\sin\phi\cos(\chi-\phi)[\cos(\chi-\phi)-\frac{d}{r}\cos\chi]B(r)\frac{d\phi}{d\eta_0}}, \tag{7}$$

where $\frac{d\phi}{d\eta_0}$ can be obtained by differentiating Eqs.(3) and (4) for η_0, when χ is determined by Eq.(5). The formula (6) is suitable for all A (r) and B (r).

Now as a special example, we consider the transparent object as a star with uniform density.* Inside the sphere, the metric is

$$B(r) = \frac{1}{4}[3(1-\frac{2MG}{R_0})^{\frac{1}{2}} - (1-\frac{2MGr^2}{R_0^3})^{\frac{1}{2}}]^2 \tag{8}$$

$$A(r) = (1-\frac{2MGr^2}{R_0^3})^{-1}, \tag{9}$$

*We are going to calculate the case of degenerate neutrino celestial object in another paper.

where M and R_0 are the total mass and radius of the object, respectively. To simplify the computation, we normalize the distance and mass with R_0, $\widetilde{M}=\frac{MG}{R_0}$, $\widetilde{d}=\frac{d}{R_0}$, $\widetilde{r}=\frac{r}{R_0}$. Then, substituting the metric (8) and (9) into Eqs.(3) and (4), $\phi(r)$ can be easily calculated by computer. One set of calculated results is shown in Figs. 2 and 3, where $\widetilde{d}$ = 0.2, $\widetilde{M}$ = 0.35.

All the light rays are bending when the source deviates from the center of the object. We can also calculate $\frac{d\phi}{d\eta_0}$ with the method of finite difference. The range we calculated is d: 0.1-0.9R_0(step 0.05), $\widetilde{M}$: 0.1-0.4(step 0.005), $\widetilde{R}$:1000R_0, η_0:0-π. Some results are shown in Fig. 4, which includes three parts with different $\widetilde{M}$(0.31,0.35,0.368). In each part there are four curves with different $\widetilde{d}$(0.60, 0.65,0.71,0.75). We get the minimal $\frac{d\phi}{d\eta_0}$ as 0.18×10^{-3}(where η_0=1.812 rad, $\widetilde{d}$=0.71,$\widetilde{M}$=0.368), $K=5\times10^2$. It can be seen from Fig.4 that the brightness amplification changes with the initial angle η_0 and the position of the source $\widetilde{d}$. Therefore, the brightness amplification certainly involves distortion.

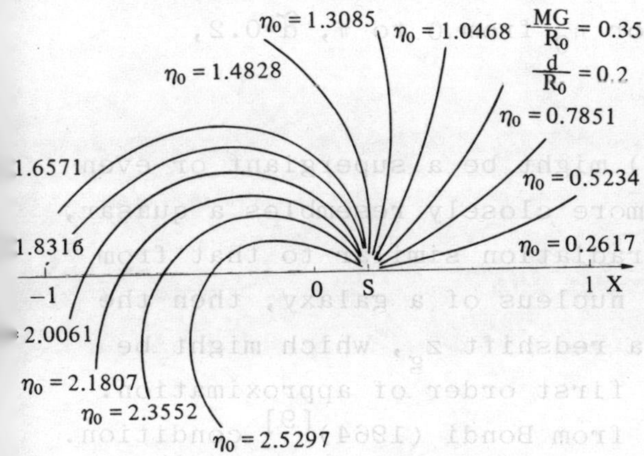

Fig. 2 The light trajectory inside the sphere, for η_0 from 0 to π $\widetilde{d}$=0.2, $\widetilde{M}$=0.35.

In addition, the relation among z, $\widetilde{M}$ and $\widetilde{d}$ is

$$1 + z = \frac{1}{B^{1/2}(d)} = \frac{2}{3(1-2\widetilde{M})^{1/2}-(1-2\widetilde{M}\widetilde{d}^2)^{1/2}} . \qquad (10)$$

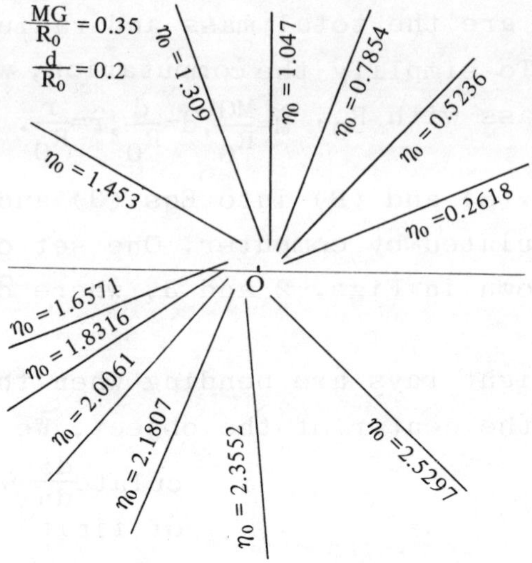

Fig. 3 The light trajectory outside the sphere, for η_0 from 0 to π, $\tilde{d}=0.2$, $\tilde{M}=0.35$.

The source (starlike object) might be a supergiant or even a nucleus of galaxy, which more closely resembles a quasar, since it emits synchrotron radiation similar to that from quasars. If the source is a nucleus of a galaxy, then the nucleus itself also causes a redshift z_g, which might be a superposition on z in the first order of approximation. The largest z_g is about 0.6 from Bondi (1964)[9] condition.

Now we can observe a brighter (several hundred times than original) and higher redshift quasistellar object (nucleus of galaxy). Furthermore, we can also observe the galaxy (belonging to the nucleus) itself. Perhaps it might explain why quasars frequently locate near galaxies. It is possible that a part of quasars could be starlike objects in neutrino objects.

Acknowledgements:

We are much obliged to Prof. Xie Xide and Prof. Zhou Shixun for their help revising our manuscript. We should also like to thank Prof. Chen-Ning Yang, Prof. Hu Hesheng, Prof. Ni Guangjiong and Dr. Xin Yuanlong for their encouragement and enlightenment.

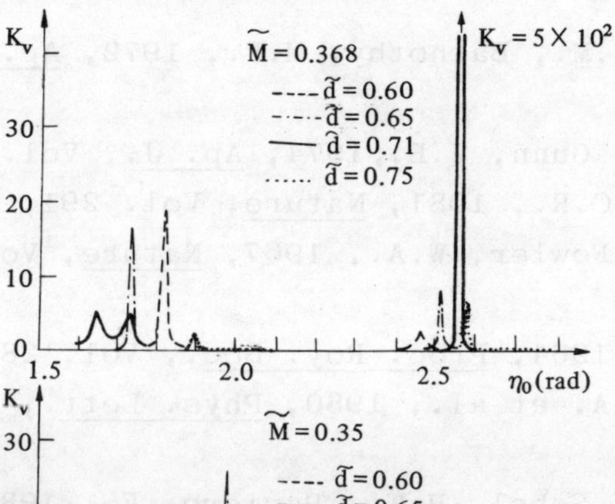

Fig. 4 K_v as a function of η_0 for $\widetilde{M}$=0.31, 0.35, 0.368; $\tilde{d}$=0.60, 0.65, 0.71, 0.75.

REFERENCES

[1] Field, G.G., Arp, A., Bahcall, J.N., 1973, "The Redshift Controversy", Benjamin, Massachusetts.

[2] Arp. H., 1980, Ann. N.Y. Acad. Sci., Vol. 336, 94.

[3] Burbidge, G., 1979, Nature, Vol. 282, 451.

[4] Hewitt, A., Burbidge, G., 1980, Ap. J. Supp., Vol. 43, 57.

[5] Barnothy, J.M., Barnothy, M.F., 1972, Ap. J., Vol. 174, 477.

[6] Gott, J.R., Gunn, J.E.,1974, Ap. J., Vol. 190, L. 105.

[7] Canizares, C.R., 1981, Nature, Vol. 291, 620.

[8] Hoyle, F., Fowler, W.A., 1967, Nature, Vol. 213, 373.

[9] Bondi, H., 1964, Proc. Roy. Soc., Vol. 282A, 303.

[10] Lubimov, V.A. et al., 1980, Phys. Lett., Vol. 94B, 266.

[11] Reines, F., Sobel, H.W., Pasierb, E., 1980, Phys. Rev. Lett., Vol. 45, 1307.

[12] Lu Tan, Fang Lizhi, 1981, ARD-81-106(to be published).

[13] Markov, M.A., 1964, Phys. Lett., Vol. 10, 122.

[14] Gao, J.G., Ruffini, R., 1981, Acta Astrophysica Sinica, Vol. 1, 19; Phys. Lett., Vol. 97B, 388.

[15] Lu Tan et al., 1981, Acta Astronomica Sinica,Vol. 22,251.

[16] Ching, C. et al., 1981, AS-ITP-81-002(to be published).

[17] Weinberg, S., 1972, "Gravitation and Cosmology", J. Wiley, New York.

A DE SITTER-FRIEDMANN PHASE TRANSITION MODEL

GUO Hanying（郭汉英）

(Institute of Theoretical Physics,
Academia Sinica, Beijing)

ZOU Zhenlong（邹振隆）

(Beijing Observatory, Academia Sinica, Beijing)

we have shown[1] that a new type of singularity-free cosmological model can be established by means of a simple finite temperature $\lambda\phi^4$ model. This model takes a series of important quantum and statistical effects into account, such as spontaneous symmetry breaking, trace anomaly and particle creation, symmetry restoration at high temperature through phase transition and so on. To begin with, the state of the universe would be a cold singularity-free anti-de Sitter one rather than a hot one. Then associated with the particle creation, the temperature would become higher and higher. When it reached a critical value T_c, a second-order phase transition would take place and the universe would transfer to a hot radiation dominated Friedmann state. The phase transition could be interpreted as the Big Bang and there would not exist singularity before it.

This contribution is mainly extracted from the authors' recent paper[1].

First, we solve the equation of motion in the T = 0 case and get an anti-de Sitter Robertson-Walker metric (RW metric) with k = -1 involving the spontaneous symmetry breaking and trace anomaly. Then we consider the particle creation and temperature effects to show how the model could undergo a de Sitter-Friedmann phase transition. Finally, we say a few words concerning the possibility of establishing a real model.

The Lagrangian of the ϕ-field taken to be the source of the classical RW metric in the model is

$$L = \frac{1}{2}\sqrt{-g}(\partial_\mu \phi \partial^\mu \phi - \frac{1}{6}R\phi^2 - \frac{1}{2}\lambda\phi^4). \tag{1}$$

Then it follows that the equations of the fields should be

$$g^{\mu\nu}\nabla_\mu\nabla_\nu\phi + \frac{1}{6}R\phi + \lambda\phi^3 = 0 \tag{2}$$

$$R_{\mu\nu} - \frac{1}{2}g_{\mu\nu}R + \Lambda g_{\mu\nu} = -8\pi\langle T_{\mu\nu}\rangle, \tag{3}$$

where $\langle T_{\mu\nu}\rangle$ is either the vacuum estimated value (VEV) (at T = 0) or the Gibbs average (at $T \neq 0$) of the metric energy-momentum tensor of the ϕ-field

$$T_{\mu\nu} = \partial_\mu\phi\partial_\nu\phi - \frac{1}{2}g_{\mu\nu}\partial^\lambda\phi\partial_\lambda\phi + \frac{1}{4}g_{\mu\nu}\lambda\phi^4$$
$$- \frac{1}{6}(\nabla_\mu\nabla_\nu - g_{\mu\nu}\Box)\phi^2 - \frac{1}{6}\phi^2(R_{\mu\nu} - \frac{1}{2}g_{\mu\nu}R). \tag{4}$$

The Gibbs average of $T_{\mu\nu}$ contains the temperature effects[2], such as energy of the black body radiation, besides the Casimir effect and trace anomaly[3,4] contributed to the VEV of $T_{\mu\nu}$ at T = 0.

We have proved that besides the unstable solution $\sigma = 0$, there exists a set of stable ones of the VEV σ of ϕ-field. The stable solutions and the scale factor a(t) of RW metric at T = 0 are given by

$$\sigma = \pm(-\frac{2}{3}\lambda^{-1}\Lambda^*)^{1/2},$$

$$a(t) = \sqrt{-\frac{3}{\Lambda^*}} \sin \sqrt{-\frac{\Lambda^*}{3}}\, t, \ k = -1, \tag{5}$$

$$\Lambda^* = 540\pi - (270^2\pi^2 - 1080\pi\Lambda)^{1/2} < 0,$$

where the contribution of pure geometric part of the trace anomaly[3] is involved. It is obvious that the stable solutions have broken down the invariance of the Lagrangian (1) under $\phi \rightarrow -\phi$ transformation, and the scalar curvature $R = 4\Lambda^*$ has played an important role.

It should be noted that the above spacetime is an anti-de Sitter one of negative constant curvature. Although $a(0) = 0$ at $t = 0$, the spacetime is still regular, i.e. there does not exist singularity, so one can analytically continue beyond the starting point. On the other hand, the metric has no horizon so that any systems of interacting particles may reach thermodynamical equilibrium states characterized by the temperature to be introduced.

Let us now consider the particle creation and the temperature effects of the model. It is usually believed that there is no particle creation for massless field on a conformally flat spacetime. However, this is not the case in our model. First, we can redefine the ϕ-field by the lift $\phi \rightarrow \phi + \sigma$ such that the VEV of new field is null but the new particle has non-zero mass $m_\phi = \frac{2}{3}\sqrt{2}|\Lambda^*|$. Secondly, some authors have recently pointed out that the self-interacting term could non-geometrically contribute to the trace anomaly and induce particle creation[4]. Although the details of such mechanisms for particle creation are not well known yet, there exists particle creation in our model. And because there is no horizon, it would be followed from the creation of ϕ-particle that the system could reach the thermodynamical equilibrium through the self-interaction and the temperature could become high enough.

It is proposed that under certain conditions the properties and the equations of the well-defined finite temperature $\lambda\phi^4$ model on the flat spacetime[5] could be extended to the thermodynamical equilibrium state of the system on the curved spacetime as long as taking the covariant derivatives and $\mu^2 - \frac{1}{6}R$ to replace the original derivatives and μ^2, which is taken to be null in our model for simplicity, respectively.

Therefore, one can show that under certain conditions the Gibbs average $\sigma(T)$ of ϕ-field should satisfy

$$g^{\mu\nu}\nabla_\mu\nabla_\nu\sigma(T) - [\lambda\sigma^2(T) + \frac{1}{6}R + \frac{1}{4}\lambda T^2]\sigma(T) = 0. \tag{6}$$

And the equation still has constant $\sigma(T)$ solutions

$$\sigma(T) = 0, \quad \sigma^2(T) = -\frac{1}{6\lambda}R - \frac{1}{4}T^2 \tag{7}$$

The second solution disappears at the temperature higher than a critical value

$$T_c = (-\frac{2}{3}\lambda^{-1}R)^{1/2}; \tag{8}$$

the state $\sigma(T) = 0$ with massless excitation quanta becomes stable and the symmetry restoration takes place at T_c, i.e. the system undergoes a second-order phase transition, and it is full of the massless excitation quanta.

The similar result can also be obtained for the effective potential at $T \neq 0$. Under certain conditions the effective potential of the system is

$$V(\sigma,T) = \frac{1}{12}R\sigma^2 + \frac{1}{4}\lambda\sigma^4 - \frac{\pi^2}{90}T^4 + \frac{m^2}{24}T^2. \tag{9}$$

Obviously, at $T \geq T_c$ the symmetry restoration takes place and $\sigma = 0$ becomes stable.

Using the effective potential and non-zero $\sigma(T)$ to solve Einstein equation, one can obtain the scale factor of RW metric at $T<T_c$

$$a(t) = \sqrt{-\frac{3}{\Lambda_T}} \sin \sqrt{-\frac{\Lambda_T}{3}}t, \ k = -1. \tag{10}$$

$$\Lambda_T^2 - 1080\pi(1+ \frac{\pi}{9} T^2)\Lambda_T - 1080\pi^2(\frac{4\pi^2}{45} + \frac{\lambda}{20})T^4 + 1080\pi\Lambda = 0.$$

This shows that the spacetime keeps itself up in the anti-de Sitter state as long as $T<T_c$.

However, the anti-de Sitter state (10) disappears at $T \geq T_c$. In fact, at $T = T_c$ the state of the universe in the model would undergo a second-order phase transition and would transfer to a hot radiation dominated Friedmann state since the spacetime is full of relativistic massless excitation quanta at the phase transition point.

It should be noticed that the model is too simple and it can not give the quantitative description of the real universe. However, the intriguing picture of the evolution in the early universe given by the model is meaningful because the main features of the model are common ones in the finite temperature spontaneous symmetry breaking gauge theory. It would be reasonable to expect that in the cosmological models based upon the standard model of weak, electromagnetic and strong interactions or based upon the GUTs (grand unified theory the important features of our model would remain.

REFERENCES

[1] Guo Hanying, Zou Zhenlong, 1982, Commun. Theor. Phys., Vol. 1, No.2.

[2] Hu B.L., 1981, Phys. Lett., Vol. 103B, 331.

[3] See for example, Gibbons, G.W., 1979, in GR, "An Einstein Centenary Survey". Hawking, S.W., Israel, W. (eds.), Cambridge.

[4] Birrell, N.D., Davies, P.C.M., 1980, Phys. Rev., Vol. D22, 322.

[5] See for example, Linde, A.D., 1979, Rep. on Prog. in Phys., Vol. 42, 389.

IX. MISCELLANEOUS

THE CERENKOV MICROWAVE LINE EMISSION OF HYDROXYL RADICAL

YOU Junhan (尤峻汉)

(Astrophysics Research Division, University of Science and Technology of China, Hefei)

LI Shouzhong (李守中)

(Department of Geophysics, Beijing University, Beijing)

I. THERMAL EQUILLIBRIUM DISTRIBUTION

Many interstellar and stellar hydroxyl sources have a variety of abnormal characteristics. Their emission lines can not interpreted by spontaneous transition. Generally, it is considered as a celestial maser, i.e. the emission lines originate from the stimulated transition.

Besides spontaneous emission and stimulated emission, Cerenkov effect can produce line emission when relativistic electrons pass through a gaseous medium[1]. In this paper, the posibility of microwave emission lines produced by Cerenkov effect is discussed using OH as a specimen. The intensity and other characteristics of OH Cerenkov microwave line emission are calculated.

The frequencies and the transition probabilities of four microwave lines between the levels of OH ground state (Fig. 1) are listed in Table 1.[2] When a relativistic electron with dimensionless velocity $\beta=v/c$ moves through a gas of refraction index n, Cerenkov radiation will occur at the frequency ν if Cerenkov radiation condition $\beta > 1/n_{\nu}$ is satisfied. Line emission near the resonance frequency will be produced by this mechanism since n_{ν} of the gas is greater than $1/\beta$ at this narrow frequency region.

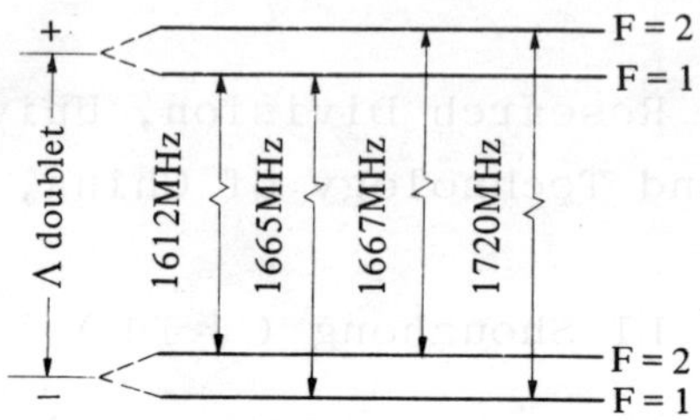

Fig.1 Levels of OH ground state.

Table 1. The main group of OH lines

Transition	Frequency [MHz] (±0.002)	Spontaneous transition probability A_{ij} [s^{-1}]
F = 2 ⟶1	1612.231	1.29×10^{-11}
F = 1 ⟶1	1665.401	7.10×10^{-11}
F = 2 ⟶2	1667.358	7.70×10^{-11}
F = 1 ⟶2	1720.533	0.94×10^{-11}

The Cerenkov radiation power produced in frequency interval $\nu \to \nu + d\nu$ can be expressed by

$$P(\nu)d\nu = \frac{4\pi^{1}e^{1}\beta_{\nu}}{c}\left(1 - \frac{1}{\beta^{2}n_{\nu}^{2}}\right)d\nu \quad . \qquad (1)$$

Suppose that the relativistic electrons have isotropic distribution of velocity, the polarization of Cerenkov line emission will disappear and spectra emission coefficient

$J^C_\nu d\nu$ can be expressed by

$$J^C_\nu d\nu = \frac{1}{4\pi}\int_{r_1}^{\infty} ne(r)dr P(\nu)d\nu = \frac{\pi e^2}{c}\,\nu d\nu \int_{r_1}^{\infty} ne(r)dr\beta\left(1-\frac{1}{\beta^2 n_\nu^2}\right). \tag{2}$$

$J^C_\nu d\nu$ is the power emitted per unit volume of gas in a given direction within an unit solid angle in the interval $d\nu$, where $\int ne(r)dr = N_e$ is density of relativistic electrons. Since, actually, $\beta \simeq 1$, then (2) becomes

$$J^C_\nu d\nu = \frac{\pi e^2}{c}\,\nu d\nu\left(1-\frac{1}{n_\nu^2}\right)N_e = 2.42\times 10^{-29} N_e \nu\left(1-\frac{1}{n_\nu^2}\right)d\nu \tag{3}$$

and $J^C_\nu d\nu$ can be obtained if the dependence of function n_ν on ν is known.

Complex refraction index $\tilde{n}_\nu$ can be calculated from the Lorentz-Lorenz formula[3]

$$\frac{\tilde{n}_\nu^2-1}{\tilde{n}_\nu^2+2} = \frac{4\pi}{3}(N_j\alpha_{ji}(\omega)+N_i\alpha_{ij}(\omega)), \tag{4}$$

where N_j, Nj are densities and $\alpha_{ij}(\omega), \alpha_{ji}(\omega)$ are polarizabilities of OH in states i and j respectively. Since we are only interested in the polarizabilities near the resonance frequencies, the polarizability in state j can be written approximately as

$$\alpha_{ji}(\omega) \simeq \frac{e^2}{\mu}\frac{f_{ji}}{(\omega_{ji}^2-\omega^2)+i\Gamma_{ji}}$$

$$= \frac{1}{(\omega_{ji}^2-\omega^2)+i\Gamma_{ji}}\frac{c^3}{8\pi^2\nu_{ji}^2}A_{ij}\left(\frac{g_i}{g_j}\right), \tag{5}$$

where f_{ji} is oscillator strength, Γ_{ji} is the damping constant, A_{ij} is spontaneous transition probability, g_i and g_j are the degeneracy parameters of states i and j respectively. We obtain

$$\tilde{n}_\nu^2 = (n_\nu + ik\nu)^2 = \frac{1+2b/[2\pi(\nu_{ji}^2-\nu^2)+i\Gamma_{ji}}{1-b/[2\pi(\nu_{ji}^2-\nu^2)+i\Gamma_{ji}} \qquad (6)$$

from Eqs. (4) and (5), where

$$b = \frac{g_i}{12\pi^2}\left(\frac{N_j}{g_j} - \frac{N_i}{g_i}\right)\frac{c^3}{\nu_{ji}^2} A_{ij}. \qquad (7)$$

n_ν^2, as a function of ν, can be deduced from Eq. (6)

$$n_\nu^2 = (A+\sqrt{A^2+9b^2\Gamma_{ji}^2\nu^2})\ /B,$$
$$k_\nu^2 = (-A+\sqrt{A^2+9b^2\Gamma_{ji}\nu^2})/B, \qquad (8)$$

where

$$A=[2\pi(\nu_{ji}^2-\nu^2)+2b][2\pi(\nu_{ji}^2-\nu^2)-b]+\Gamma_{ji}^2\nu^2,$$
$$B=2\{[2\pi(\nu_{ji}^2-\nu^2)-b]^2+\Gamma_{ji}^2\nu^2\}. \qquad (9)$$

The values of b and damping constant Γ_{ij} under a given density N_{OH} and temperature T can be calculated as ν_{ij} and A_{ij} in Table 1 are known. Thus, the values of n_ν^2 as a function of ν can be calculated from Eq. (8). Substituting these values of n_ν^2 into Eq. (3), the variation of J_ν^C with ν can be obtained. The typical curves of n_ν^2 and J_ν^C are shown in Fig. 2 and Fig. 3 respectively. The total emission coefficient is given by

$$J^C=4\pi\int_{\nu_{min}}^{\nu_{max}} J_\nu^C d\nu=3.04\times10^{-28}\times N_e\int_{\nu_{min}}^{\nu_{max}} \nu(1-\frac{1}{n_\nu^2})d\nu. \qquad (10)$$

The values of the integral $\int\nu(1-1/n_\nu^2)d\nu$ for four OH lines are listed in Table 2, in which we assume γ=200(β=0.9999875, i.e. $n_\nu^2\geq1.000025$ in the frequency interval $\nu_{min}\to\nu_{max}$).

The characteristics of Cerenkov microwave line emission are:

a) The shape of the line is asymmetric.

b) The width of the line ($\nu_{max} - \nu_{min}$) depends on the energy of electrons and the population difference per unit volume of OH.

c) The total emission coefficient $J^C \propto N_e$ (cf.[3]). Calculation shows that in a L.T.E. gas of T=10K, J^C will be comparable with the intensity of spontaneous emission J^{sp} when $N_e=10-10^2 cm^{-3}$. In a gas of nonthermal equilibrium, J^C can be even greater than J^{sp} for 1 or 2 order of magnitude. This means that when N_e in the molecular cloud is sufficiently large, Cerenkov microwave line emission is probably one of the mechanisms of the celestial maser emission.

d) The intensity ratio between Cerenkov lines is the same as that of spontaneous emission lines in the case of L.T.E. gas, and will change as the population is deviated from L. T.E..

e) Cerenkov lines will show polarization to a certain degree according to the anisotropy of the velocity distribution of relativistic electrons.

f) The direction of radiation determined by Cerenkov angle $\theta_c=\cos^{-1}(1/\beta n_\nu)$, varies with n_ν within a large range.

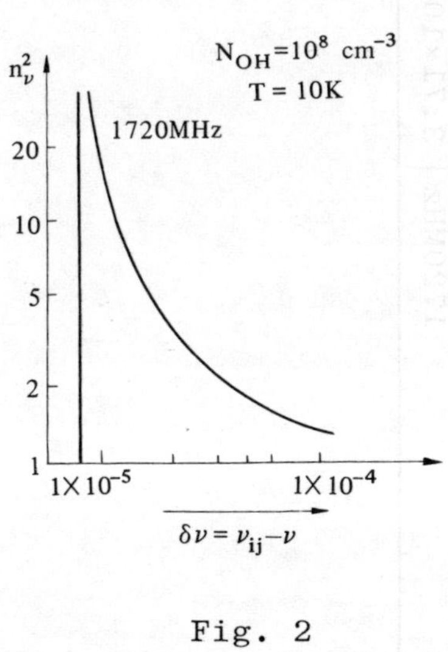

Fig. 2

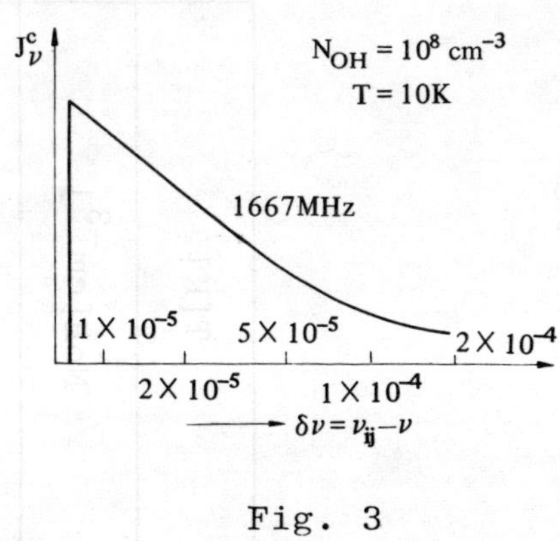

Fig. 3

Table 2

T[K]		10			100		
NoH[cm^{-3}]		10^5	10^8	10^{14}	10^5	10^8	10^{14}
$\int_{\nu_{min}}^{\nu_{max}} \nu(1-\frac{1}{n_\nu^2})d\nu$	1612MHz	3.26×10^2	3.25×10^5	3.24×10^{11}	2.7×10	3.25×10^4	3.24×10^{10}
	1665MHz	1.73×10^3	1.73×10^6	1.73×10^{12}	1.75×10^2	1.73×10^5	1.73×10^{11}
	1667MHz	3.12×10^3	3.12×10^6	3.12×10^{12}	3.08×10^2	3.12×10^5	3.12×10^{11}
	1720MHz	3.71×10^2	3.69×10^5	3.69×10^{11}	3.1×10	3.70×10^4	3.69×10^{10}

II. NONTHERMAL EQUILIBRIUM DISTRIBUTION

The microwave emission lines of most OH molecular sources have obvious nonthermal characterstics[4]. These sources are ususlly interpreted as a celestial maser. Various pumping models have been suggested[5] but none of them has been successful. The principal difficulty in celestial maser theory is evident insufficiency of pumping energy. For instance, the luminosity of maser lines radiated by several nonthermal OH sources were listed by Litvak[6] and all have the magnitude of 10^{44}–10^{45} microwave photons per second in spite of the total number of the pumping photons supplied by associated celestial bodies is only 10^{44}–10^{46} photons per second. Evans et al[7]. gave a similar interrogation based on their data of 11 OH sources of Turners class I[8].

The other difficulty comes from various inevitable suppressing effects on celestial maser, such as the nonhomogeneity and the unstability of the population inversion in the maser regions. This unstability may caused by the disturbance waves existed in plasma or by the self-terminating effect[9].

Another nonthermal microwave line emission mechanism, i.e. the Cerenkov microwave line emission is suggested in this paper (see part I.). If a certain pumping model make a population inversion in OH source, and sufficient relativistic electrons move through this gas, the Cerenkov microwave emission lines of OH with anomalous intensity ratio would be produced.

With this emission mechanism, the 1720 MHz line of anomalous OH sources of Turner's class IIa (physically associated with SNR) can be interpreted satisfactorily.

(1) Population and Rate Equations

Based on the fact that there is only 1720 MHz line in Turner's class IIa sources, the population may be affirmed to satisfy the following inequality:

$$N_4/g_4 > N_1/g_1;\ N_4/g_4 \leqslant N_2/g_2;\ N_3/g_3 \leqslant N_2/g_2 \text{ and } N_1/g_1,$$

which can achieve in certain pumping models. For example, in Litvak's model[10] the population is

$$N_4/g_4 = N_2/g_2; N_3/g_3 = N_1/g_1;\ N_4/g_4 > N_1/g_1. \tag{11}$$

The rate equations may be written as

$$\begin{aligned}\dot{N}_i &= -\frac{4\pi}{c} I(B_{ij}N_i - B_{ji}N_j) - A_{ij}N_i - C_{ij}N_i - \Gamma N_i + C_{ji}N_j + R_i(N_{OH} - N_{ij})\\ \dot{N}_j &= \frac{4}{c} I(B_{ij}N_i - B_{ji}N_j) + A_{ij}N_i + C_{ij}N_i - \Gamma N_j - C_{ji}N_j + R_j(N_{OH} - N_{ij}),\end{aligned} \tag{12}$$

where i and j are the symbols of upper and lower levels respectively, B and A are Einstein coefficients, C_{ij} and C_{ji} are the downward and upward collision rates, N_{OH} is the total number density of OH molecules, R_i and R_j are the pump rates per molecule into states i and j. The decay constant Γ represents the sum of various effects which take a molecule away from a maser level i or j to any other levels and is assumed to be the same for both maser levels[11]. Eq. (12) reduces to

$$\begin{aligned}\dot{N}_i &= R_i(N_{OH} - N_{ij}) - N_i(\Gamma + C_{ij} + A_{ij})\\ \dot{N}_j &= R_j(N_{OH} - N_{ij}) + N_i(A_{ij} + C_{ij}) - N_j(\Gamma + C_{ji})\end{aligned} \tag{13}$$

considering the physical conditions of the cloud and the negligible terms.

The dynamical solution of the population difference

$\delta N(t) = N_4/g_4 - N_1/g_1$ can be solved from Eq.(13). It shows that the possibility of inversion is decided not only by pumping rate, but also by the density N_{OH} and the temperature T of the maser gas. Considering the nonhomogeneity of N_{OH} and T in the cloud, the nonhomogeneous inversion is naturally obtained. The net effect of the nonhomogeneity and instability is that the radiation of maser line will be suppressed.

(2) Cerenkov Line Emission Coefficient

In spite of the fact that maser radiation is suppressed, the Cerenkov line emission is still effective. Both the regions of $\delta N>0$ and $\delta N<0$ will produce Cerenkov line radiation with a negligible frequency difference.

1. Value of the number density of OH. The 1720 MHz line of IIa source comes from some small, bright spots ($T_B \simeq 10^{12}$K, size $\sim 10^{14}$ cm) spreading in HII regions of SNR[5]. This probably implies that they are small compact regions. A upper limit of hydrogen density N_H given by Litvak[6] is $N_H \simeq 10^{13}$ cm^{-3}, and $N_{OH}/N_H \lesssim 10^{-5}-10^{-6}$[12], so $N_{OH} \simeq 10^8$ cm^{-3} seems reasonable. From Eq.(11) the extreme value of $|\delta N|$ (for 1720 MHz and 1612 MHz lines) attains to $0.1N_{OH}$.

2. Cerenkov line emission coefficient J^C. As N_{OH} and δN is known, J^C can be calculated in accordance with the procedure mentioned in part I. Emission coefficient

$$J^C = 2.42\times10^{-29}\times N_e \int_{\nu_{min}}^{\nu_{max}} \nu\left(1-\frac{1}{n_\nu^2}\right)d\nu \,(\text{erg cm}^{-3}\ \text{s}^{-1}\ \text{sr}^{-1}).$$

Because the typical value of Lorentz factor γ is 10^3 (see Ref.[13]), Cerenkov radiation will occur in the frequency interval $\nu_{max}\rightarrow\nu_{min}$ where $n_\nu \geqslant 1/\sqrt{1-\gamma^{-2}} \cong 1.000005$ is satisfied. Calculation shows that for $N_{OH}=10^8$ cm^{-3}, we obtain

$$J^C_{1720}=1.49\times10^{-21}N_e,$$
$$J^C_{1612}=1.31\times10^{-21}N_e. \tag{14}$$

and $J^C_{1665}=J^C_{1667}=0$ due to the fact that $\delta N=0$ for 1665 MHz and 1667 MHz transitions.

(3) Radiative Transfer Equation

Suppose the radiation field in the source is isotropic and the energy density of the field is $u_\nu=4\pi I_\nu/c$. The radiative transfer equation for OH molecular cloud is

$$\frac{\partial I_\nu}{\partial x}\delta\nu=h\nu\frac{I_\nu}{c}(B_{ij}N_i-B_{ji}N_j)\phi(\nu)\delta\nu+\frac{h\nu}{4\pi}A_{ij}N_i\phi(\nu)\delta\nu+J^C_\nu\delta\nu. \tag{15}$$

The terms in the right side of Eq. (15) represent the stimulated absorption energy, spontaneous emission energy and Cerenkov emission energy respectively. $\phi(\nu)$ is the normalized profile function.

Integration of Eq. (15) over frequencies yields

$$\frac{\partial I}{\partial x}=(b_{ij}N_i-b_{ji}N_j)I+a_{ij}N_i+J^C/\Delta\nu, \tag{16}$$

where

$$I=\int I_\nu\phi(\nu)\delta\nu=\frac{\int I_\nu\delta\nu}{\Delta\nu},\quad b_{ij}=\frac{h\nu B_{ij}}{c\Delta\nu}=b_{ji}g_j/g_i,$$
$$a_{ij}=\frac{h\nu A_{ij}}{4\pi\Delta\nu}.$$

As Cerenkov term J^C is greater than spontaneous term, Eq. (16) reduces to

$$\frac{\partial I}{\partial x}=K_\alpha I+K_\beta, \tag{17}$$

where

$$K_\alpha = g_i b_{ij} \delta N, \; K_\beta = J^C/\Delta\nu.$$

The general solution of Eq. (17) is

$$I = (I_0 + K_\beta/K_\alpha)e^{K_\alpha x} - K_\beta/K_\alpha. \quad (18)$$

For the population of class IIa sources, K_α of 1612 MHz line is negative, optical depth $K_\alpha x = 2.7\times10^8 >> 1$, so the Cerenkov emission of 1612 MHz will be absorbed and can not be observed.

On the other hand, for 1720 MHz transition, the mean absorption coefficient $\bar{K}_\alpha^{1720} = 0$ owing to the nonhomogeneity and instability of population inversion. For $N_{OH} = 10^8\ cm^{-3}$, $\Delta\nu = 2$ KHz and $x = 10^{14}$ cm, $K_\beta^{1720} \simeq 3\times10^{-25} N_e$ and we obtain the intensity of Cerenkov radiation

$$I^{1720} = 3\times10^{-11} N_e (erg\ cm^{-2}\ s^{-1}\ sr^{-1}). \quad (19)$$

As compared with observational value of 1720 MHz line

$$I_{obs}^{1720} = \frac{2\nu^2}{c^2} kT_B \simeq 9\times10^{-7} (erg\ cm^{-2}\ s^{-1}\ sr^{-1}), \quad (20)$$

it is concluded that the nonthermal emission of 1720 MHz line can be interpreted by Cerenkov effect providing a density of relativistic electrons $N_e = 10^4 cm^{-3}$. As stated above, 1665 MHz and 1667 MHz lines can not be produced by Cerenkov effect and 1612 MHz emission is absorbed, so the observational characteristics of class IIa sources are interpreted satisfactorily. It is pointd out that this nonthermal line radiation mechanism may be a promising way out of the difficulty in the celestial maser theory.

REFERENCES

[1] You Junhan, Cheng Fuhua, 1980, Acta Physica Sinica, Vol. 29, 927.

[2] Cook, A.H., 1977, "Celestial Masers", Cambridge Univ. Press.

[3] Condon, E.U., Odishaw, H. (eds.), 1958, "Handbook of Physics", McGraw-Hill, New York.

[4] Robinson, B.J., McGee, R.X., 1967, Ann. Rev. Astron. Astrophys., Vol. 5, 183.

[5] ter Haar, D., Pelling, M.A., 1974, Rep. Prog. Phys., Vol. 37, 481.

[6] Litvak, M.M., 1972, in "Atoms and molecules in astrophysics", Carson, T.R., Roberts, M.J. (eds.), Academic press, 201.

[7] Evans, N.L. et al., 1979, Ap. J., Vol. 227, 450.

[8] Turner, B.E., 1970, J. Roy. Astron. Soc. Canada, Vol. 64, 221 and 282.

[9] Guo Guangcan, 1980, Journal of University of Science and Technology of China, Vol.10, No.3, 22 (in Chinese).

[10] Litvak, M.M., 1969, Ap. J., Vol. 156, 471.

[11] Goldreich, P., 1975, in "Physique atomique et moleculaire et matiere interstellaire", course 8, Roger Bolian (ed.), North-Holland publishing company press.

[12] Goss, W.M., 1968, Ap. J. Suppl., Vol. 15, 131.

[13] Saggion, A., 1975, Astron. Astrophys., Vol. 44, 285.

CORRELATIONS IN THE SYNCHROTRON-SELF COMPTON SPECTRA

YOU Junhan(尤峻汉)
(Astrophysics Research Division, University of Science and Technology of China, Hefei)

XIE Guangzhong(谢光中), BAO Mengxien(鲍孟贤), LI Kaihua(李开华)

(Yunnan Observatory, Academia Sinica, Kunming)

I. INTRODUCTION

In recent years, the X-ray radiation has been observed in some synchrotron radio sources[1-3], such as the radio source MK 501. The radio and X-ray radiation are all considered to have originated from the contribution of relativistic electrons in this source, i.e. from the "synchrotron-self Compton effect". The radio emissions are produced by the synchrotron process, then relativistic electrons collide with the radio photons produced by themselves, resulting in the production of high energy photons. The high frequency parts will enter into the X-ray band. The synchrotron-self Compton effect has been investigated by some authors[4-7], but the discussions on the correlations between the synchrotron radiation at radio frequency band and the Compton radiation in the ultra-violet to X-ray band are not yet complete and sufficient.

A series of definite relations between the synchrotron

and Compton spectra are given in this paper. Analysis of these correlations will give some semi-quantitative criteria for the observational tests of the correctness of the synchrotron-self Compton effects. The quasi-simultaneous observation technique in various wave bands (radio, infra-red, visible, ultra-violet up to X-ray) developed in recent years[8] makes it possible to test these correlations.

II. BASIC FORMULAE

(1) The Synchrotron Radio Radiation

If the magnetic field strength in a source is B, and the density of relativistic electron with isotropic distribution is

$$N(\gamma)\, d\gamma = N_0\, \gamma^{-n}\, d\gamma. \quad (\gamma_1 \leq \gamma \leq \gamma_2), \tag{1}$$

then the spectral emissivity is[8]

$$J_\nu = C\, N_0\, B^{\frac{n+1}{2}}\, \nu^{-\frac{n-1}{2}}\,. \tag{2}$$

The power law spectrum(2) holds in the interval(ν_a,ν_b), where the lower and upper limits of frequencies ν_a and ν_b are given by[9]

$$\nu_a \simeq \frac{3}{2}\, \gamma_1^2\, (\frac{eB}{2\pi mc}), \quad \nu_b \simeq \frac{3}{2}\, \gamma_2^2\, (\frac{eB}{2\pi mc}). \tag{3}$$

For the optically thin and homogeneous radio sources (the optically thick case will be discussed later), the intensity is $I(\nu_i) \simeq J(\nu_i)L$, where L is the dimension of a source. For the isotropic sources, the density of radiation field is

$$U_{ph}(\nu_i) = \frac{4\pi}{c}\, I(\nu_i) \simeq \frac{4\pi}{c}\, LJ(\nu_i), \tag{4}$$

so we obtain the number density of radio photons in the interval $\nu_i - \nu_i + d\nu_i$:

$$n_{ph}(\nu_i)d\nu_i = n_0 \nu_i^{-p}\, d\nu_i, \tag{5}$$

where

$$n_0 = \frac{4\pi L}{hc} C N_0 B^{\frac{n+1}{2}}, \quad P = \frac{n+1}{2}.$$

(2) The Compton Radiation

The Compton spectral emissivity is

$$J(\nu) = 8\pi r_0^2\, hcG$$

$$= 8\pi r_0^2 hc \int_\gamma \int_{\nu_i} N(\gamma)\, d\gamma\, f\left(\frac{\nu}{4\gamma^2\nu_i}\right) n_{ph}(\nu_i)\, d\nu_i,$$

where r_0 is the classical electron radius and the function f(x) in the integrand has the form

$$f(x) = x + 2x^2 \ln x + x^2 - 2x^3 \quad (0 \leqslant x \leqslant 1)$$
$$f(x) = 0 \ (x>1), \quad x \equiv \frac{\nu}{4\gamma^2\nu_i}. \tag{6}$$

The band of frequency of the Compton radiation can be obtained as follows. When an electron with energy γ collides with a photon with energy $h\nu_i$, the maximum scattering frequency will be $4\gamma^2\nu_i$, and the most probable frequency will be $0.61\times 4\gamma^2\nu_i \simeq 2.4\gamma^2\nu_i$. In the quasi-monochromatic approximation, it is considered that the scattering frequency equals $\sim 2.4\gamma^2\nu_i$. Hence the photons with frequency equal to that of the lower limit come from the collisions between electrons with lowest energy γ_1 and radio photons with the lowest frequency ν_a. In the same way, the photons with frequency equal to that of the upper limit come from the collisions between electrons with γ_2 and photons with ν_b. So we obtain the wave band of the Compton spectrum

$$2.4\ \gamma_1^2\ \nu_a \leqslant \nu \leqslant 2.4\,\gamma_2^2\ \nu_b. \tag{7}$$

In order to find the Compton spectrum (6), at first we complete the integration for ν_i, i.e. for the variable $x \equiv \frac{\nu}{4\gamma^2\nu_i}$. The lower limit of ν_i should be taken as the larger one between ν_a and $\frac{\nu}{4\gamma^2}$, i.e. $x_{min} = 1$ or $\frac{\nu}{4\gamma^2\nu_a}$ (this comes from the property of function $f(x)$, when $x \equiv \frac{\nu}{4\gamma^2\nu_i}$, $f(x) = 0$). Therefore the calculation of Compton spectral emissivity (6) should be completed in three wave bands, according to the value of x_{min}.

a) $\nu \geqslant 4\gamma_2^2\nu_a$ (high frequency part of the Compton spectrum). Here for any value of γ (but $\gamma_1 \leqslant \gamma \leqslant \gamma_2$) we always have $\nu \geqslant 4\gamma^2\nu_a$, therefore $\nu_{i\,min} = \frac{\nu}{4\gamma^2}$ (or $x_{min} = 1$). Combining $\nu \geqslant 4\gamma_2^2\nu_a$ with Eq. (7), the wave band of high frequency part of the Compton spectrum is

$$4\,\gamma_2^2\ \nu_a \leqslant \nu \leqslant 2.4\,\gamma_2^2\ \nu_b. \tag{8}$$

b) $\nu \leqslant 4\gamma_1^2\ \nu_a$ (low frequency part). Here for any γ ($\gamma_1 < \gamma \leqslant \gamma_2$) we have $\nu < 4\gamma^2\nu_a$, and $\nu_{i\,min} = \nu_a$ (or $x_{min} = \frac{\nu}{4\gamma^2\nu_a}$). But from Eq. (7) we see that the low frequency part $\nu \leqslant 4\gamma_1^2\nu_a$ is very close to the lower cut-off frequency ($\sim 2.4\,\gamma_1^2\nu_a$). The calculation of the low frequency Compton spectrum is not significant and it is omitted in this paper.

c) $4\gamma_1^2\nu_a \leqslant \nu < 4\gamma_2^2\nu_a$ (middle frequency part). Choosing the value $\nu_{i\,min}$ (or x_{min}) is difficult. From $\nu = 4\gamma^2\nu_a$, we can determine a critical value of γ as $\gamma_c = \frac{1}{2}\sqrt{\frac{\nu}{\nu_a}}$, when

$\gamma < \gamma_c$, $\nu_{i\,min} = \frac{\nu}{4\gamma^2}$ (or $x_{min}=1$); when $\gamma > \gamma_c$, $\nu_{i\,min} = \nu_a$ (or $x_{min} = \frac{\nu}{4\gamma^2\nu_a}$). Therefore, in Eq. (6) the integral range of γ should be divided into two parts, i.e. $\gamma_1 - \gamma_c$ and $\gamma_c - \gamma_2$, and the results is complex. But if we compare the high frequency band Eq. (8) and the middle frequency band $4\gamma_1^2\nu_a - 4\gamma_2^2\nu_a$, and use the values of ν_a and ν_b of Eq. (3), we obtain that the ratio of the intervals of high and middle frequencies is $\sim \gamma_2^2/\gamma_1^2 >> 1$. So the most important part in Compton radiation is only the high frequency band, and the middle frequency radiation is only a small proportion. If we use the spectrum at high frequency band instead of the entire Compton spectrum in the whole wave band Eq. (7), the error will be negligible.

Letting $x_{min}=1$(high frequency band), and substituting it into Eq. (6), for the high frequency Compton spectrum we obtain

$$J_\nu^{Comp} = 2^{2p+1}\pi r_0^2\, hC\, n_0 N_0 [\frac{1}{P} + \frac{P-1}{(p+1)^2} - \frac{2}{p+2}] \ln(\frac{\gamma_2}{\gamma_1})\, \nu^{-(p-1)}. \tag{9}$$

Eq. (9)shows a power law Compton spectrum with a spectral index $\alpha = p-1 = (n-1)/2$, the same as the index in the synchrotron radio spectrum.

Although the importance of the middle frequency Compton radiation is small, in order to show the variation of the spectral shape we still give the approximate result of the middle frequency spectrum

$$J_\nu^{Comp} \simeq 2^{2p+1}\pi r_0^2\, hC\, n_0 N_0 [\frac{1}{p} + \frac{p-1}{(p+1)^2} - \frac{2}{p+2}] \ln\left(\frac{\sqrt{\frac{\nu}{\nu_a}}}{2\gamma_1}\right) \nu^{-(p-1)}$$

$$+\frac{2\pi r_0^2 h\, C\, n_0 N_0}{p(n+1)}\ \nu_a^{-p}[2^{n+1}(\frac{\nu a}{\nu})^{\frac{n+1}{2}} - \gamma_2^{-(n+1)}]\nu. \qquad (10)$$

Eq. (10) shows that the middle frequency Compton spectrum has deviated from the power law shape $\sim \nu^{-(p-1)}$ and presents a flattened shape.

(3) The Spectral Luminosities P_ν and Flux Densities S_ν of Synchrotron and Compton Radiations, and Redshift Correction

For a homogeneous and optically thin spherical source with radius L, the spectral luminosity can be simply evaluated as $P_\nu \simeq \frac{4\pi}{3} L^3 J_\nu$. From Eq. (2), the luminosity of synchrotron radio radiation is

$$P_\nu^{sy} = (\frac{4\pi}{3} L^3)\, C\, N_0 B^{\frac{n+1}{2}} \nu^{-\frac{n-1}{2}}\ (\mathrm{erg\ s^{-1}\ Hz^{-1}\ Sr^{-1}}). \qquad (11)$$

From Eq. (9), for the luminosity of high frequency Compton radiation we obtain

$$P_\nu^{Comp} = (\frac{4\pi}{3} L^3) 2^{2P+1} \pi r_0^2\, h\, C\, n_0\, N_0 [\frac{1}{p} - \frac{2}{p+2} + \frac{p-1}{(p+1)^2}]$$

$$\cdot \ln(\frac{\gamma_2}{\gamma_1}) \nu^{-(p-1)} (\mathrm{erg\ S^{-1}\ Hz^{-1}\ sr^{-1}}). \qquad (12)$$

In order to derive the flux density S_ν, we must consider the Doppler effect. If the redshift of the source is z and its recession velocity is $\beta = \frac{v}{c}$, the relation between the emission frequency ν_e and the observed frequency ν_0 is

$$\nu_0 = \nu_e \frac{1-\beta}{\sqrt{1-\beta^2}}, \qquad (13)$$

and

$$1 + z = \frac{\nu_e}{\nu_0} = \sqrt{\frac{1+\beta}{1-\beta}} . \qquad (14)$$

If the source is at distance D_1 from the observer, and the number of emitted photons per second in $\nu_e - \nu_e + d\nu_e$ is $N(\nu_e)\, d\nu_e$, then the flux density in $\nu_0 - \nu_0 + d\nu_0$ is

$$S(\nu_0) d\nu_0 = \frac{N(\nu_e) d\nu_e}{4\pi D^2} \left(\frac{1-\beta}{\sqrt{1-\beta^2}} \right) h\nu_0 ,$$

where the factor $\left(\frac{1-\beta}{\sqrt{1-\beta^2}} \right)$ comes from the Doppler effect of the received number of photons which is the same as the Doppler effect of frequency. Because of $d\nu_e = d\nu_0 \frac{\sqrt{1-\beta^2}}{1-\beta}$,

and $\frac{h\nu_e N(\nu_e)\, d\nu_e}{4\pi} \equiv P(\nu_e)\, d\nu_e$ is the spectral luminosity of the source in $\nu_e - \nu_e + d\nu_e$, thus

$$S(\nu_0) = \frac{P(\nu_e)}{D_1^2} \left(\frac{1-\beta}{\sqrt{1-\beta^2}} \right) . \qquad (15)$$

For the synchrotron radio source, $P^{sy}(\nu_e)$ is given by Eq. (11). Replacing ν in Eq. (11) by ν_e we get the synchrotron flux density

$$S^{sy}(\nu_0) = \frac{1}{D_1^2} \left(\frac{1-\beta}{\sqrt{1-\beta^2}} \right) \left(\frac{4\pi}{3} L^3 \right) C N_0 B^{\frac{n+1}{2}} \nu_e^{-\frac{n-1}{2}} .$$

Replacing ν_e by ν_0, β by Z, we have

$$S^{sy}(\nu_0) = \frac{(1+Z)^{-\frac{n+1}{2}}}{D_1^2} \left(\frac{4\pi}{3} L^3 \right) C N_0 B^{\frac{n+1}{2}} \nu_0^{-\frac{n-1}{2}} . \qquad (16)$$

The Compton spectral luminosity $P^{Comp}(\nu_0)$ is given by

Eq. (12), and we obtain

$$S^{Comp}(\nu_0) = \frac{2^{n+3}}{D_1^2}(1+Z)^{-\frac{n+1}{2}}\left(\frac{4\pi}{3}L^3\right)\pi r_0^2\left[\frac{1}{n+1} - \frac{2}{n+5} + \frac{n-1}{(n+3)^2}\right]$$

$$\cdot \ln\left(\frac{\gamma_2}{\gamma_1}\right)(4\pi L)\, C\, N_0^2\, B^{\frac{n+1}{2}}\, \nu_0^{-\frac{n-1}{2}} \quad . \tag{17}$$

The derivation of the formulae for the middle frequency Compton spectrum is omitted in the above, because it is only a small proportion in the Compton radiation.

Considering the redshift influence, the frequency bands of radiations would be also changed. For the synchrotron radiation, Eq. (3) will be replaced by

$$\nu_{a0} \leqslant \nu_0 \leqslant \nu_{b0}, \tag{18}$$

where the lowest observed frequency is $\nu_{ao} \equiv \frac{\nu_a}{1+z} \simeq \frac{3}{2}\gamma_1^2 \cdot \left(\frac{eB}{2\pi mc}\right)\frac{1}{1+z}$, and the highest one is $\nu_{bo} \equiv \frac{\nu_b}{1+z} \simeq \frac{3}{2}\gamma_2^2\left(\frac{eB}{2\pi mc}\right)\frac{1}{1+z}$.

In the same way, the Compton radiation band Eq. (7) will be replaced by

$$2.4\,\gamma_1^2\,\nu_{ao} \leqslant \gamma_0 \leqslant 2.4\,\gamma_2^2\,\nu_{bo}. \tag{19}$$

III. THE CORRELATIONS BETWEEN THE RADIATIONS IN TWO WAVE BANDS

From the basic formulae mentioned above, we obtain the correlations of the synchrotron and Compton radiations as follows:

(1) Correlations of the Spectral Indices

According to Eq. (16) the spectral index of synchrotron radiation is $q = (n - 1)/2$ and from Eq. (17) the spectral index of Compton radiation in the high frequency band is $\alpha = p - 1 = (n - 1)/2$. Therefore,

$$q = \alpha = (n - 1)/2. \tag{20}$$

This is an important observable characteristic of synchrotron-self Compton effect.

But for the middle frequency band of Compton spectrum, it is obviously deviated from the power law shape and presents a flattening characteristic (see Eq.(10)).

(2) Correlation of the Frequency Bands

According to Eq. (18), the wave band of radio synchrotron radiation is $\nu_{ao} \lesssim \nu_0 \lesssim \nu_{bo}$. From Eq. (19) the ultra-violet-X-ray frequency band of Compton spectrum is $2.4\,\gamma_1^2\,\nu_{ao} \lesssim \nu_0 \lesssim 2.4\,\gamma_2^2\,\nu_{bo}$, i.e.

$$\nu_{0\,max}^{\ Comp} = 2.4\,\gamma_2^2\,\nu_{bo}, \quad \nu_{0\,min}^{\ Comp} = 2.4\,\gamma_1^2\,\nu_{ao}. \tag{21}$$

This is another observable characteristic of the synchrotron-self Compton effect. From Eq. (21) we have

$$\nu_{0\,max}^{\ Comp} \Big/ \nu_{0\,min}^{\ Comp} = \left(\frac{\gamma_2^2}{\gamma_1^2}\right)\left(\frac{\nu_{bo}}{\nu_{ao}}\right) = \left(\frac{\nu_{bo}}{\nu_{ao}}\right)^2. \tag{22}$$

(3) Correlation of the Energies of Synchrotron and Compton Radiations

From Eqs. (16) and (17), for the integral flux we obtain

$$S^{sy} = \int_{\nu_{ao}}^{\nu_{bo}} S^{sy}(\nu_0)d\nu_0$$

$$= \frac{2}{D_1^2}(1+z)^{-\frac{n-1}{2}}(\frac{4\pi}{3}L^3)C\,No\,B^{\frac{n+1}{2}}(3-n)^{-1}\left[\nu_{bo}^{\frac{3-n}{2}} - \nu_{ao}^{\frac{3-n}{2}}\right], \quad (23)$$

$$S^{Comp} = \int_{\nu_{0\,min}^{Comp}}^{\nu_{0\,max}^{Comp}} S^{Comp}(\nu_0)d\nu_0$$

$$= \frac{2^{n+4}}{D_1^2}(1+z)^{-\frac{n-1}{2}}(\frac{4\pi}{3}L^3)\pi r_0^2$$

$$\cdot\left[\frac{1}{n+1} - \frac{2}{n+5} + \frac{n-1}{(n+3)^2}\right]\ln(\frac{\gamma_2}{\gamma_1})(4\pi L)C\,N_0^2 B^{\frac{n+1}{2}}(3-n)^{-1}$$

$$\cdot\left[(2.4\,\gamma_2^2\,\nu_{bc})^{\frac{3-n}{2}} - (2.4\,\gamma_1^2\,\nu_{ao})^{\frac{3-n}{2}}\right]. \quad (24)$$

From these, we can obtain the flux ratio $\frac{S^{Comp}}{S^{sy}}$. Considering $\nu_{bo} >> \nu_{ao}$, $\gamma_2 >> \gamma_1$, and $\gamma_0 \simeq 2.8\times 10^{-13}$cm, we have

$$\frac{S^{Comp}}{S^{sy}} \simeq 7.74\times 10^{-25}\times 2^{n+4} L N_0\left[\frac{1}{n+1} - \frac{2}{n+5} + \frac{n-1}{(n+3)^2}\right]\ln\left(\frac{\nu_{bo}}{\nu_{ao}}\right)\left(2.4\,\gamma_2^2\right)^{\frac{3-n}{2}} \quad (25)$$

Eq. (25) shows that the larger the values of L and N_0, the stronger is the synchrotron-self Compton effect.

(4) Correlation Between the Peak Frequencies

From Eq. (25) we see that the dimension of the source L and the density coefficient N_0 of relativistic electrons must be sufficiently large to produce obvious synchrotron-self Compton effect. But if L and N_0 are large, the synchrotron-self absorption would not be negligible. This makes the source an optically thick one at the lower frequency band of radio radiation, so that, the power law radio spectrum does not hold in the entire radio band (ν_{ao}, ν_{bo}) and the radio flux will fall down at the lower frequency edge by the absorption. Hence, we expect that the Compton spectrum will also present a turn over at its lower frequency edge. The values of the two peak frequencies of the synchrotron and Compton radiations are relative to each other. This correlation emerges in optically thick source.

If we denote the peak frequency of synchrotron radio radiation by $\nu_{po}^{sy}(\nu_{po}^{sy} > \nu_{ao})$, and that of Compton radiation by ν_{po}^{Comp}, the correlation between ν_{po}^{sy} and ν_{po}^{Comp} can be obtained as follows.

For an optically thick source, the most probable frequency of radio photon is $\nu_p^{sy} = (1 + z)\, \nu_{po}^{sy}$. In the quasi-monochromatic approximation, the radio source can be considered as a monochromatic radiation field with frequency ν_p^{sy}. For the relativistic electrons with power law energy spectrum Eq. (1), the most probable electrons are those with the lowest energy γ_1. The collisions between electrons with γ_1 and radio photons with γ_p^{sy} will produce monochromatic Compton radiation with a frequency $\sim 2.4\ \gamma_1^2\ \nu_p$. Although the Compton scattered photons with different frequency $\sim 2.4\ \gamma^2\ \nu_p^{sy}$ can also be produced by the electrons with energy γ, due to the power of Compton radiation $\propto \gamma^2$, and the energy spectrum distribution $N(\gamma) \propto \gamma^{-n}$, the contribution of electrons with energy γ to the Compton radiation will be proportional to γ^{-n+2}. Usually $n > 2$, so the Compton radiation from the electrons with lowest energy γ_1 is the strongest. Hence, the peak frequency of Compton spectrum is $\nu_p^{Comp} \simeq 2.4\ \gamma_1^2\ \nu_p$. Replacing it by observed peak frequency, we obtain

$$\nu_{po}^{Comp} \simeq 2.4\ \gamma_1^2\ \nu_{po}^{sy}. \tag{26}$$

Eq. (26) shows the correlation between ν_{po}^{sy} and ν_{po}^{Comp}.

Because of the existence of ν_{po}^{sy} and ν_{po}^{Comp} for a optically thick radio source, it is necessary to make some corrections for the correlation formulae (21)-(25), which are satisfied for an optically thin source. To make the corrections, we need only to replace the lower limit of radio frequency ν_{ao} by ν_{po}^{sy}, and replace (21) by (26); correspondingly the ν_{min}^{Comp} would be replaced by ν_{po}^{Comp}.

(5) Correlation of Variations of Flux Density

For a variable synchrotron radio source the Compton radiation in X-ray band is also variable. The synchronism of variations is also an important observable characteristic of the synchrotron-self Compton effect.

REFERENCES

[1] Baity, W.A. et al., 1975, Ap. J., Vol. 199, L5.

[2] Mushotzky, R.F., 1977, Nature, Vol. 265, 225.

[3] Ives, J.C. et al., 1976, Ap. J., Vol. 207, L159.

[4] Hoyle, F., Burbidge, G., 1966, Nature, Vol. 209, 751.

[5] Jone's, T.W., O'Dell, S.L., Stein, W.A., 1974, Ap. J., Vol. 188, 353.

[6] Jone's, T.W., O'Dell, S.L., Stein, W.A., 1974, Ap. J., Vol. 192, 261.

[7] Marscher, L., 1977, Ap. J., Vol. 216, 244.

[8] Kondo, Y. et al., 1981, Ap. J., Vol. 243, 690.
[9] Ginzburg, V.L., Syrovatskii, 1965, Ann. Rev. Astron. Astrophys., Vol. 3, 297.

MOTION AND RADIATION OF ELECTRONS IN INHOMOGENEOUS MAGNETIC FIELD

AN Shuyuan(安树元), LUO Liaofu(罗辽复)
(Department of Physics, Neimenggu University, Huhehaote)

The motion and radiation of electrons in the neighbourhood of neutral sheet are discussed. A simplified model of field configuration is proposed. Assume that the type of field is

$$H_x = H_y = 0,\ H_z = \begin{cases} -B\varepsilon(y) & \varepsilon(y) = \dfrac{y}{|y|} \quad (|y| > \dfrac{\ell}{2}) \\ 0\ , & (|y| \leqslant \dfrac{\ell}{2}), \end{cases} \tag{1}$$

or

$$H_x = H_y = 0,\ H_z = -B\varepsilon(y), \qquad \varepsilon(y) = \frac{y}{|y|} \quad (y \neq 0). \tag{2}$$

For the field (2) the relativistic Hamiltonian is

$$\begin{aligned} \hat{H} &= c\ \vec{\alpha}\cdot\vec{\pi} + \beta m\ c^2, \\ \vec{\pi} &= \vec{p} + \frac{e}{c}\ \vec{A}\ , \\ A_y &= A_z = 0,\ A_x = B\ |y|. \end{aligned} \tag{3}$$

The eigenvalue of the energy:

$$E = m c^2[1+(\frac{P}{mc})^2+2n \frac{B}{B_r}]^{\frac{1}{2}} \quad (n=0,1,2,\ldots) \tag{4}$$

$$B_r = \frac{m^2c^3}{e\pi} = 4.41\times10^{13} \text{ gauss.}$$

The corresponding eigenfunctions:

$$w = (\hat{H} + E)\, y$$

$$Y_{n,+1,q} = \begin{pmatrix} \psi_{n-1,q} \\ 0 \\ 0 \\ 0 \end{pmatrix} (y < 0), \quad \begin{pmatrix} \psi'_{n,q} \\ 0 \\ 0 \\ 0 \end{pmatrix} (y > 0)$$

$$Y_{n,-1,q} = \begin{pmatrix} 0 \\ \psi_{n,q} \\ 0 \\ 0 \end{pmatrix} (y < 0), \quad \begin{pmatrix} 0 \\ \psi'_{n-1,q} \\ 0 \\ 0 \end{pmatrix} (y > 0) \tag{5}$$

$$\psi_{n,q} = N_n \exp\left\{ \frac{i}{\hbar}(pz+qx) - \frac{1}{2}(\frac{y-a}{\lambda})^2 \right\} H_n(\frac{y-a}{\lambda})$$

$$\psi'_{n,q} = N'_n \exp\left\{ \frac{i}{\hbar}(pz+qx) - \frac{1}{2}(\frac{y+a}{\lambda})^2 \right\} H_n(\frac{y+a}{\lambda})$$

$$(a = \frac{c}{eB}\, q, \qquad \lambda = (\frac{\hbar c}{eB})^{\frac{1}{2}}).$$

(p,q are the eigenvalues of π_2 and P_x respectively). By use of the condition of continuity at $y = 0$ we obtain

$$H_n^2\ (\frac{a}{\lambda}) - 2n\ H_{n-1}^2(\frac{a}{\lambda}) = 0. \tag{6}$$

We have proved that there are 2n discrete zeros of polynomials $H_n^2\ (\xi) - anH_{n-1}\ (\xi)$ distributed symmetrically from $\xi= 0$ to $\xi=(2n+1)^{\frac{1}{2}}$ and found all of them[1]. In terms of $\xi_r^{(n)}$ one obtains the quantization of a and q

$$a_{n,r} = \xi_r^{(n)}\lambda, \quad r = \pm1,\ldots\pm n, \tag{7}$$

$$q_{n,r} = (\frac{e\hbar B}{c})^{\frac{1}{2}}\ \xi_r^{(n)}$$

The similar results can be deduced for more realistic configuration of field (1). Thus we predict that the momentum of charged particles flying from the interface of inhomogeneous field takes discrete eigenvalues.

By use of Schwinger's method[2], the general formulae of spectral and angular distributions of radiation by an electron in magnetic field (1) or (2) are obtained

$$P(\psi,\chi,\omega,\omega_0')=\sum_{n=1}^{\infty}\delta(\omega-n\omega_0^*)P_n(\psi,\chi,\omega_0')$$

$$\omega_0^*=\left(\frac{1}{\omega_0'}-\frac{2\beta\sin\frac{\pi\omega_0}{2\omega_0}\cos\psi\cos\chi}{\pi\,\omega_0}\right)^{-1} \tag{8}$$

$$P_n(\psi,\chi,\omega_0')=\frac{2e^2}{\pi^3 c}\frac{n^2\omega_0^{*3}\omega_0'}{\omega_0^2}\Bigg\{\beta^2\Bigg[\sum_{k'=-\infty}^{\infty}J_{k'}\left(\frac{n\omega_0^*\beta\cos\psi}{\omega_0}\right)$$

$$\times\frac{\sin\frac{\pi\omega_0}{2\omega_0'}\left(\frac{n\omega_0^*}{\omega_0}-k'\right)}{\frac{n\omega_0^*}{\omega_0}-k'}\sin\zeta\Bigg]^2$$

$$+\tan^2\psi\Bigg[\sum_{k'=-\infty}^{\infty}J_{k'}\left(\frac{n\omega_0^*\beta\cos\psi}{\omega_0}\right)$$

$$\times\frac{\sin\frac{\pi\omega_0}{2\omega_0'}\left(\frac{n\omega_0^*}{\omega_0}-k'\right)}{\frac{n\omega_0^*}{\omega_0}-k'}\cos\zeta\Bigg]^2\Bigg\}$$

$$\zeta=\frac{n\pi}{2}-k'\chi+\frac{n\omega_0^*\beta\cos\psi}{\omega_0}\cos\frac{\pi\omega_0}{2\omega_0'}\sin\chi, \tag{9}$$

where ψ refers to the angle between radiating direction $\vec{n}$ and X = y plane, χ refers to the azimuth angle, ω_0 is the frequency of electron in homogeneous field, $\omega_0=\frac{eB_0}{mc}(1-\beta^2)^{\frac{1}{2}}$

and ω_0' — the frequency of y-coordinate in inhomogeneous field (2).

For synchrotron radiation when $\frac{\omega_0'}{\omega_0} >> 1$ the radiation concentrates in the direction $\psi^2 << 1$, $\chi^2 << 1$ and a simplified formula is deduced

$$P_n(\psi,\chi,\omega_0') \simeq \frac{4e^2\omega_0^2}{\pi^3 c}(1-\beta^2+\psi^2+\chi^2)^{-3}\left\{\frac{(\frac{\omega_0'}{\omega_0})^2(\psi^2+\chi^2)\sin^2 n\pi\eta}{\eta^2}\right.$$

$$\left. + \left(\frac{\pi\sin n\pi\eta}{2\eta} - \frac{1-\cos n\pi\eta}{n\eta^2}\right)^2\right\} \qquad (10)$$

$$\left(\frac{\omega_0'}{\omega_0} >> 1, |\chi| \lesssim \sqrt{1-\beta\cos\psi} - \frac{\pi\omega_0}{2\omega_0'},\ n\omega_0^* \lesssim \omega_0\left(\frac{\pi\omega_0}{2\omega_0'} + |\chi|^{-3}\right)\right),$$

$$\eta = \frac{\omega_0^*}{\omega_0'}(1-\beta\cos\psi) \simeq \frac{1-\beta^2+\psi^2}{1-\beta^2+\psi^2+\chi^2}.$$

It is approximately equal to

$$P_n(\psi,\chi,\omega_0') \simeq \frac{e^2\omega_0^2}{\pi^3 c}(1-\beta)^{-3}\frac{1-(-1)^n}{n^2} \qquad (11)$$

for $\frac{\omega'}{\omega_0} >> (1-\beta^2)^{-\frac{1}{2}}$ and $\psi^2+\chi^2 \sim 0$. The synchrotron radiation spectrum is essentially discrete due to the rapid transition of circulation center along the direction of the discontinuity of field. In many cases the fundamental is dominant, thus we have

$$P_1 = \int P_1(\psi,\chi,\omega_0')\, d\Omega$$
$$\simeq 3.1\times 10^{-3}\ B^2 \left(\frac{E}{mc^2}\right)^2 \text{ eV s}^{-1}, \qquad (12)$$

with $\omega = \frac{\omega_0'}{1-\beta}$. It is remarked that the emitted line is linear polarized as in the homogeneous field.

For cyclotron radiation, to the order of β^2, Eq. (9) can be approximated

$$P_n(\psi,\chi,\omega_0') = \frac{2}{\pi^3}\frac{e^2\omega_0'^2}{c}\frac{\beta^2(\frac{n\omega_0'}{\omega_0})^2}{[1-(\frac{n\omega_0'}{\omega_0})^2]^2}\begin{cases}(1-\cos^2\psi\cos^2\chi)\sin^2\frac{\pi\omega_0}{2\omega_0'} & (n \text{ even}).\\ (1-\cos^2\psi\sin^2\chi)\cos^2\frac{\pi\omega_0}{2\omega_0'} & (n \text{ odd}).\end{cases} \tag{13}$$

$$(\frac{\omega_0'}{\omega_0}\beta << \frac{1}{n})$$

After angular integration it reads

$$P_n(\omega_0') \frac{16e^2\omega_0'^2}{3\pi^2c}\frac{\beta^2(\frac{n\omega_0'}{\omega_0})^2}{[1-(\frac{n\omega_0'}{\omega_0})^2]^2}\begin{cases}\sin^2\frac{\pi\omega_0}{2\omega_0'} & (n \text{ even}),\\ \cos^2\frac{\pi\omega_0}{2\omega_0'} & (n \text{ odd}).\end{cases} \tag{14}$$

On account of the frequent collisions between electrons and plasmas in neutral sheet. Eqs. (13) and (14) should be averaged with respect to ω_0' and lead to

$$P(\psi,\chi,\omega) = \frac{e^2\beta^2}{\pi^3 c}\frac{\omega_0^3}{\omega^2}(\frac{\omega_0}{\omega}+1)^{-2}(\frac{\omega_0}{\omega}-1)^{-2}\sum_{n=1}^{M}\frac{\sin^2\frac{n\pi}{2}(\frac{\omega_0}{\omega}-1)}{n}$$

$$\times[1-\cos^2\psi\cos^2(\frac{n\pi}{2}-\chi)], \tag{15}$$

$$P(\omega) = \frac{8}{3}\frac{e^2\beta^2}{\pi^2 c}\frac{\omega_0^3}{\omega^2}(\frac{\omega_0}{\omega}+1)^{-2}(\frac{\omega_0}{\omega}-1)^{-2}\sum_{n=1}^{M}\frac{\sin^2\frac{n\pi}{2}(\frac{\omega_0}{\omega}-1)}{n}, \tag{16}$$

$$M = \max\{n < \frac{2\omega}{\omega_0}\}.$$

This shows that the profile of the spectrum of cyclotron radiation is a zigzag line from $\frac{\omega_0}{2}$ and descends as ω^{-2} for large ω. As for the angular distribution, when $|\psi| \sim \frac{\pi}{2}$

the radiation is χ- isotropic and as $|\psi|$ decrease from $\frac{\pi}{2}$ the χ- isotropy is broken gradually. The radiation is linear-polarized at $\psi \sim 0$ and elliptic polarized in other directions.

It is not difficult to estimate the absorption of the radiation by the plasma. For synchrotron radiation the main effect is collision absorption. It can be neglected unless the wave propagates in neutral sheet for a long distance. For cyclotron radiation the main effect is resonance absorption which only occurs at frequency $\omega_s = s\omega_0 (s = 1,2)$. For example, in solar flare the absorption coefficient is of the order of 10^{-7} cm^{-1}.

The authors are grateful to Qu Qinyue, Lu Tan and Peng Qiuhe for their discussions.

REFERENCES

[1] An Shuyuan, 1981, Acta Scientiarum Naturalium Universitatis Intramongolicae, Vol. 12, No. 2, 55 (in Chinese).

[2] Schwinger, J., 1949, Phys. Rev., Vol. 75, 1912.

THE END =

THE BEGINNING